Ultrastructure of the connective tissue matrix

ELECTRON MICROSCOPY IN BIOLOGY AND MEDICINE

Current Topics in Ultrastructural Research

SERIES EDITOR: P.M. MOTTA

Already published in this series

Motta, P.M. (ed.): Ultrastructure of Endocrine Cells and Tissues.
ISBN: 0-89838-568-7.

Van Blerkom, J. and Motta, P.M. (eds.): Ultrastructure of Reproduction: Gametogenesis, Fertilization, and Embryogenesis. ISBN: 0-89838-572-5.

Series Editor
P.M. MOTTA, Department of Anatomy, Faculty of Medicine, University of Rome, Viale R. Elena 289, 00161 Rome, Italy
Advisory Scientific Committee
D.J. ALLEN (Toledo, Ohio, USA) / A. AMSTERDAM (Rehovot, Israel) / P.M. ANDREWS (Washington, DC, USA) / L. BJERSING (Umea, Sweden) / I. BUCKLEY (Canberra, Australia) / F. CARAMIA (Rome, Italy) / A. COIMBRA (Porto, Portugal) / I. DICULESCU (Bucharest, Romania) / L.J.A. DIDIO (Toledo, Ohio, USA) / M. DVORÁK (Brno, Czechoslovakia) / H.D. FAHIMI (Heidelberg, FRG) / H.V. FERNÁNDEZ-MORÁN (Chicago, Ill., USA) / T. FUJITA (Niigata, Japan) / E. KLIKA (Prague, Czechoslovakia) / L.C.U. JUNQUEIRA (São Paulo, Brazil) / R.G. KESSEL (Iowa City, Iowa, USA) / B.L. MUNGER (Hersey, Pa., USA) / O. NILSSON (Uppsala, Sweden) / K.R. PORTER (Boulder, Colo., USA) / J.A.G. RHODIN (Tampa, Fla., USA) / K. SMETANA (Prague, Czechoslovakia) / L.A. STAEHELIN (Boulder, Colo., USA) / K. TANAKA (Yonago, Japan) / K. TANIKAWA (Kurume, Japan) / I. TÖRÖ (Budapest, Hungary) / J. VAN BLERKOM (Boulder, Colo., USA)

Ultrastructure of the Connective Tissue Matrix

Edited by

A. RUGGERI, M.D.

Department of Anatomy, Faculty of Medicine
University of Bologna, Bologna, Italy

and

P.M. MOTTA, M.D., Ph.D.

Department of Anatomy, Faculty of Medicine
University of Rome, Rome, Italy

1984 **MARTINUS NIJHOFF PUBLISHERS**
a member of the KLUWER ACADEMIC PUBLISHERS GROUP
BOSTON / THE HAGUE / DORDRECHT / LANCASTER

Distributors

for the United States and Canada: Kluwer Boston, Inc., 190 Old
Derby Street, Hingham, MA 02043, USA
for all other countries: Kluwer Academic Publishers Group, Distribution Center, P.O. Box 322, 3300 AH Dordrecht, The
Netherlands

Library of Congress Cataloging in Publication Data

Main entry under title:

Ultrastructure of the connective tissue matrix.

    (Electron microscopy in biology and medicine)
    Includes index.
    1. Connective tissues. 2. Ground substance (Anatomy)
3. Ultrastructure (Biology) I. Ruggeri, A. (Alessandro)
II. Motta, Pietro. III. Series. [DNLM: 1. Connective
tissue--Ultrastructure. 2. Collagen. 3. Proteoglycans.
4. Elastin. 5. Microscopy, Electron--Methods.
Q8 532.5.C7 U47]
QM563.U47 1984   611'.74   83-17326

ISBN-13: 978-1-4612-9789-5   e-ISBN-13: 978-1-4613-2831-5
DOI: 10.1007/978-1-4613-2831-5

Copyright

Preface

In recent years, the techniques of electron microscopy have developed so widely and rapidly that they now cover the fields of research once the unique apanage of sister research techniques such as biochemistry, physiology, immunology, X-ray diffraction, etc. It is now possible to reach molecular and submolecular levels, making this technique indispensable in every type of research. Electron microscopy alone often provides enough information to solve given problems.

In the field of the connective tissue matrix, knowledge of the molecular structure of collagen, proteoglycans and elastin and their interaction has been to a large extent elucidated by electron microscopy. The field over which electron microscopy ranges in the investigation of the connective tissue matrix is so wide that the aim of this volume is to collect the main ultrastructural acquisitions disseminated in various journals and monographs in one book.

The intent of this volume is to: (a) integrate different and new microscopic methods and review the results of such an integrative approach; (b) present a comprehensive ultrastructural account of selected aspects of the field; (c) point out gaps or controversial topics in our knowledge; (d) outline pertinent future research and expansion of the subject.

The chapters of this volume, prepared by recognized authorities in the field, briefly present traditional information on the topic, but mainly describe the very new trends on the subject, showing with the help of a rich and valuable selection of micrographs the contribution that these integrated submicroscopic techniques have produced in the field. It is hoped that this book will represent a valuable help to specialists concerned with the normal and pathological structure of the connective matrix in embryology, development, adult life and aging.

We wish to express our sincere thanks to the contributors of the volume and to all the members of the advisory scientific committee for having enthusiastically and patiently responded to our numerous requests during the preparation of the volume.

We also wish to carry out the desire of the contributors, and of the Italian Group for Calcified Tissue Research, in dedicating this volume to Prof. Rodolfo Amprino on his seventieth birthday. This dedication is an acknowledgement of his fundamental contribution to the study of bone development and of his constant and exemplary commitment in stimulating and coordinating research in the field of connective tissue.

Editors

Contents

List of contributors

Benazzo, Franco, Clinica Ortopedica, Università, 27100 Pavia, Italy

Bezerra, M.S.F., Laboratório de Biologia Celular, Faculdade de Medicina da USP, Av. Dr. Arnaldo 455, 01246 São Paulo, Brazil

Bonucci, Ermanno, Istituto di Anatomia Patologica, Policlinico Umberto I, Viale Regina Elena, 324, 00161 Roma, Italy

Chapman, John A., Department of Medical Biophysics, University of Manchester Medical School, Manchester M13 9PT, United Kingdom

Craig, Alan S., Applied Biochemistry Division, DSIR, Palmerston North, New Zealand

Fornieri, Claudio, Istituto di Patologia Generale, Via Campi, 287, 41100 Modena, Italy

Hulmes, David J.S., Department of Medical Biophysics, University of Manchester Medical School, Oxford Road, Manchester, M13 9PT, United Kingdom

Junqueira, Luiz C.U., Laboratório de Biologia Celular, Faculdade de Medicina da USP, Av. Dr. Arnaldo 455, 01246 São Paulo, Brazil

Kádár, Anna, 2nd Central Electron Microscope Laboratory, 2nd Department of Pathology, Semmelweis Medical University, Budapest IX., Üllöi-út 93, Hungary

Marchini, Maurizio, Istituto di Anatomia Umana Normale, Via Irnerio, 48, 40126 Bologna, Italy

Montes, Gregorio S., Laboratório de Biologia Celular, Faculdade de Medicina da USP, Av. Dr. Arnaldo 455, 01246 São Paulo, Brazil

Motta, Pietro M., Università di Roma, Istituto di Anatomia Umana Normale, Viale Regina Elena, 289, 00161 Roma, Italy

Parry, David A.D., Department of Chemistry, Biochemistry and Biophysics, Massey University, Palmerston North, New Zealand

Pasquali-Ronchetti, Ivonne, Istituto di Patologia Generale, Via Campi, 287, 41100 Modena, Italy

Reale, Enrico, Medizinische Hochschule Hannover, Abteilung Elektronenmikroskopie, Karl-Wiechert-Allee 9, D-3000 Hannover 61, Federal Republic of Germany

Ruggeri, Alessandro, Università di Bologna, Istituto di Anatomia Umana Normale, Via Irnerio, 48, 40126 Bologna, Italy

Serafini-Fracassini, Augusto, Department of Biochemistry and Microbiology, University of St. Andrews, St. Andrews, Fife KY16 9AL, United Kingdom

Thyberg, C. Johan O., Department of Histology, Karolinska Institutet, P.O. Box 60400, S-104 01 Stockholm, Sweden

CHAPTER 1

Electron microscopy of the collagen fibril

JOHN A. CHAPMAN and DAVID J.S. HULMES

1. Introduction

1.1. Identification of collagen

Collagen is identified by those properties that stem from the predominantly triple-chain helical structure of its molecules. A prerequisite for the formation of this triple helix is a Gly-X-Y repeating tripeptide unit in the amino acid sequence of the three chains, where X and Y can be any amino acids but are often the imino acids proline and hydroxyproline. This sequence, with glycine in every third position and with an unusual abundance of hydroxyproline, forms the basis for the chemical identification of collagen (for review, see 1). An unambiguous physical identification is provided by X-ray diffraction; the helix parameters established by high-angle X-ray scattering are unique to collagen (2).

Although electron microscopists have for many years identified collagen by its appearance as long unbranched banded fibrils with a characteristic periodicity of 60 – 70 nm (see, for example, Figures 3–8, it is now known that not all collagens exist in this form. Types I, II and III collagen all form periodic-structured fibrils, but type IV collagen molecules in basement membranes do not occur in the fibrillar form and aggregate instead as a mat-like network; less is known about the distribution and structure of other collagen types. Thus although character-istically banded fibrils are unmistakably collagen, this is not the sole criterion for the identification of mature collagen in the electron microscope.

Nevertheless, we shall be concerned here only with collagen in the fibrillar form and, in particular, with the molecular basis underlying its periodicity and the intraperiod band pattern. Type I collagen, the main constituent of tendon, skin, bone and vessel walls, has been studied for longer than other types, and more is known about its structure at all levels. It is the source of material for most of the experimental data presented in this chapter. The band patterns of type II and type III collagen fibrils differ only slightly from the type I pattern, implying a broadly similar axial arrangement of molecules in each of these fibril types.

1.2. The collagen molecule

A variety of physical and physiochemical studies, including direct visualisation of individual molecules in the electron microscope, have demonstrated the rod-like nature of the collagen molecule. Molecules of type I collagen have a length, L, which is slightly less than 300 nm and a diameter, $2r$, of about 1.4 nm. The rod is neither rigid nor randomly flexible but appears to possess an intermediate level of semi-flexibility which probably varies along its length (3, 4). The three helically wound polypeptide chains ('α chains') which make up the rod each comprise about 1,000 amino acid residues. The details of the three-dimensional structure of the triple helix, established by X-ray diffraction and using known bond lengths and bond angles, have been reviewed elsewhere (2). For the interpretation of the intraperiod band pattern of the collagen fibril, we shall be less concerned with the structure in three dimensions than with the projection of that structure on to the molecular axis. From our point of view the parameter that matters is the residue-to-residue spacing, h, in an axial direc-tion. This is known (from the position of the merid-ional reflection in the X-ray pattern) to be close to 0.29 nm (2). The known values for the D-periodicity (67 nm in rat tail tendon) and the number of residues in a D-period ($= 234$) show that the mean value of h is, more accurately, 0.286 nm. As the meridional reflection is diffuse, the assumption that the residues are uniformly spaced throughout the triple-helical body of the molecule is not strictly valid and the residue-to-residue spacing can be expected to vary slightly along the molecule.

Ruggeri, A and Motta, PM (eds): Ultrastructure of the connective tissue matrix. ISBN-13:978-1-4612-9789-5
© 1984, Martinus Nijhoff Publishers, Boston, The Hague, Dordrecht, Lancaster.

1.3. Amino acid sequence

The three α-chains are non-identical in type I collagen, where the molecule comprises two identical α1(I) chains and one α2(I) chain, but are identical in type II and in type III. The amino acid sequences of the α1(I) and α2(I) chains are known (5); so too are the sequences of the α1(III) chain and most of the α1(II) chain. Species differences occur, but a substantial measure of homology between species exists. Of the 1,055 residues in the α1(I) chain of calf skin collagen, 1,014 occur in the repeating Gly-X-Y triplets essential for triple-helical packing. The N-terminal 16 residues and C-terminal 25 residues do not have glycine in every third position and exist in a less regular conformation. N-terminal and C-terminal extrahelical peptides occur at the ends of the 1,029-residue-long α2(I) chain but are shorter.

Some of these features are illustrated in Figure 1 which shows the amino acid sequences in the three α-chains at the N-end of a type I collagen molecule from calf skin. Roughly 5 per cent of a complete molecule appears in the figure. The N-terminal extrahelical peptides are on the left and the numbering of residues begins with the first glycine in the triple-helical part of the molecule. No attempt has been made to indicate the coiling of the chains. Although the residue-to-residue spacing, h, has been shown as constant throughout, this is far from being the case in the extrahelical peptides; their conformations, probably folded, are still in doubt. The formation of the triple helix requires the three chains to be mutually staggered by one residue to accommodate the glycines close to the central axis of the molecule and to allow the X and Y side-chains to project outwards. The position of the α2 chain with respect to the two α1 chains in type I collagen is not yet known with certainty; here the order α1–α2–α1 has been assumed. Throughout the sequences, the charged residues Arg, Lys, Asp, Glu are printed in bold capitals (this will later be used to mimic the effect of heavy metal staining); the residues His, Hyl, Glx (i.e. Glu or Gln), where the staining behaviour is less certain, are in small capital letters; all other residues are

shown in italic. It will be noted that there is a marked tendency for charged residues to occur in groups (underlined), separated by stretches devoid of charge.

The complete sequence of type I collagen appears in Figure 13 a,b,c,d. About 80 per cent of the data are from calf skin; where calf skin data were not available, data from rat skin or (for the α2 chain only) chick skin have been used instead.

This knowledge of the amino acid sequences of the three α chains and the essentially one-dimensional nature of the collagen molecule, with its near-constant axial separation between residues in all but 3 per cent of its tertiary structure, now permit the direct correlation of structural data obtained by electron microscopy with chemical sequence data. In this respect collagen provides a valuable model system for studying the chemical basis of ultrastructure and the action of heavy metal stains and other reagents on a protein.

1.4. D-periodicity in fibrils

It has been recognised for some time that the periodic structure in collagen fibrils arises because the molecules are assembled in parallel array and are mutually staggered (i.e. axially displaced with respect to one another) by integral multiples of a common distance, D (6–9). Low-angle X-ray diffraction of hydrated, slightly stretched fibrils suggests that D is close to 67 nm in the native state in rat tail tendon collagen (10, 11), although the same technique has yielded slightly different values in other tissues, with the D-period in skin significantly shorter than in tendon (11–13). Electron microscopy, inevitably of dehydrated specimens, usually gives lower values, commonly around 64 nm in fibrils deposited on a supporting film from suspension. In embedded and sectioned material, values over a wide range can be encountered, presumably as a result of mechanical stresses imposed during sectioning.

The relative axial relationships between molecules in a fibril are illustrated in Figure 2. It is to be

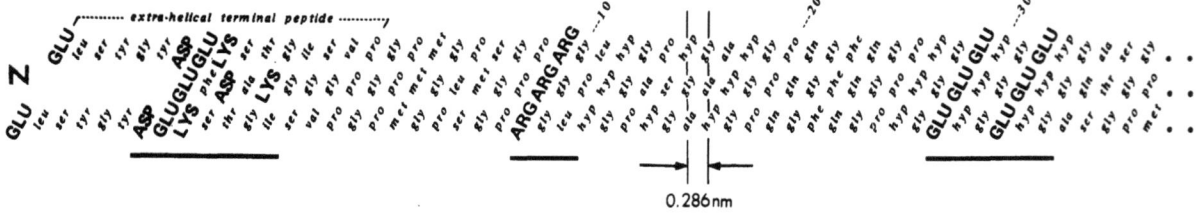

Figure 1. The triple-chain sequence of type I calf skin collagen at the N-end of the molecule. Bold capitals indicate charged residues.

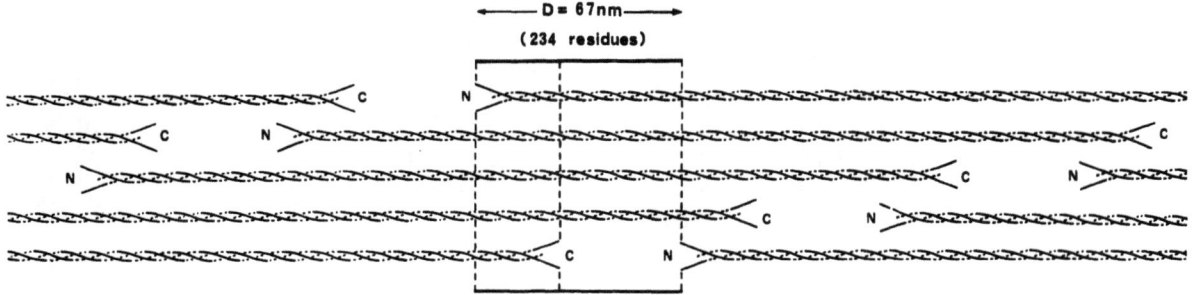

Figure 2. The regular staggering of molecules in a collagen fibril (the 'Hodge-Petruska' packing arrangement (9), first proposed by Tomlin (6)).

remembered that this is merely a diagrammatic representation in two dimensions and cannot show all possible stagger relationships in three dimensions (which could include not only the 1D and 4D staggered contacts shown here but also 0D, 2D and 3D staggers betwwen adjoining molecules). The figure shows, in essence, the predicted positions of molecules projected on to the fibril axis.

As the ratio of molecular length to D-stagger is non-integral (the values quoted here give L/D = 4.48), each D-period can be seen from Figure 2 to be divided into two roughly equal zones, an 'overlap' zone which includes the N- and C-ends of molecules, and a 'gap' zone which does not. For every five molecular segments in an overlap, there are only four in a gap zone, and the protein density in this zone can be expected to be roughly 4/5 that in the overlap zone. The ratio L/D = 4.48 implies that the axial extent of the overlap zone should be 0.48D but measurements on electron micrographs (see Section 3.5) indicate an axial extent closer to 0.40D. The difference is probably due to a condensed or folded conformation of the extrahelical terminal peptides in the dehydrated fibril, leading to a broadening of the gap.

2. The collagen fibril in the electron microscope

2.1. Selection and preparation

The earliest methods for the ultrastructural study of biological fibres involved dispersing or homogenising the tissue and depositing a small amount, suitably diluted, on a supporting film. This procedure is well suited to tendon (e.g. from rat tail) which is predominantly collagen but is less applicable to tissue in which extra-fibrillar material tends to obscure fibrils. The examination of intact tissues, with the aim of preserving the form of tissue components

and the spatial relationships between them, came later with the development of embedding and ultra-thin sectioning techniques. The adjoining chapter (14) is largely concerned with data obtained in this way. Newer techniques which permit the examination of surfaces (e.g. scanning electron microscopy, freeze-fracturing) are sometimes used as an alternative approach to the study of tissue components *in situ* (Chapter 4).

A further selective procedure applicable to collagen makes use of the property that collagen extracted from connective tissue (usually young) by weak acid or neutral salt solution can be made to reconstitute into native-type fibrils which exhibit an intra-period band pattern indistinguishable from that in directly extracted native fibrils. As reconstituted fibrils can readily be prepared from purified solutions, yielding fibrils free of contaminants and with clearly defined staining patterns, they have been widely used for high-resolution studies of intra-period structure. Most of the work described here, correlating staining patterns with amino acid sequence data, was carried out on reconstituted fibrils usually from citric acid or acetic acid extracted calf skin.

The reconstitution of fibrils from solution has received a good deal of attention, from physical chemists as well as electron microscopists. Although reconstitution *in vitro* superficially resembles growth *in vivo* it is now recognised that many other factors may operate *in vivo*. This has prompted fresh approaches to the selection of material for electron microscopic and other studies of fibril formation. As described later (Section 5.2), new information has emerged from the direct examination of macromolecular aggregates in the extracellular milieu around cultured fibroblasts actively synthesising collagen (15). More recently still, attempts have been made to prepare collagen fibrils from newly synthesised collagen precursor molecules (i.e. pro-

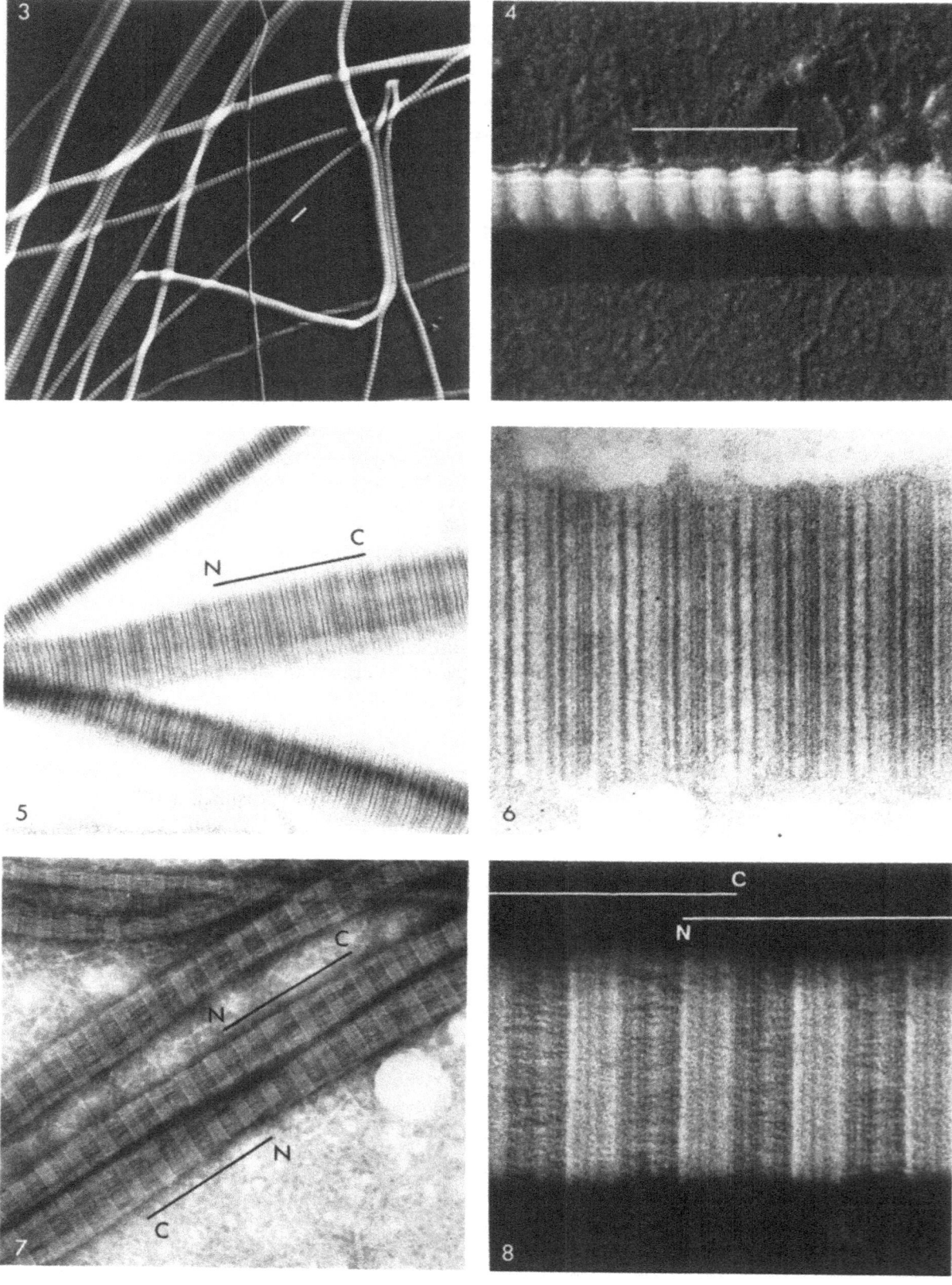

collagen) by enzymic removal of the propeptides *in vitro* (16).

2.2. *Enhancement of image constrast*

The elemental composition of most biological materials is such that the scattering of electrons differs little from one part of a specimen to another and contrast in the electron-optical image is therefore weak. There are various ways in which heavy metal atoms, which scatter electrons strongly, can be incorporated in the specimen to introduce contrast. Which of these contrasting methods has been used will influence markedly the appearance of a collagen fibril in the electron microscope.

In the early days of electron microscopy, shadow-casting was widely employed and gave useful information about the form of fibrils and their growth behaviour. Collagen fibrils, deposited from suspension on a supporting film and coated *in vacuo* with a thin layer of heavy metal at a small angle to the supporting film, display a striking 'gas mask tubing' appearance (Figures 3, 4). This appearance is, of course, due to the different mass thickness of the overlap and gap zones in the fibril and differential shrinkage after dehydration. The periodically changing surface contours of the fibril are accentuated by the shadow effect, particularly when the direction of evaporation of the heavy metal is other than at right angles to the fibril axis. In the conventional shadow-casting technique applied to objects as large as fibrils, resolution is limited to some tens of nm by aggregation effects in the evaporated metal layer and only rarely is it possible to detect any intraperiod surface substructure other than that arising from the gap-overlap mass distribution. Heavy metal shadowing can nevertheless be used to visualise individual collagen molecules deposited on a clean flat substrate (such as freshly cleaved mica); visualisation is improved by using rotary shadowing to coat the collagen molecules from all directions at a glancing angle.

The most extensively used contrasting technique in biological electron microscopy has always been staining with solutions of heavy metal salts. Indeed, it is the only technique normally applicable to sectioned material. The word 'staining' is used here to mean the procedure in which, after exposure of the specimen to the heavy metal salt solution, unreacted staining solution is removed by washing, leaving only those staining ions or molecules which have reacted specifically with the specimen (Figures 5, 6). We shall also refer to this procedure as 'positive staining' when there is possibility of confusion with the 'negative staining' technique (see below). Commonly used stains are uranyl and leads salts (so-called 'cationic stains') and salts in which the heavy metal is in the anion (e.g. phosphotungstate). A detailed interpretation of the image of the stain distribution in fibrils is considered in Section 3.

In negative staining (or 'negative contrasting') only excess heavy metal staining solution is drained off, and no attempt is made to remove unbound stain by washing. The result is to leave a thin layer of dried stain in and around the specimen, outlining it and filling internal voids with the electron-dense contrasting medium. The internal structure that can be revealed depends on the size of the heavy metal staining molecule and the extent to which it can penetrate. In collagen fibrils the interstices in the gap zones are large enough to be readily penetrable, giving the characteristic alternation of dark (gap) and light (overlap) zones along the negatively stained fibrils (Figures 7, 8). Oddly, one of the first electron micrographs of a negatively stained fibril to be published (18) appeared before the technique itself was recognised and described. Several years elapsed before the technique was intentionally applied to collagen (19, 20).

2.3. *Positive and negative staining patterns*

When collagen fibrils are exposed to solutions of heavy metal salts, up to twelve staining bands per period can be distinguished (21). Band patterns in positively stained fibrils appear in Figures 5, 6, 10b. The asymmetric nature of the pattern (i.e. absence of mirror symmetry) implies a polarisation in direction of the molecular units in the fibril. The labelling of the bands in a D-period, shown in Figure 9a, follows the conventional notation in which groups of bands are denoted by letters and each band within a

---

Figures 3, 4. Collagen fibrils from human Achilles tendon, shadowed with gold-palladium. The periodicity is ~65 nm. The line represents the length of a collagen molecule (~300 nm).

Figures 5, 6. Calf skin collagen fibrils reconstituted from solution, positively stained with phosphotungstic acid (PTA) and uranyl acetate (UA). The periodicity is ~65 nm. In Figure 5, the line shows the length and polarity of a collagen molecule.

Figures 7, 8. Negatively stained calf skin collagen fibrils (Figure 7: PTA; Figure 8: lithium tungstate). The periodicity is ~65 nm. The lines indicate collagen molecules.

6

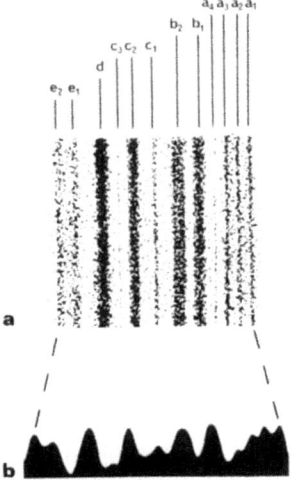

Figure 9. Labelling of the bands in a single D-period of a positively stained fibril (Figure 9a), and, below, a microdensitometric trace of the stain distribution (Figure 9b).

group is distinguished by a numerical suffix. An alternative, wholly numerical, notation has been proposed (22) but has not found wide acceptance, probably because the old notation is too well established. The four closely spaced **a** bands in the pattern have a mean centre-to-centre spacing of 3.4 nm and provide a useful test of pattern resolution. All staining patterns here are shown with the collagen molecules directed so that their N-ends point to the left (which results in a reversed alphabetical order for the lettering of the bands).

The pattern is readily observable in stained sections of embedded connective tissues in regions where collagen fibrils are sectioned longitudinally, although the section thickness is usually sufficient to reduce specimen resolution in the image to a level where the four **a** bands cannot be distinguished. As noted earlier (Section 2.1) the most clearly resolved band patterns are obtainable from reconstituted fibrils (Figure 10b). This is probably because fibrils prepared by reconstitution *in vitro* are less compact than those formed *in vivo* and flatten on the supporting film on drying, giving thinner specimens in which higher resolutions are attainable; distortion is often present, however, and fields of view with such clearly defined bands as those in Figure 10b are infrequent. It is not unusual to observe a slight curving of the bands, as in Figures 17 and 18; there is a marked tendency for this to be greatest in band c_2 and for the curvature to be directed with its concave side to the right (i.e. in the N→C direction). It is most readily seen in micrographs by observing the pattern at a glancing angle. The frequent occurrence of this

effect suggests that it is not a random distortion but reflects a basis structural feature in the fibril. As the N-ends of molecules terminate at the c_2 band, the implication is that these N-ends are under tensile strain. This strain cannot be uniform across the fibril but must be greatest in its peripheral regions.

The D- staggered arrangement of molecules in a fibril was first elucidated by comparing positive staining patterns from fibrils with those from segment long spacing (SLS) collagen. Superposition of four SLS patterns, mutually staggered by D, was shown to give a pattern resembling that from a fibril (7,23). At the time it was thought that the molecular length, L, was 4D and the proposed molecular array was therefore referred to as 'quarter-staggered'.

The negative staining pattern, by locating the positions of the ends of the molecules, gives a more complete picture. The conspicuous subdivision of each D-period into a stain excluding (overlap) zone and a stain-accessible (gap) zone reveals the non-integral relationship between L and D in which L is approximately 4.4D (9). Figure 10a, b shows the negative and positive staining patterns compared; the D-staggered molecular array is indicated above. It is evident from this figure that the N-end of the molecule (at the left margin of each overlap zone) lies in the vicinity of the c_2 band in a fibril and that the C-end occurs close to the a_3 band.

The pattern from a negatively stained fibril appears to have a dual character. Although the principal contrast effects come from the gap-overlap subdivision of each D-period, superimposed on this pattern of broad light and dark zones is a finer band pattern similar to that obtained after positive staining. As suggested by others (9, 24) negative staining patterns would seem to retain an element of positive staining as well (see also Section 3.4).

3. The chemical basis of the staining pattern

3.1. Introduction

In the image of any stained biological specimen, the electron microscopist is viewing the products of a chemical reaction. Knowledge of the nature and extent of this reaction and of the factors that influence it should lead to a greater understanding of the significance of the image. As soluble reactants and products are removed in the positive staining procedure, only products bound to the specimen will contribute to the image. In positive staining, therefore, the important question is: what are the chemical groups on the specimen involved in this binding reaction?

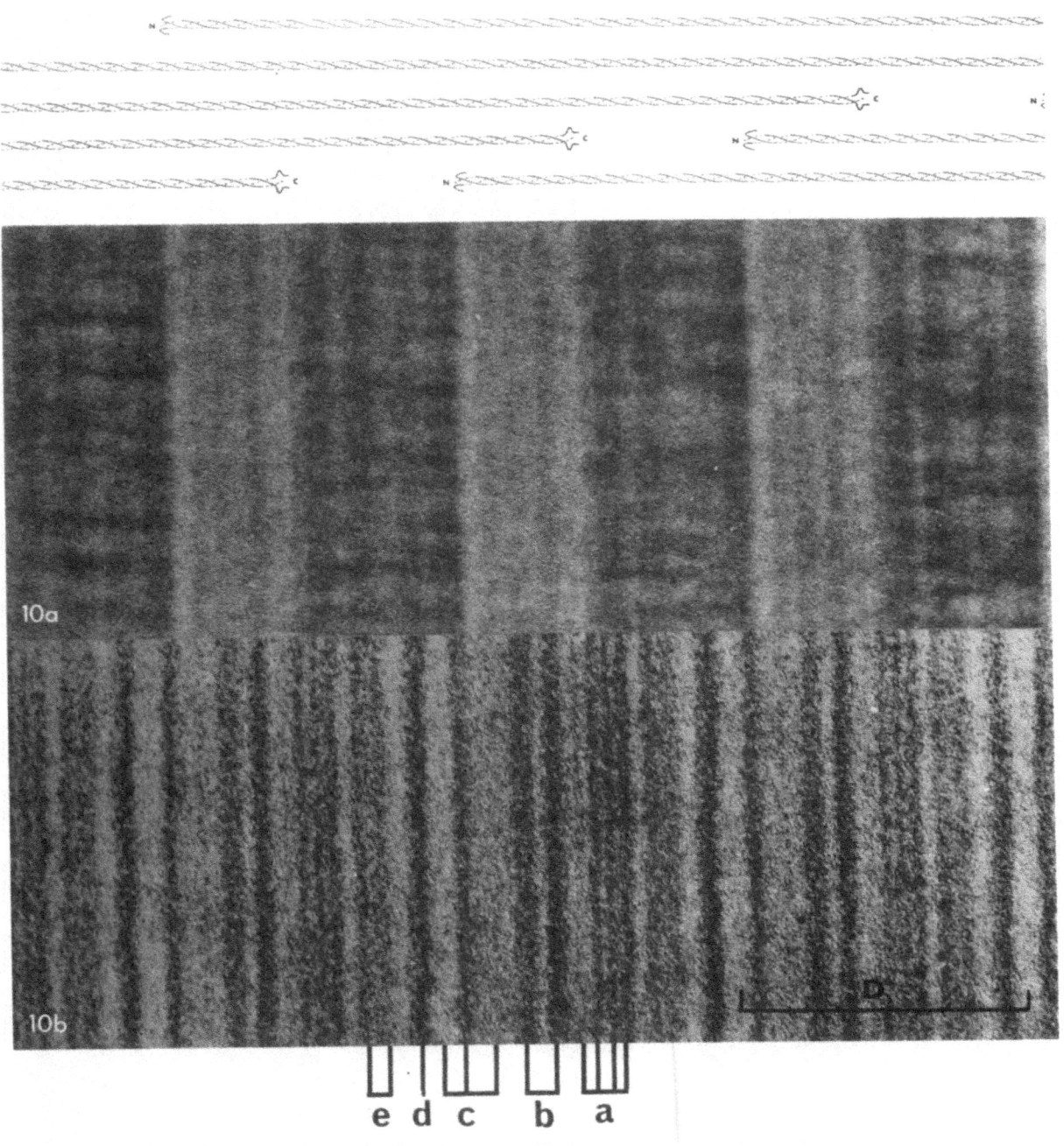

Figure 10. Matching of the negative staining pattern (Figure 10a) with the positive staining pattern (Figure 10b). The corresponding array of D-staggered molecules appears above, suitably aligned.

The characteristic band pattern exhibited by collagen stained with heavy metal salts has long been thought to be due to the uptake of the heavy metal ions on charged amino acid side groups along the collagen molecules (7,8). With current knowledge of the amino acid sequences of the polypeptide chains of collagen, this is most clearly demonstrated by a direct comparison of the band pattern from stained SLS collagen with the molecular charge distribution predicted by the sequence data (see Section 4.1 and Figure 11a, b). For type I collagen the molecular charge distribution is the summation of two α1

8

charge distributions and one α2 charge distribution. The substantial measure of agreement between the location of charged residues in the summed 2α1 + α2 charge distribution and the location of dark bands in the SLS staining pattern means that this pattern can legitimately be described as the 'molecular staining pattern' (21) It is also evident that discrete staining bands exist because the charged amino acid residues tend to occur in groups, separated by regions sparsely populated with charged residues.

3.2. Comparison of the fibril staining pattern with sequence data

The relationship between the fibril staining pattern and the SLS-derived molecular pattern, first established in 1960 in the classic studies of Hodge and Schmitt (7) and Kühn et al. (23) and extended three years later by Hodge and Petruska (9) into the 'gap-overlap model' of axial molecular packing (Figure 2), can also be viewed in the light of the chemical sequence data. We note first, from Figure 11b, c, that the bands in the aperiodic SLS pattern can, for the most part, be matched in position, although not in intensity, with bands in the periodic fibril pattern. This result shows that a considerable measure of charge–charge association must occur when collagen molecules assemble into fibrils; equally, apolar regions sparse in charge will tend to associate with other apolar regions.

An essential step in the interpretation of the fibril staining patterns in terms of sequence data is the accurate determination of the relative axial positioning of the assembled molecules. We need to know how an amino acid residue on one molecule is positioned with respect to amino acid residues on neighbouring staggered molecules. This requires a precise evaluation of D in terms of residue spacings. An accuracy rather better than one residue spacing is called for, bearing in mind that an error of one residue in D will lead to a mismatching of four residue spacings between molecules in 4D-staggered contact. The accuracy with which h, the axial residue-to-residue spacing, can be established by X-ray diffraction is only sufficient to establish the number of residue spacings in D to within 2 – 3 spacings (i.e.

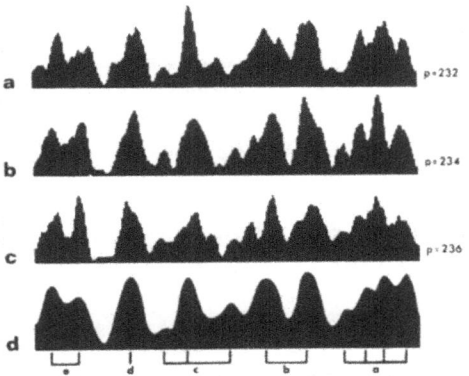

Figure 12. Predicted charge distributions in the fibril, assuming that the number of residues in a D-period is 232, 234, 236 (Figure 12a, 12b, 12c). Distributions are based on α1 and α2 sequence data and have been 'smoothed' to assist comparison with the observed fibril staining pattern. Highest correlation with the stain intensity distribution (Figure 12d) occurs when a D-period comprises 234 residue spacings.

D must be in the range 231 – 236 residue spacings)

D can be found with greater accuracy than this by comparing the fibril staining pattern with the pattern predicted by the sequence. This is done for a range of values of D, seeking that value which gives best agreement. Figure 9b shows the distribution of stain intensity in a single D-period of doubly stained collagen. This intensity distribution was obtained by microdensitometry of patterns such as that shown in Figure 10b. The axial distribution of charge in a D-period (for a chosen value of D) can be predicted from sequence data by summing five D-staggered molecular charge distributions (four in the gap zone). The result of this summation is shown in Figure 12a,b,c for D equal to 232, 234 and 236 residue spacings respectively. To assist comparison with the staining pattern, in which image resolution is unlikely to be better than 2.5 nm, the predicted charge distributions have in each case been 'smoothed' by 7 residue spacings (i.e. convolved with a smoothing function of this width). Below, in Figure 12d, appears the stain intensity distribution of Figure 9b for comparison with the smoothed charge distribution. This comparison, particularly in the region of the four closely spaced **a** bands, suggests that agreement is best when D is equal to 234 residue

Figure 11. The black lines in Figure 11a show the locations of charged residues in the central part of a type I collagen molecule, as predicted by α1 and α2 sequence data. Figure 11b is the band pattern from SLS collagen (stained with PTA and UA), matched with the charge distribution over this central region. (A comparison covering the whole molecule appears in Figure 22.) Figure 11c is the fibril staining pattern, aligned with the SLS pattern and showing how SLS and fibril bands match in position but not intensity. The numbering of the SLS bands is indicated above Figure 11a; bands 9/10, 23, 36, 49 are those contributing to the **d** line in the fibril.

10

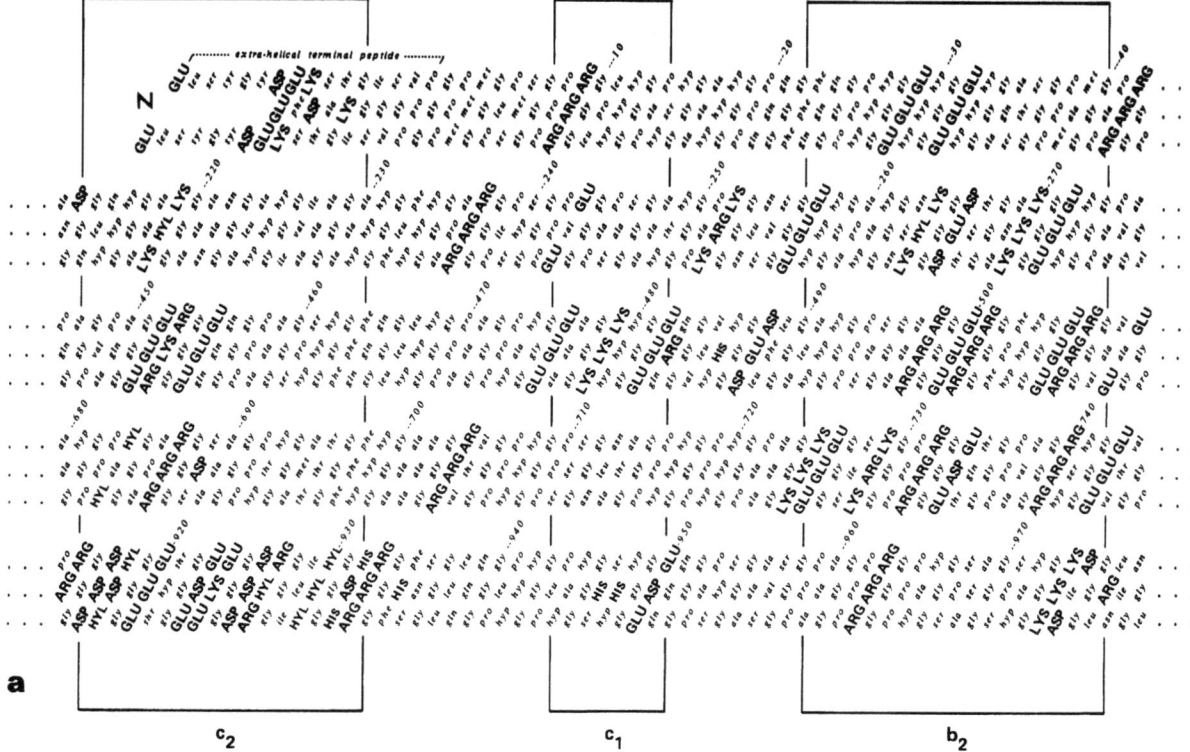

Figure 13 a, b, c, d (extending over four pages). Triple-chain molecular sequences regularly staggered by 234 residues to simulate the axial arrangement of molecules in a fibril of type I calf skin collagen. In effect, the complete display is an enlargement of the central rectangle of Figure 2, now showing axial intermolecular relationships at the amino acid level throughout a D-period. The overlap zone appears in Figures 13a and b, the gap zone in Figures 13c and d. The numbering of the residues begins with the first Gly in the triple helix and does not include residues in the extra-helical terminal peptides. Regions along the D-period with the greatest concentrations of charged residues (in bold capitals) are enclosed within rectangles. These rectangles correspond with bands in the fibril staining pattern, identified by the letters below.

spacings. In practice, the comparison was made objectively by a computer-aided correlation procedure, seeking greatest correlation between electron-optical data and sequence data (25). In addition, the stain intensity distribution used for the analysis was computer-averaged from up to 50 microdensitometric recordings to reduce 'noise' in the pattern. This correlation procedure gives an optimal value of D = 234.2 ± 0.5 residue spacings (26). Uniform spacing of the residues in the triple-helical body of the molecule is assumed; the correlation is improved however by supposing that the extrahelical terminal peptide regions are in a more contracted conformation.

An independent derivation of the number of residues in a D-period can be obtained from sequence data alone. Collagen molecules assemble into fibrils as a result of specific intermolecular attractive forces in which electrostatic interactions between charged amino acid residues of opposite sign, and hydro-

phobic interactions between the larger non-polar amino acid side-chains, play a dominant role. It has already been noted that the matching of bands in the SLS and fibril staining patterns indicates that charge–charge association and apolar–apolar association must occur on assembly. An analysis of possible electrostatic and hydrophobic interactions between two identical but staggered molecular sequences should therefore give greatest scores for the number of possible interactions when the stagger is D or integral multiples thereof. This analysis, first carried out by Hulmes et al (27) and later confirmed by others, shows that maxima in the interaction scores occur when D equals 234, 2D equals 468 residue spacings, etc., with an error not greater than one residue spacing. The result is in good agreement with that obtained from the direct comparison of the fibril staining pattern with the charge distribution predicted by the sequence information.

With the value of D established, the side-to-side

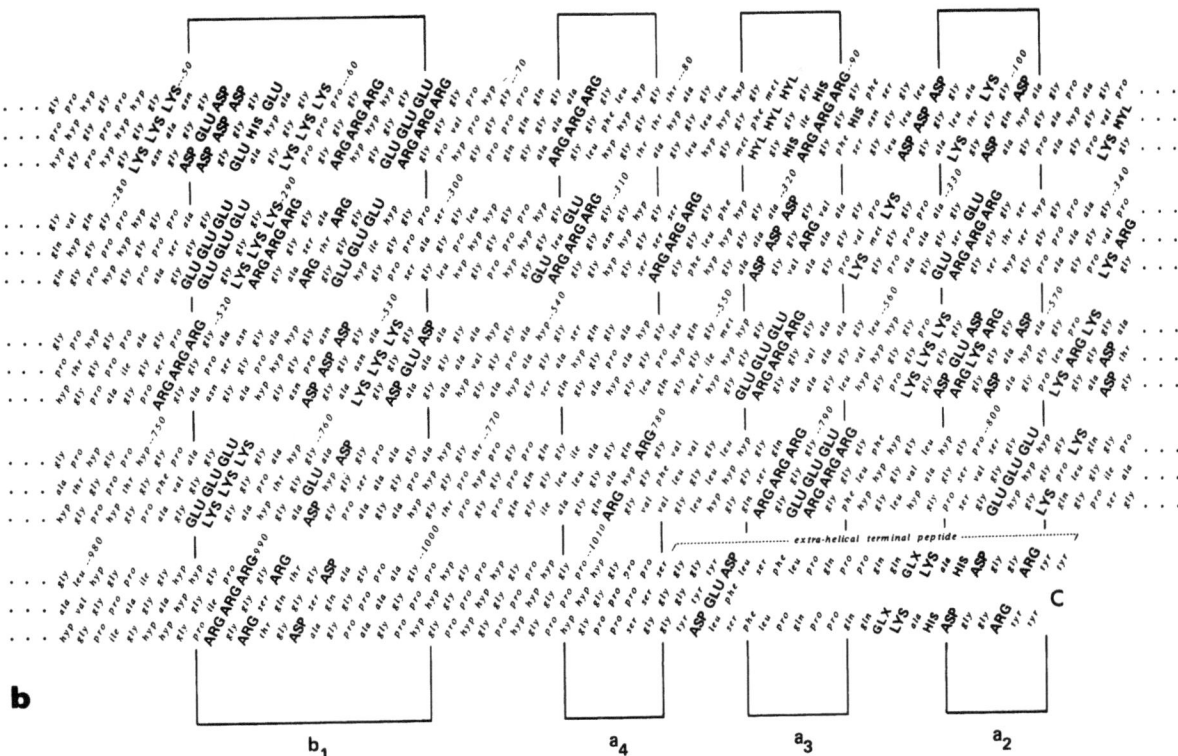

b

b₁   a₄   a₃   a₂

relationships between the molecular sequences in a collagen fibril can be displayed (28). This display is shown in Figure 13 a, b, c, d stretching across four pages. Those parts of five molecular sequences falling within a single D-period are shown. As the second molecular sequence starts where the first leaves off, the complete molecular sequence appears in the whole display. Although there is no structural requirement for an integral value, D has been taken to be 234 residue spacings exactly. The data are for type I α1 and α2 collagen, mainly (80%) from calf skin, with gaps in the calf skin data made up by using rat skin or chick skin data. The sources of the sequence data are given in a footnote in reference 28. The three chains in a molecule are mutually staggered by one residue spacing; as in Figure 1, the order α1–α2–α1 has been assumed for the three chains, and charged residues are distinguished by being printed in bold capitals. The residues are shown as uniformly spaced throughout, although this is certainly far from true in the end regions of the molecule (i.e. the extrahelical terminal peptides – see Section 3.5.). Even in the triple-helical body of the molecule , devations of a few per cent are possible.

The tendency for charged residues to associate is evident from the greater concentration of bold print in certain parts of the display. These greater concen-

trations, enclosed in rectangles, correspond closely with the observed bands in the fibril staining pattern. The bands are identified by the letters at the bottoms of the rectangles. The correspondence can be demonstrated more convincingly by direct visual comparison. Figure 14 illustrates how a number of displays can be put together to form an extended array of staggered molecular sequences representing 'Hodge-Petruska' molecular packing in a fibril. When, as in Figure 15, this array is aligned with the fibril staining pattern, the matching of the charge-rich regions in the array with the dark bands in the pattern is immediately apparent.

3.3. 'Anionic' and 'cationic' staining of collagen

This comparison of sequence data and electron optical data provides direct evidence that the band pattern in stained collagen fibrils reflects the axial distribution of charged amino acid residues along a fibril. It also shows that the staining reaction is essentially electrostatic in character. The comparison was however made using a collagen fibril doubly stained with phosphotungstic acid (PTA) and uranyl acetate (UA) where all charged residues may be supposed to bind one or other of the heavy metal ions. When

12

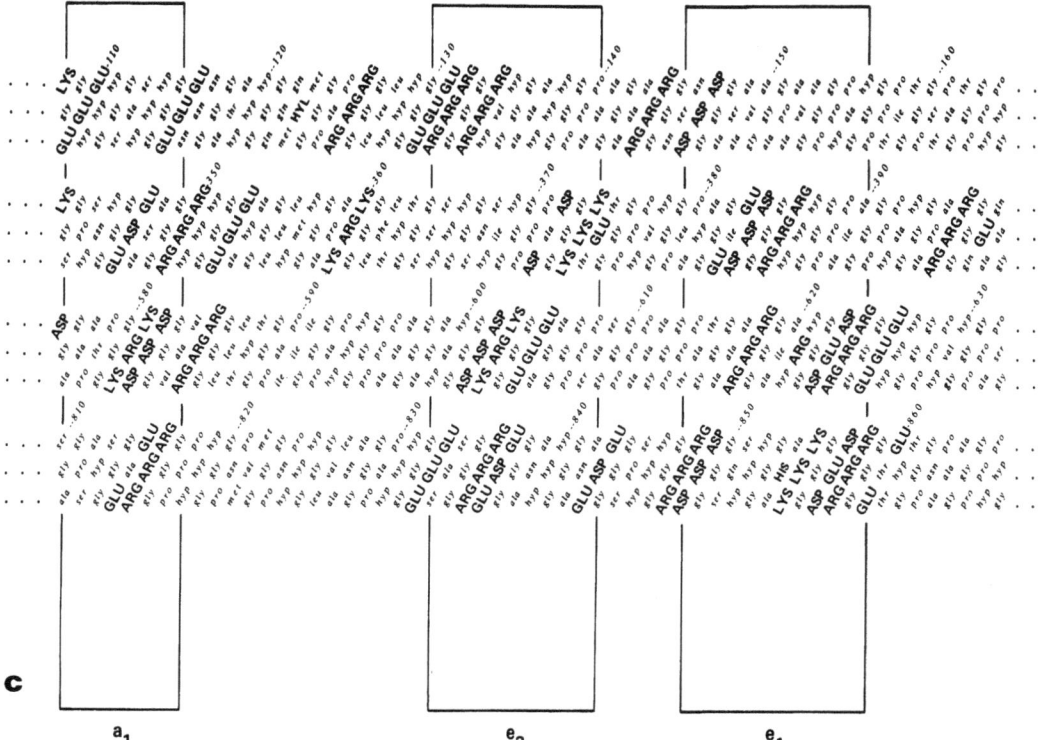

Figure 13c.

collagen is singly stained with a heavy metal salt in solution, the distribution of an 'anionic' stain such as PTA might be expected to reflect only the occurrence of positively charged residues, with the opposite staining behaviour occurring when a supposedly 'cationic' stain such as UA is used. This was indeed the view taken by early workers (7, 8); uptake of phosphotungstate ions was assumed to occur on the positively charged side-chains of arginine, lysine, hydroxylysine and perhaps the few histidines along the collagen molecule, whereas uranyl ions were taken to be cationic, reacting only with negatively charged aspartic acid and glutamic acid residues. A detailed comparison with sequence data shows, however, that this is an oversimplified view and that the band patterns from singly stained collagen cannot always be interpreted in such as straightforward way.

In the case of staining with PTA, and probably also with other commonly used stains where the heavy metal is in the anion, the distribution of stain intensity in the fibril band pattern (Figure 17) does indeed correlate closely with the predicted distribution of positive charge along the fibril (29). Little or no uptake occurs on negatively charged residues. The binding groups in the collagen are clearly the

ε-amino groups of lysine and hydroxylysine, the guanidino groups of arginine and, presumably, the imidazole groups of histidine. With its relative abundance of lysine and arginine, it is not suprising that collagen takes up PTA more readily than other proteins.

Uranyl salts exhibit a much more complex staining behaviour (30). The band pattern from a fibril stained with UA (Figure 18) does not yield the expected high correlation with the axial distribution of negatively charged residues. Rather, the correlation is better if it is supposed that binding of uranyl ions occurs on charges of both sign. This surprising result arises because uranyl acetate is a weak electrolyte in aqueous solution, and acetate ions are for the most part associated with the uranyl ions. The charged uranyl ions which are present in solution exist not as a single species (UO_2^{2+}) but as a variety of complexes, anionic as well as cationic. These participate in binding reactions with all the charged residues on the collagen molecule, and the relative uptake on negatively charged and positively charged side-chains will depend on concentration and pH. No conditions exist for which UA staining reflects only the negative charge distribution. Uranyl nitrate, which dissociates strongly in aqueous solution, be-

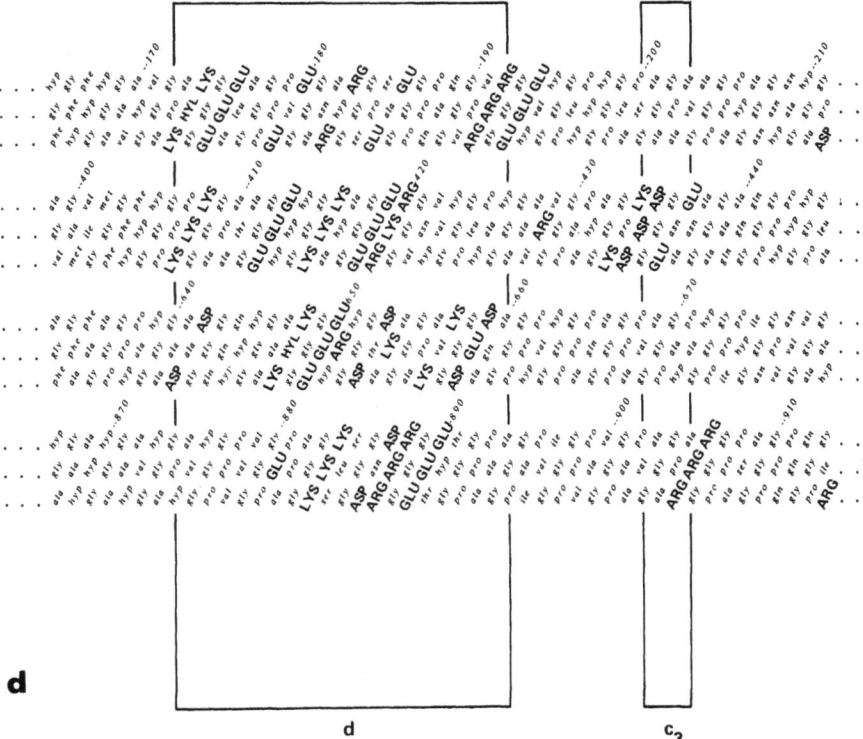

Figure 13d.

haves quite differently. Unlike the acetate, it is taken up predominantly by negative charges on the collagen; as in the case of PTA, this reaction is sensitive to the presence of phosphate and other ions.

Although, therefore, the fibril band pattern from collagen stained with PTA (and ammonium tungstate) can reasonably be ascribed to anionic staining and to reveal the location in the specimen of the positively charged residues of arginine, lysine, hydroxylysine and possibly histidine, the converse is not true for fibrils stained with UA. Here the staining is only partly cationic, and it has a strong, usually dominant, anionic component. Uranyl nitrate, which is usually regarded as a less effective stain than UA, can however bind predominantly to the negatively charged side-chains of aspartic and glutamic acids under suitable conditions. Figure 16 shows a comparison between densitometric traces from singly stained collagen fibrils and predicted charge distributions. The comparison is limited to that part of a D-period in the vicinity of the a bands. The intense a_3 band in the ammonium tungstate-stained fibril, compared with the relative weakness of the same band after uranyl nitrate staining, can be seen to be due to a marked excess of positive charge in this band.

3.4. Interpretation of the negative staining pattern

The gap-overlap alternation in stain accessibility resulting from the D-staggering of 4.4D-long molecules is the dominant feature of the collagen fibril negative staining pattern. Other smaller-scale features are, however, present and the pattern exhibits local variations in stain distribution superimposed on the broad light and dark overlap and gap zones. These finer details, which depend on the staining conditions, have yet to be as fully explained in chemical terms as the banding of the positive staining pattern. Similarities to the positive staining pattern

Figure 14. An extended array of D-staggered molecular sequences, simulating 'Hodge-Petruska' packing in a fibril, can be formed by assembling together many copies of Figure 13. Only charged residues are shown here.

Figure 15. (a) The fibril staining pattern (PTA + UA) matched with (b) the extended array of charged residues in a fibril (i.e. that predicted by the assembly of Figure 14). The comparison is best made with the page tilted.

are obvious (Figure 10 a, b), with local maxima in the negative stain intensity distribution matching in position the dark bands obtained after positive staining. Nevertheless, relative intensities differ in the two patterns, and interbands (charge-sparse regions showing no stain uptake after positive staining) vary in their behaviour towards the negative stain. This

variation is more apparent when SLS collagen, rather than native D-periodic fibrils, is negatively stained.

What are the factors that determine the small-scale deposition of negative stain along a collagen fibril or segment (or, for that matter, around any protein), when the image is observed at high resolu-

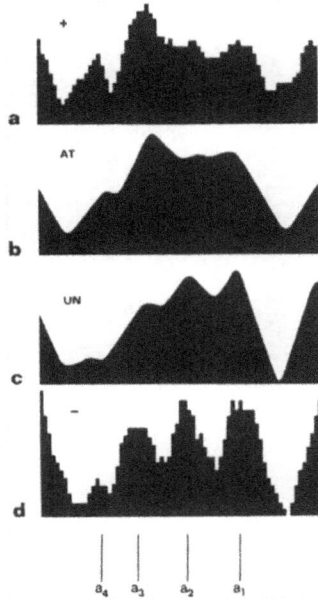

a_4 a_3 a_2 a_1

Figure 16. Ammonium tungstate is taken up by positively charged side-chains in the collagen fibril (29); uranyl acetate reacts with all charges, but uranyl nitrate can, under certain conditions, be taken up predominantly by negatively charged side-chains (30). Figures 16a and d show the predicted distributions of positive charge and negative charge along the fibril in the vinicity of the **a** bands. Figures 16b and c are averaged microdensitometric traces from AT-stained and UN-stained fibrils in the **a** band region. The charge distributions, which are smoothed to facilitate comparison, assume a 50% contraction of the C-ends of the α-chains.

tion? We use the expression 'high resolution' here in a relative sense to mean the highest resolution normally attainable with conventional transmission electron microscopic techniques on biological specimens. The resolution limit is usually set by electron-induced radiation damage in the specimen and is rarely better than about 2 nm unless exceptional precautions are taken. It has already been seen, from the observed agreement between electron-optical and chemical data, that resolutions of this order of magnitude can be reached in images of positively stained collagen fibrils. Although a similar resolution is probably present after negative staining, the image is less readily interpreted because a number of factors may contribute to the local distribution of stain and these are less well understood. It is reasonable to suppose that these factors will include (a) the volumes of the amino acid side-chains projecting from the tightly coiled body of the collagen molecule,(b) the hydrophobicity of the non-polar side-chains and (c) the ionic character of the polar side-chains. The first two will have a stain-excluding effect, the last stain-attracting. It is generally as-

sumed that negative staining involves some measure of simultaneous positive staining (9, 24), implying that the last factor makes a significant contribution. The uptake of staining ions on charged side-chains, known to occur in solution, could well result in a greater concentration of stain in these regions after drying down. Conversely, hydrophobicity could lead to a reduced stain concentration in its immediate vicinity.

Katayama and Nonomura deny that simultaneous positive staining occurs and suggest that the principal factor determining the finer details of the collagen negative staining pattern is the 'bulkiness' of side-chains, defined as the ratio of the volume of a side-chain to its length (31). This makes long side-chains less 'bulky' than shorter ones of similar volume. A recent study of SLS-collagen negatively stained with uranyl nitrate or phosphotungstic acid shows that this is largely true for the small uranyl ion (where longer side-chains can project through the film of negative stain) but may not be true for the large phosphotungstate ion (where side-chain volume and the distribution of positive charge determine the deposition of stain) (32).

3.5. *The ends of the molecules*

The short extrahelical terminal peptides, sometimes referred to as 'telopeptides', constitute less than 4 per cent of the complete collagen molecule. Yet they play a key role in fibrillogenesis and are crucial in the initial steps of the assembly process *in vitro*. Their removal by enzyme digestion drastically alters fibril initiation and growth, and has a profound effect on final fibril morphology (33–39). Chemical modification such as iodination of the tyrosine residues, present in the telopeptides but absent elsewhere, also leads to a marked alteration in fibril-forming ability (40, 41). It is also significant that the oxidisable lysyl residues, which provide the potential sites for the formation of covalent crosslinks, are located in the telopeptides; in type I collagen α1 chains these lysyl residues occur roughly in the middle of each extra-helical peptide. It is now evident that a detailed knowledge of the conformations and interaction behaviour of the telopeptides is an essential step towards a better understanding of many aspects of the fibril assembly process. As the telopeptides constitute such a small part of a collagen fibril, X-ray diffraction has hitherto been unable to furnish any high resolution structural data, and much of our existing information comes from electron micro-scopic studies in combination with sequence data,

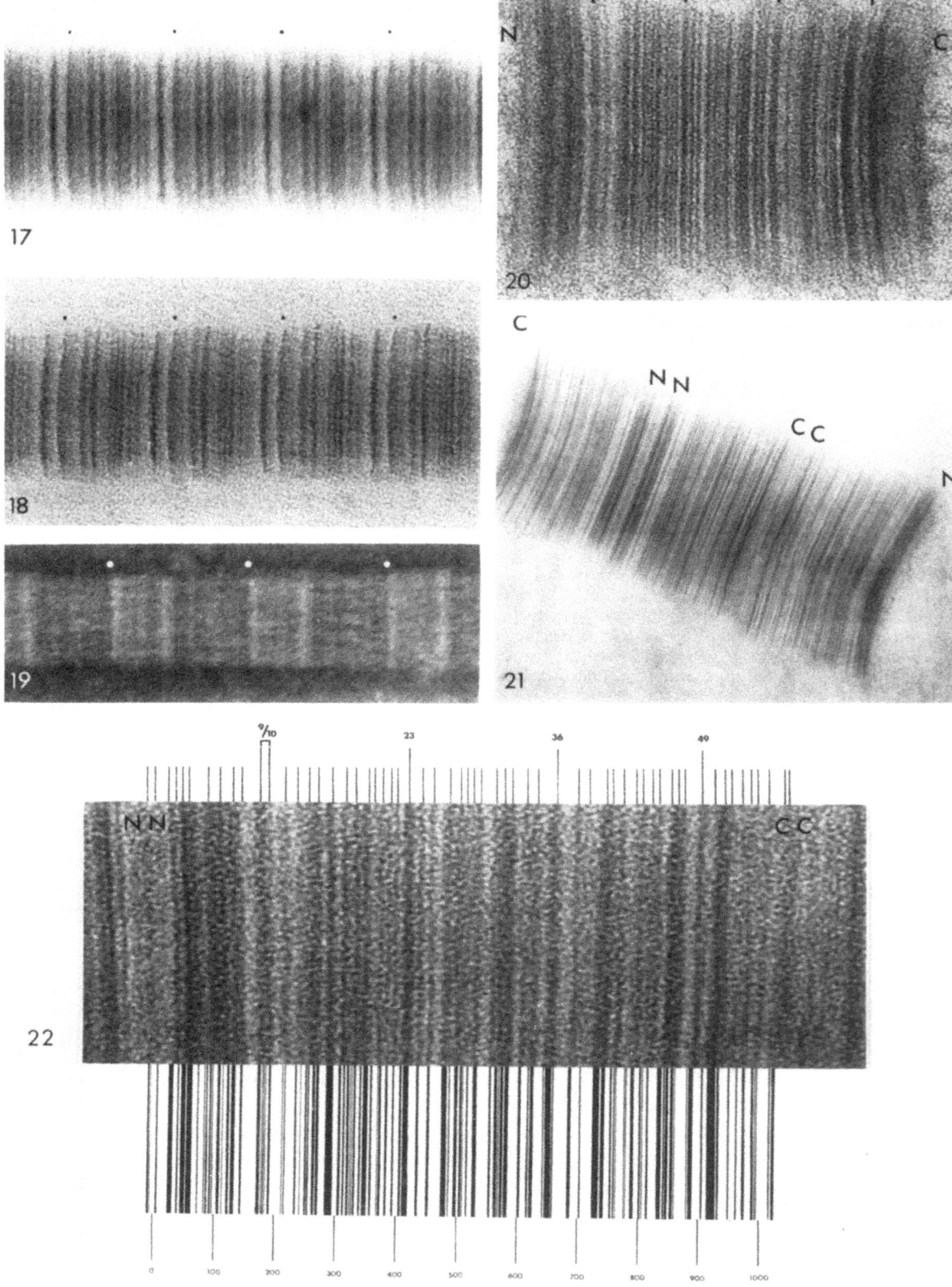

supplemented by predictions based on model building.

In the absence of any certain knowledge of their conformations, the N-terminal and C-terminal extrahelical peptides in the four-page display of Figure 13 are shown in an extended form in which the residue-to-residue spacing is, for convenience, taken to be the same as that existing in the triple-helical body of the molecule. If this were so, the N- end would terminate, as shown, well into the c_2 band in the fibril and the C-end would extend into and beyond the a_2 band. The fraction of a D-period occupied by the overlap zone would then be almost 0.51. The negative staining pattern provides clear evidence that this is not the case in fibrils prepared for electron microscopy. The overlap/D-period ratio is closer to 0.40; an abrupt change in contrast occurs at the right hand edge of the c_2 band, and the less well-defined contrast step in the a_3–a_2 region falls more to the left of the a_2 band (Figure 10a,b). Negative staining, moreover, frequently shows prominent white (stain excluding) bands at the edges of the overlap zones, pointing to a greater protein density in these regions. The effect is particularly marked in fibrils of smaller diameter (Figure 19). Electron microscopy indicates, therefore, that the two ends are in a more condensed conformation, with the extrahelical peptides contracted or folded in such a way that the mean residue-to-residue spacing is considerably less than the triple-helical spacing of 0.286 nm.

Recent electron-optical studies of SLS dimers (two SLS segments abutting at their ends) and of the effects of prior residue removal by exopeptidases have now provided additional information about the C-terminal extrahelical region (42). End-to-end dimerisation is common in SLS specimens prepared from undigested and enzyme-digested collagen: N–N and C–C dimer junctions can be seen in Figure 21. Figure 23b is a typical microdensitometric trace obtained from a C–C junction in SLS from undigested collagen, with the numbering of the staining bands following the notation introduced by Bruns and Gross (43). These bands can be identified with groups of charged residues and Figure 23a shows this identification in the sequence at the C-end of the molecule for bands 55–58; the extreme bands 57 and 58 are located in the extrahelical terminal peptides whereas bands numbered 56 and less occur in the triple helix. Combining the microdensitometric data and the sequence data now allows the axial extent of the C-terminal extrahelical region to be estimated. A plot of staining band location against sequence position yields the expected slope of 0.286 nm per residue in the triple-helical region but only half this value in the extrahelical region (42). As the distribution of stain intensity corresponding to bands 57 and 58 on either side of the C–C dimer junction precludes any significant overlap of the peptides, it may be concluded that each C-terminal α1 peptide exists in a condensed conformation and occupies a total axial distance which is roughly half that occupied by the same number of residues in a triple-helically wound chain.

Carboxypeptidase in glucose buffer can be shown to remove from the collagen all but a few residues in the C-terminal extrahelical peptides (42). When SLS segments are prepared from collagen digested in this way, the extreme staining band, number 58, is seen from Figure 23c to be eliminated entirely by the enzyme. This carboxypeptidase treatment can be used to confirm that the C-ends are in a condensed conformation in the fibril. When fibrils are prepared from carboxypeptidase-digested collagen, the effect on the staining pattern is shown in the microdensitometric traces of Figure 24. Only the a_3 band is reduced in intensity by the enzyme and the a_2 band is unaffected. Reference to Figure 13b shows that the a_2 band would have been reduced in intensity if the C-ends had been in extended conformations in the fibril.

Contraction of the extrahelical terminal peptides could be due, in part at least, to the dehydration inevitable in specimen preparation for electron microscopy. There is, however, evidence from other sources that the residues in the ends occupy a smaller axial extent than elsewhere, even before dehydration. Low-angle X-ray diffraction from moist tendon collagen fibrils yields many orders in a meridional

Figure 17. Reconstituted calf skin collagen fibril stained with ammonium tungstate (1% aqeous solution, pH 3.2).
Figure 18. Reconstituted calf skin collagen fibril stained with uranyl acetate (c = 10^{-5} M, pH 4.4). As in Figures 17 and 19, the positions of the c_2 bands are indicated by dots.
Figure 19. A rat tail tendon collagen fibril, negatively stained with phosphotungstate (1%, pH 7.2).
Figure 20. A single SLS segment from calf skin type I collagen, positively stained with PTA and UA. Bands 9/10, 23, 36, 49 (all contributing to the fibril **d** line) are indicated by short lines.
Figure 21. Polymeric SLS with N–N and C–C end-to-end junctions, similarly stained.
Figure 22. An enlargement of the central part of the previous figure, identifying the SLS bands with the (α1 + α2)-derived charge distribution over the complete molecule. The numbering of the SLS bands appears above while the residues are numbered below.

18

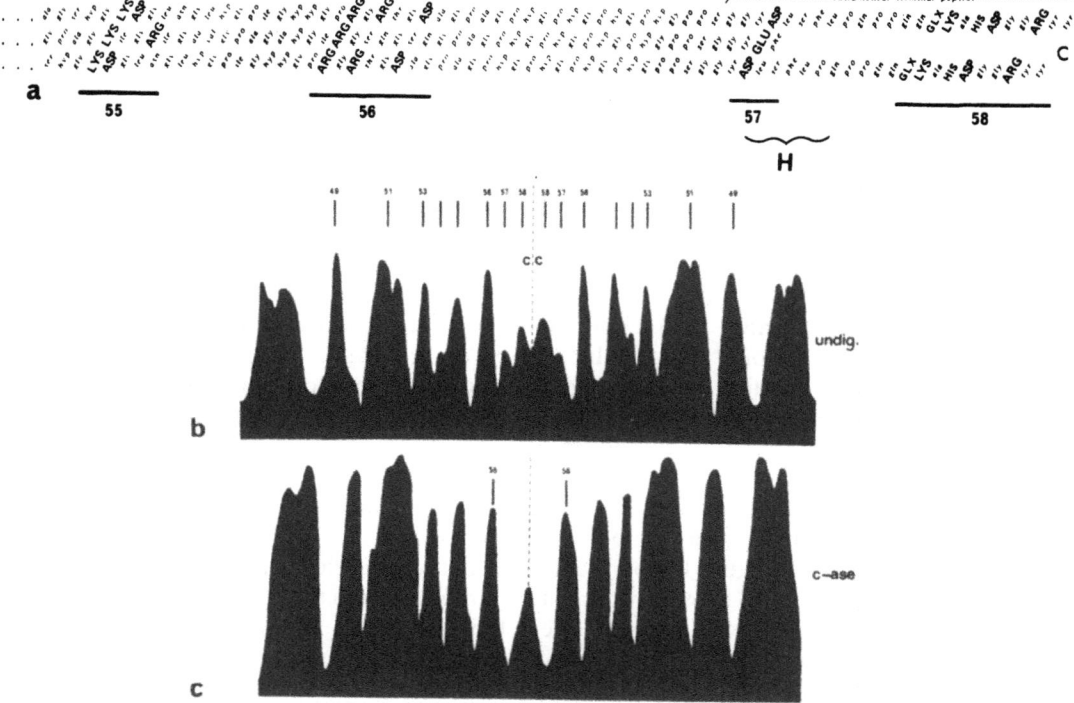

Figure 23. (a) SLS bands 57 and 58 are located in the C-terminal extrahelical peptide region. (b) These bands can be detected in a microdensitometric trace of the stain intensity distribution in a C–C dimer on either side of the dimer junction; their positions indicate that the C-terminal peptides are contracted in SLS. (c) Prior carboxypeptidase digestion removes band 58.

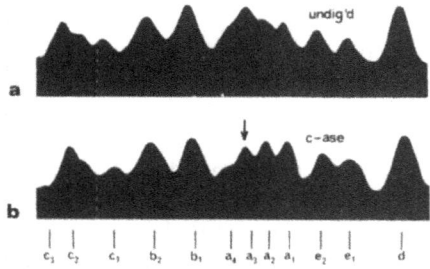

Figure 24. Fibrils reconstituted from carboxypeptidase-digested collagen show a reduced stain intensity of the a_3 band: (a) undigested; (b) digested. If the C-terminal extra-helical peptides were in an extended conformation in the fibril, as Figure 13b suggests, the loss of the charged residues at the end of these peptides would affect the a_2 band more than the a_3.

direction of reflections from the D-periodic 67-nm repeating unit. From the intensities of these reflections it has been inferred that the distribution of density within a D-period approximates to a simple step function in which the fraction occupied by the denser overlap zone is about 0.46. A more detailed picture emerges when sequence data are utilised in the interpretation of the X-ray data. Knowing the volumes of individual amino acid residues, the sequence data can be used to predict meridional intensities. Predicted intensities calculated in this way are found to be very sensitive to the conformation of the telopeptides and best agreement with intensities actually observed occurs when the telopeptides are taken to be contracted, rather than extended (44). A considerable measure of agreement between prediction and observation is obtained when the residue-to-residue translation in the N-terminal peptides is assumed to be 70 per cent of that in the triple helix and, in the C-terminal peptides, 50 per cent.

These condensed conformations are consistent with models of telopeptide secondary structure predicted from their amino acid sequences. The rules for predicting protein secondary structure from sequence data have only a limited reliability, but they do point to a hairpin-like folding of the N-telopeptide, putting the potential crosslink-forming lysyl residue at the far end of the loop (45). A feature of the C-telopeptide, near to its point of attachment to the triple helix, is a high concentration of hydrophobic residues ('H' in Figure 23a). A contraction of this part of the telopeptide chain into a 'hydrophobic cluster' is energetically favoured, optimising both intramolecular interactions and intermolecular interactions, notably those between molecules mutually staggered by 4-D periods (42). Not only does this contraction favour interactions involving residues at the far end of the telopeptide, including the positioning of the oxidisable lysyl residue for crosslink formation, but it also explains certain small-scale features of the negative staining image of collagen fibrils and SLS segments. Negatively stained SLS C-C dimers display a prominent white line on each side of the dimer junction; pepsin digestion, which removes the outer half of the telopeptide, causes the two lines to merge into one, while carboxypeptidase, which removes most of the telopeptide residues, eliminates the white region altogether. The white line can now be seen to be due to the stain-excluding behaviour of the condensed cluster of hydrophobic residues at the base of the C-telopeptide.

4. Polymorphic forms of collagen

4.1. Segments

Collagen molecules can aggregate into ordered structures in a number of quite different ways depending on environmental factors such as pH and the presence of other molecules and ions. This polymorphism has been extensively characterized by electron microscopy. The segment long spacing (SLS) form, first observed by Schmitt and co-workers when they added nucleotides to collagen preparations, was an important factor in the experimental observations which led to the explanation of the D-periodicity of fibrils in terms of a regular array of D-staggered molecules (11–13). Since then it has played a significant part in establishing the order of the cyanogen bromide peptides in the α1 and α2 chains of collagen, in studying procollagen peptides and in determining the sites of cleavage by collagenases and the extent of degradation by exopeptidases (for references see 46, 47).

SLS is a non-fibrous form of collagen in which the molecules assemble in non-staggered array with like features aligned in accurate transverse register. Although the length of a single segment can be identified with the molecular length, dehydration and distortion tend to shorten segment lengths to some extent. Each segment is polarised, with molecules all pointing in the same direction. Exceptionally, symmetrical segments in which molecules point in both directions can be formed (48). A typical polarised segment from undigested collagen is shown in Figure 20. A symmetrical segment occurs in Figure 29.

The usual method of preparation is precipitation in the presence of adenosinetriphosphoric acid (ATP) at low pH (46). The adenosine is not an essential part of the inducing agent and triphosphoric acid (TP), which has the advantage of being more stable than ATP, is equally effective (49). The features required of the inducing molecule are, apparently, a small size and several negative charges on its surface; perdisulphuric acid will induce SLS formation although not as effectively as ATP or TP. Just why such molecules should act in this way is not entirely clear. The most likely explanation is that it is a combination of two effects. At low pH, below the pK values of the carboxyl groups of the aspartic and glutamic acid residues on the collagen, the negative charges on the carboxyls will be largely suppressed and positive charges on the collagen will predominate; a small, multiply negatively charged molecule, such as ATP or TP, may then be expected to bring about charge neutralisation by bridging the positive charges on adjoining collagen molecules (8, 49). The greatest number of these electrostatic bridging interactions occur when adjoining molecules align in transverse register. The neutralisation of opposing charges of like sign (which would otherwise tend to impede zero-stagger alignment) will have a second, possibly greater effect; it will allow like regions rich in large non-polar side-chains to fall into register throughout the lengths of adjoining zero-staggered molecules. In this way, mutual alignment in precise transverse register can also be expected to favour direct hydrophobic interactions between the molecules (50).

The nature of an SLS segment means that its image in the electron microscope is, effectively, a laterally extended picture of a single collagen molecule. As described earlier (Section 3.1) the distribution of dark bands in the SLS staining pattern shows good agreement with the distribution of charged residues along the collagen molecule (Figure 11a,b and Figure 22). The 'rules' governing the uptake of 'cationic' or 'anionic' heavy metal stains on SLS are not quite the same as those which apply when fibrils are stained, as account has to be taken of the presence of the negatively charged ATP or TP in the segments and also of the low pH at which staining is carried out; indirect uptake of positively charged staining ions on positively charged residues will occur via the ATP or TP (30). Nevertheless SLS collagen can be a useful 'test specimen' for studying possible staining procedures which might allow the visualisation of specific amino acids in a protein. It has also been used to demonstrate the specific staining of sugars (51).

The joining of SLS segments end to end to form dimers or oligomers (as in Figures 21 and 22) not only establishes the position of the ends of the molecules (33, 42) but also gives information on the location of groups of charged or hydrophobic residues in the telopeptides and on the extent of enzyme digestion at the ends (see Section 3.5). Exopeptidase digestion leads to a change in the aggregation properties of collagen molecules in SLS formation; prolonged treatment with pepsin (which removes residues from both extrahelical terminal peptides) results not only in dimeric, trimeric and tetrameric SLS but also in aggregates of extreme width, so much so that they can appear as long banded tapes (33). The appearance of these tapes suggests that little distortion is present and that the rod-like triple-helical molecules are closely parallel throughout. When SLS is formed from non-digested collagen (i.e. when the extrahelical telopeptides are present) tapes are not formed and distortion is usually apparent as a non-uniform curvature of the segment, presumably because the larger volume-per-residue ratio in the telopeptides prevents close packing of the molecules.

SLS aggregates can now be prepared from all the different collagen types. The band patterns, although similar, are not identical and the variations between them reflect the sequence differences between the types.

Until recently, SLS has been generally regarded as an artifically produced form of collagen, restricted to experimental situations where its formation *in vitro* has been brought about by specific inducing agents. Only a few isolated instances of naturally occurring, characteristically banded SLS segments in tissues were known (52–54). In Section 5.2 it will be seen that this view must now be revised with the discovery of SLS-like segments in a wide range of extracellular situations where newly secreted procollagen (the higher molecular-weight precursor of collagen with peptide extensions at both ends of the molecule) is being enzymatically processed into collagen (15, 55).

4.2. Symmetrically banded fibrils

The fibrous long spacing (FLS) polymorphic form of collagen was first observed in acid extracts of rat and calf skins and of rat tail tendon, probably with polyanionic impurities present. Its precipitation from collagen solutions *in vitro* in the presence of chondroitin sulphate (or certain other glycosaminoglycans) has since been described by numerous workers.

As its name suggests, FLS is a fibrous aggregate of

collagen with a periodicity greater than the D-periodicity of normal native fibrils. Its intraperiod band pattern is strikingly different from the native-type pattern and is symmetrical rather than polarised, indicating that molecules are not unidirectionally arrayed as in native fibrils but point, in roughly equal numbers, in both directions. A variety of patterns and periodicities can be observed, often in the same preparation (Figures 25–28).

Precipitation of FLS from collagen solution *in vitro* requires a low initial pH and the presence of a suitable polyanionic inducing agent in adequate concentration; small ions need to be removed from the initially structureless precipitate by dialysis before periodic-structured FLS aggregates will form (56). As with SLS formation, the inducing molecule has to be negatively charged, but, unlike SLS, it must be a large, apparently flexible, molecule, such as chondroitin sulphate, with a multiplicity of negative charges on its surface. Polynucleotides will not induce FLS formation and only a limited range of glycosaminoglycans and plant polysaccharides will do so.

In many preparations, the most commonly occurring form, referred to as FLS I, has a periodicity of about 250 nm and a narrow (46 nm) overlap zone, with a molecule spanning two overlap zones and the intervening gap zone (Figure 25). A more recently discovered form, FLS IV, has a smaller periodicity of about 170 nm but a broader (127 nm) overlap zone (Figure 28); as before a molecule spans two overlaps and one gap but the stagger relationships are different (48). FLS II and FLS III, appearing in Figures 26 and 27, are intermediate forms in which more than one type of intermolecular stagger must be present. The stagger relationships in the four forms are summarized diagrammatically in Figure 32. Fibrils in which the stagger relationship changes from one type to another along the length of a single fibril are not uncommon. Different precipitating conditions or relative concentrations fail to favour one form of FLS over another.

The principles governing the aggregation of molecules in FLS are probably similar to those for SLS. In both cases the negatively charged inducing molecule will participate in charge-neutralising bridging reactions between positively charged residues on adjoining collagen molecules. The larger size of the polyanionic inducing molecule required for FLS formation implies that positively charged residues can be bridged over a greater range and that only the broad features of the positive charge distribution will be significant. Doyle and co-workers have suggested that the crucial factor is the distribution of *unpaired*

21

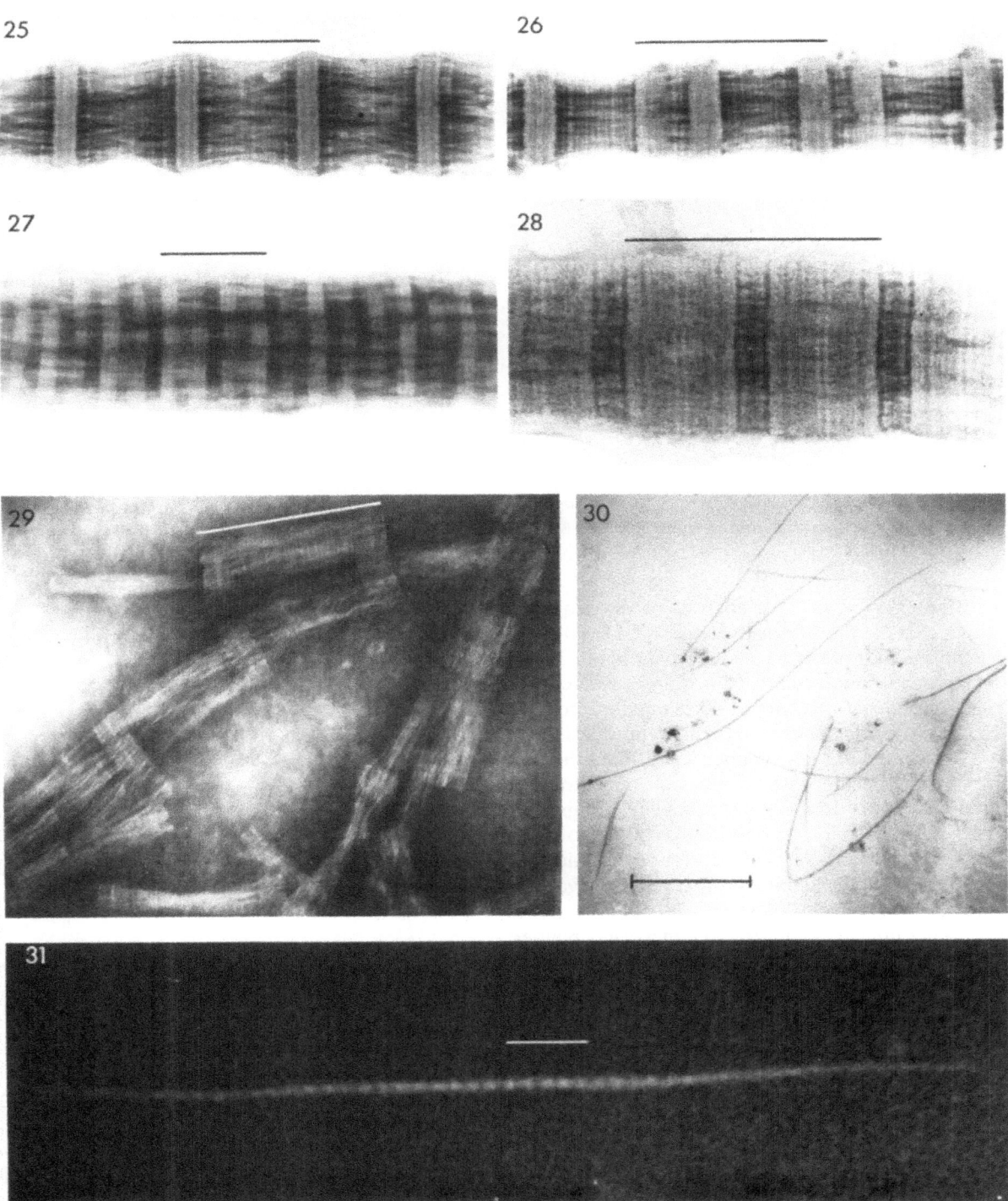

Figures 25–28. Fibrous long spacing (FLS) polymorphic forms of collagen: Figure 25, FLS I; Figure 26, FLS II; Figure 27, FLS III; Figure 28, FLS IV. In each case, the length of a molecule is shown by a horizontal line.

Figure 29. Incompletely-formed FLS fibrils (mainly FLS I) in a negatively stained preparation. The fibrils appear to form by the association of symmetrically banded segments. The line, adjacent to one of these segments, is the molecular length.

Figure 30. Early native fibrils reconstituting from solution. The solution was sampled before any rise in turbidity had occured. Fibrils are positively stained. The marker = 5 μm.

Figure 31. Dark field image of an unstained early fibril, suitable for mass measurements. The line indicates the molecular length, 300 nm.

22

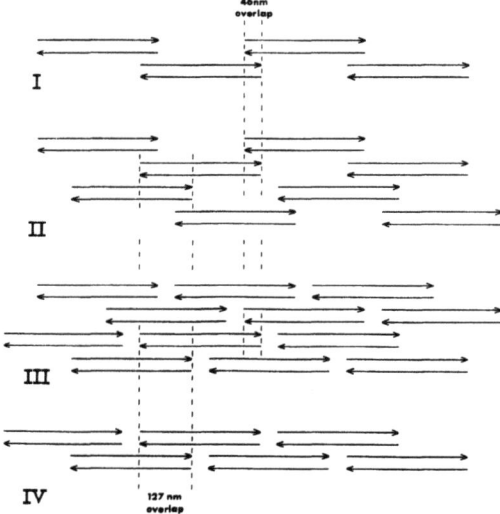

Figure 32. Axial interrelationship between molecules in the various polymorphic forms of fibrous long-spacing collagen, FLS I, II, II, and IV. Each arrow represents one molecule, length 300 nm.

positive charge (50). An analysis of the sequence shows that, of the 87 positively charged residues in the helical part of the α1 chain, all but about 35 can be paired intramolecularly (to within 3 residue spacings) with negative charges. The unpaired positive charges (over 70 per cent of them arginine residues) are distributed roughly symmetrically about the centre of the molecule, with the greatest concentrations occurring towards the two ends. The bridging of these positive-charge-rich regions by polyanions can account for the antiparallel association of molecules and the staggered alignments observed in the various polymorphic forms of FLS.

It is not unusual to observe symmetrically banded segments in preparations of incompletely formed FLS fibrils (as in Figure 29), suggesting that antiparallel in-register association of polyanion-linked molecules takes place before staggered associations.

Numerous reports have described the occurrence of long spacing fibrils *in vivo*, either as naturally occurring structures or, more usually, in pathological situations, notably in nervous tissues and related tumours where the structures, frequently tactoidal in outline, appear as roughly aligned fine filaments with conspicuous but diffuse dark bands occurring every 100–150 nm (for references see 57). These and related structures have been variously described as FLS collagen or as 'Luse bodies'. Although their occurrence in close proximity to normal D-periodic collagen fibrils is suggestive, the evidence that they are indeed collagen is far from conclusive. A recent

study of the effect of collagenase on the formation of similar long spacing fibrous aggregates in cultures of rat skin indicates that the aggregates form in the presence of elevated levels of endogenous collagenase; there is some evidence to suggest that they could be derived from degraded type III collagen from reticular fibres attacked by the enzyme (57). In this and other cases the lack of intraperiod banding precludes unambiguous identification by a direct comparison with the known banding patterns found in FLS collagens precipitated *in vitro*, and it would seem prudent to regard the nature and possible significance of these peculiar structures as unknown until a more positive identification can be achieved using, for example, labelled antibodies specific for genetically distinct collagen types.

D-periodic fibrils can, under exceptional circumstances, exhibit a symmetrical banding pattern. Factors which favour the formation of such fibrils are slow precipitation at an initially low pH (~4.5) and in a relatively high ionic strength salt solution (50, 58–60); D-periodic symmetric fibrils or tactoids will also form after pepsin treatment of the collagen (33). The banding pattern can be shown to be a superposition of two antiparallel native-type patterns with a specific stagger between them. Different staggers generate different types of D-periodic symmetric fibrils; up to now four types, designated DPS I, II, III and VI have been observed (50, 59, 60). Hydrophobic interactions appear to play a key role in the formation of these structures; the distribution of large hydrophobic residues along the collagen molecule exhibits approximate planes of mirror symmetry and this would allow antiparallel packing when electrostatic interactions are suppresssed (50).

4.3 Obliquely banded fibrils

A remarkable polymorphic form of collagen in which the banding is obliquely inclined to the fibril (or, more usually, tactoid) axis has been described by numbers of workers (see, for example, 34, 59, 61, 62). It has been observed most frequently in reconstituted cartilage (type II) collagen, precipitated from solution under conditions similar to those giving rise to D-periodic symmetric fibrils, although it has also been reported in preparations of reconstituted type I collagen fibrils (34). The oblique banding pattern is actually a step pattern, shown to be due to a regular staggering, by 9–10 nm, of normally banded, D-periodic subfibrils. The overall structure is ribbon-like (rather than helical), and the angle of inclination of the banding with respect to the tactoid axis depends on the size of the subfibrils. The diameter of the subfibrils is variable; it is commonly in the range 10–20 nm, but optical diffraction of electron micrographs reveals instances in which the D-periodic subfibrils have diameters of only 4 nm (62). It is not yet known why these obliquely banded structures form in this way. Nor have the experimental conditions required for their formation been clearly defined.

5. Fibril growth *in vitro*

5.1. *Growth of reconstituted fibrils*

The discovery that fibrils will spontaneously self-assemble from solutions of extracted collagen when the pH, temperature and ionic strength are adjusted to physiological values led to the idea that this provided an explanation of fibril formation in vivo. It was thought that the rod-like collagen molecules were secreted from the cell in monomeric form and that fibril growth proceeded extracellulary in a manner essentially identical to that which occurs when fibrils reconstitute *in vitro*. We now know that this is far from being the whole story and that the process is much more complicated than this. As we discuss in the next section, it is now apparent that cell-mediated factors and other components of the extracellular milieu must play an important role in exerting control over the process *in vivo*. Nor is it even certain that isolated monomers exist as an intermediate species outside the cell.

Nevertheless, the concept of self-assembly as a description of the underlying mechanism is still a useful one, even though other factors do operate *in vivo*. Self-assembly occurs because the molecules 'recognise' one another and form specific associations by virtue of the chemical nature of their surfaces. We have seen how the origin of the D-periodicity of fibrils can be traced to the distribution of hydrophobic and charged amino acid side chains along the molecule (27). In this respect collagen resembles other multiunit protein systems which assemble as a result of specific associations between subunits. Additional information may be supplied *in vivo*, but the basic information needed for the formation of D- periodic fibrils clearly resides in the collagen amino acid sequence and is an intrinsic property of collagen *per se*. For this reason alone, the assembly process warrants study in isolation from the cell.

Assembly *in vitro* of type I collagen fibrils has been studied for many years using, as starting material, soluble collagen extracted by various weak acid or neutral salt solutions from rat tail tendon, rat skin (normal and lathyritic), calf skin and many other connective tissues (see, for example, 17). Soluble collagen prepared in any one of these ways (other than from lathyritic animals) almost invariably displays some degree of covalent crosslinking, inter- as well as intramolecular, and will therefore be heterogeneous in composition and only partly monomeric. Nor can it be assumed that the terminal extrahelical peptides are intact unless precautions have been taken to avoid attack by tissue proteases. Between investigators therefore, the starting material for reconstitution studies may vary considerably.

Raising the temperature of the solution (commonly to 20–35°C, remembering that denaturation of the collagen monomer occurs close to 37°C) and increasing the pH (to neutrality) will initiate the reconstitution of fibrils from the solution. Electron microscopy shows that a range of precipitating conditions will yield D-periodic fibrils as the end product (17). Optimal conditions that will give fibrils similar to those observed *in vivo* have now been established (60). If the conditions do not deviate too widely from these, the banding pattern appears identical to that in native fibrils formed *in vivo*. Under more extreme conditions, polymorphic or non-banded aggregates form. Reconstituted fibrils tend to be noticeably less compact than those formed *in vivo* and to flatten on a supporting film when prepared for electron microscopy. Individual fibrils are usually fairly uniform in diameter; the distribution of the diameters can be moderately sharp or very broad, depending on the reconstitution procedure (17). In transverse cross section, fibril outlines often deviate considerably from circularity.

It is the intermediate stages in the reconstitution process that have commanded greatest attention. As fibrils precipitate, the solution becomes increasingly turbid, and the process can be followed by measurements of this turbidity, plotting the optical density of the solution against time. These plots reveal the existence of a distinct lag period, in which no increase in optical density occurs, followed by sigmoidal growth to a final plateau (17, 63). The rate of the fibril formation (taken as the reciprocal of the time, $t_{1/2}$, to reach half-maximal turbidity) is proportional to the concentration of collagen in the solution, suggesting that turbidity can be regarded as a measure of growth by accretion (60). The lag period has been interpreted in various ways. A long-held view is that it represents a nucleation phase, similar to that which occurs in the precipitation of inorganic compounds from solution (36, 63–65). The demonstration that a collagen solution, heated and cooled during the lag period, can exhibit thermal memory (i.e. it retains a history of temperature changes) has been cited as evidence for the presence of nuclei (17, 37). Others have now interpreted the data in terms of a more complex multistep model, possibly involving conformational changes in the extrahelical end regions of the collagen molecule (38, 39, 66), or stabilisation of early aggregates by the formation of cross-linking aldehyde adducts (67). Studies of the earliest aggregates by inelastic light scattering favour initial steps that lead to 'linear' rather than 'lateral' growth, giving rise to thin filaments containing no more than a few molecules in cross-section (4, 68, 69). Increase in turbidity does not occur until lateral association of these filaments takes place.

The nature of the assembly during the turbidity lag period and immediately afterwards would seem to present a problem amenable to electron microscopy. In practice, however, the difficulties are great. Long thin filaments have been observed by numbers of workers, but it is impossible to be sure that some further aggregation has not occurred during the final stage of drying down on a supporting film. Specimen preparation and sampling techniques can have a profound effect on the appearance of collagen aggregates. What is more, quite minor differences in the experimental procedures used to initiate fibril assembly in vitro have a dramatic effect on the subsequent assembly process. Some workers start with a cold (0°–4° C) collagen solution adjusted to neutral pH (~pH 7), where it remains for some time, and then initiate precipitation by raising the temperature (38–41, 60, 66, 70). Others start with an acidic (~pH 4) collagen solution equilibrated at the required temperature (20°–35° C) and initiate fibril formation by adding a prewarmed buffer solution to adjust the system to pH 7 (17, 42, 71, 72). The two procedures yield similar fibrils in the final gel, but fibril formation proceeds via dissimilar pathways. Not only are the turbidimetric curves different (63), but the aggregates which form during the intermediate stages show marked differences in morphology (73). In the second procedure (initation by pH neutralisation after prewarming) early fibrils showing D-periodic banding can be detected at low concentration during the lag period (72) at a time when little more than long thin filaments, 2–4 nm in diameter, are present after the first procedure (thermal initiation after pH neutralisation) (66). Figure 30 shows these early fibrils (i.e. fibrils with both ends visible) at the end of the lag period, immediately before the rise in turbidity. Subsequent growth in the two cases occurs via different intermediate aggregates (70, 73),

and it is very clear that generalisations about the nature of the fibril assembly process cannot be made from observations which rely wholly on one set of precipitation conditions.

Measurement of the diameters of collagen fibrils reconstituted in vitro shows that the final distribution of diameters is established long before all the collagen has precipitated. In one investigation (using the 'initation by pH neutralization' method) the final diameter distribution was reached when only one quarter had precipitated, as judged by the optical density of the solution (71). By this stage fibrils ends are rare and new fibrils are not being formed so it is evident that the remaining three quarters of the collagen still in solution must accrete preferentially on to the growing ends of fibrils rather than on to the central maximum-diameter region. A reduction in the accretion of new material on to the surface of the fibril as its diameter increases implies the existence of some kind of intrinsic self-limiting growth mechanism, even in vitro.

This has now been confirmed by mass measurements in the electron microscope. Distortion during drying makes direct estimates of fibril diameters inaccurate and a more quantitative approach is provided by dark field electron microscopy, used to measure mass distributions in unstained specimens (Figure 31). Cross-sectional masses of fibrils (per unit length) can be determined with an accuracy of a few molecular masses (per unit length). The method not only confirms that accretion on to the fibril surface decreases as the number of molecules in the fibril cross-section increases but also shows that the onset of this self-limiting growth process can occur at quite an early stage, when the number of molecules in a fibril cross-section is less than 100 (72). The two ends of an early fibril (reconstituted by the 'pH neutralization' method) do not exhibit identical mass/length distributions; the rate of change of mass with axial position at the more sharply pointed N-end is roughly half that at the blunter C-end. It is quite possible that the two ends are growing at different rates. There is some evidence of this from electron microscope autoradiographic studies of fibrils reconstituted (following thermal initiation) from solutions initially containing ^{125}I-labelled collagen; the location of the radioactive label shows that the fibrils lengthen more rapidly at their N-ends (41).

5.2. Growth in tissue culture

The precise role of the connective tissue cell in controlling the formation of extracellular collagen fibrils has been debated for many years. It was recognised at an early stage that some means must exist to prevent premature fibril formation before the collagen molecules had reached their final destination outside the cell. Biosynthesis of collagen, it was postulated, might not be completed until after secretion, intracellular collagen molecules could be associated with an inhibitor of fibrillogenesis released into the extracellular milieu, or collagen molecules might be secreted in precursor form and then enzymatically processed to 'fibril-assembly-competent' molecules.

Early electron microscopic studies of fibrillogenesis in tissue culture used tissue explants or isolated cells cultured on formvar-coated glass coverslips; after incubation, specimens were fixed, dried and heavy-metal shadowed for electron microscopy to reveal the formation of characteristically banded D-periodic collagen fibrils. In this way extracellular fibril formation was shown to occur in close proximity to the cell surface. Using simultaneous measurement of hydroxyproline production, Fitton Jackson and Smith noted a distinct time lag between secretion of a hydroxyproline-rich material and fibril formation and suggested that the hydroxyproline-rich material was a precursor which subsequently became transformed into collagen fibril (74).

Later studies used fixation,embedding and ultrathin sectioning techniques. Combining as before, measurements of hydroxyproline production with electron microscopy, Goldberg and Green detected long non-periodic filaments, roughly 8.5 nm in diameter, coinciding with the initial onset of hydroxyproline production (75). D-periodic fibrils, roughly 40 nm in diameter, formed later. Although collagen synthesis continued to take place in these cultures (shown by continued hydroxyproline production and an increase in the number of D-periodic fibrils), no significant increase in fibril diameters above this maximum value of about 40 nm occurred. Here then was clear evidence of the existence of a mechanism limiting fibril diameters in the presence of a continued supply of collagen, similar to that described earlier in the growth of reconstituted fibrils (71).

We now know that the formation of collagen fibrils by cells (*in vivo* or in tissue culture *in vitro*) is much more complex than was originally suspected and that the protein undergoes extensive processing after synthesis. A host of post-translation modifications, intracellular and extracellular, takes place, and the initially synthesised collagen precursor, procollagen, possesses extension peptides at both the N- and C-ends (for review see 76). The extracellular post-translational modifications include removal of these end-peptides by specific procollagen proteases to allow fibril formation to proceed.

The higher molecular weight precursor of collagen was discovered simultaneously in 1971 in tissue culture and in the hyperextensible skin of cattle suffering from the hereditary defect, dermatosparaxis (see 76). Early concepts of a precursor with only an N-terminal extension soon gave way to the realisation that propeptide extensions occurred at both ends, giving a total molecular weight for procollagen of approximately 450,000. In normal tissues, the two propeptides are eventually removed by specific N- and C-terminal procollagen proteases, but in dermatosparactic tissues the N-terminal protease is defective. D-periodic fibrils can still form, but they are thin with a loose, twisted appearance in longitudinal section and a bizarre spoke-like outline in transverse section (77, 78). Immunoelectronmicroscopic staining with ferritin-labelled antibodies demonstrates a D-periodic distribution of N-propeptide in dermatosparactic fibrils (78). Even in normal skin, the thinner collagen fibrils can show a D-periodic uptake of ferritin-labelled antibody directed against the N-terminal extension propeptides (79). This uptake is abolished if the antibodies are blocked by isolated N-propeptide or if the tissues are pretreated with N-terminal procollagen protease, so there is little doubt that N-propeptide extensions are not removed as rapidly as was originally thought and can persist on the surface of newly formed fibrils.

Further lateral growth may require the removal of these N-terminal propeptides from the fibril surface. If this is the case, the N-terminal protease could have a regulatory role in fibril growth (79). Lateral accretion would cease if the enzyme were not released or if its activity were inhibited. It is significant that reconstitution of fibrils from pN-collagen (i.e. procollagen without the C-terminal extension) gives smaller diameter fibrils (as determined by scanning electron microscopy) than those reconstituted from equal concentrations of collagen solutions under identical buffer conditions (80). Conversely, addition of purified N- and C-terminal procollagen proteases to procollagen at physiological pH, ionic strength and temperature leads to the formation of unusually wide (up to 1 μm) D-periodic tactoids (16). These observations support the idea that fibril diameters *in vivo* could be regulated by the relative rates of processing of the N- and C-terminal propeptides.

The question of the form in which procollagen is secreted from the cell has recently been the subject of some controversy. Although it is generally agreed that the presence of both the N- and C-terminal

26

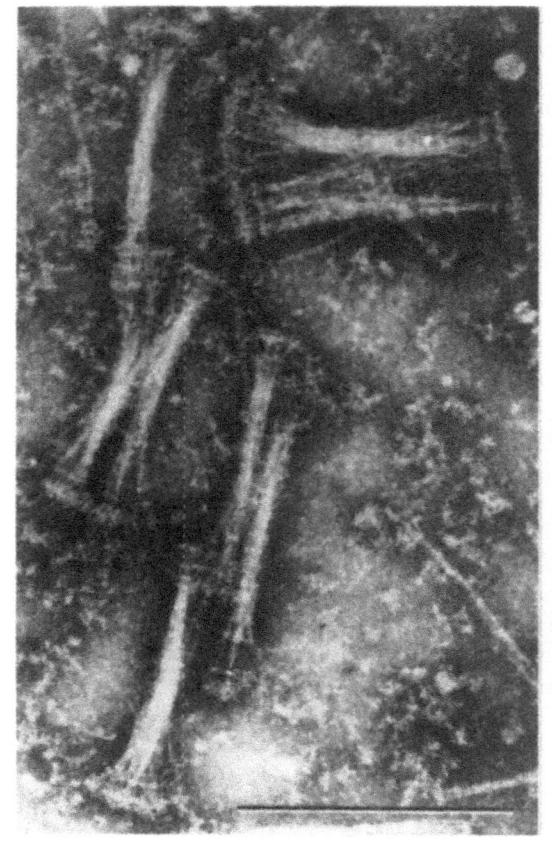

33

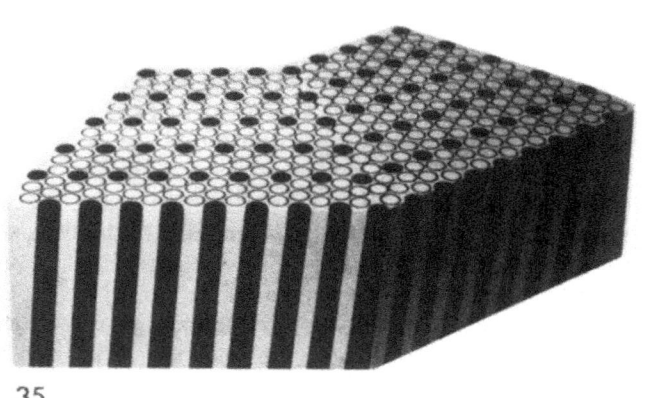

35

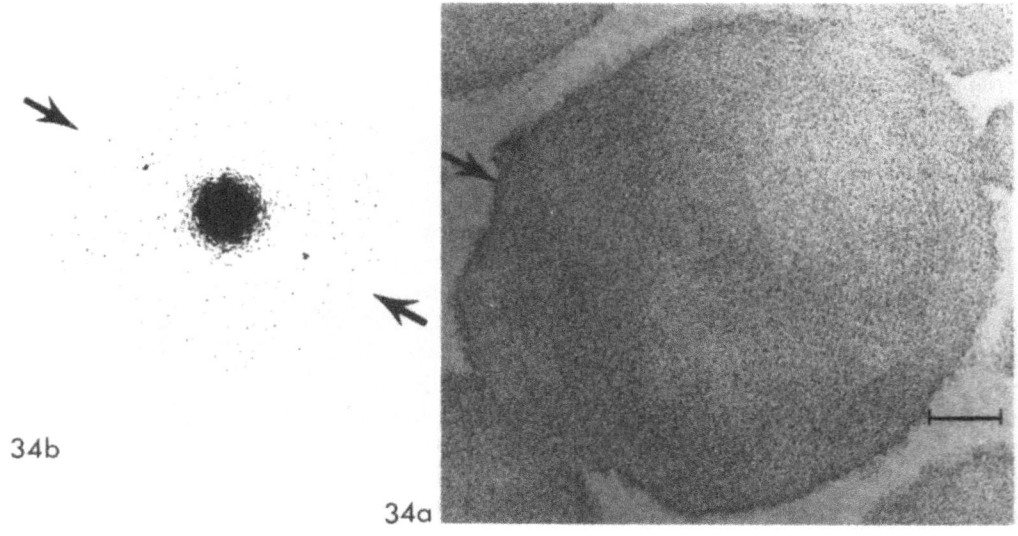

34b

34a

precursor peptides inhibits the aggregation of procollagen into D-periodic fibrils, numerous observations of intracellular secretory vacuoles containing naturally occurring SLS-like aggregates, often loosely filamentous in structure, suggest that procollagen molecules do assemble in non-staggered in-register array prior to export (15, 55, 81, 82). The procollagenous nature of these secretory SLS packages has recently been verified by immunoelectron-microscopy (82). To determine the fate of procollagen in the extracellular environment, Bruns and co-workers examined fibroblast culture media from a wide range of tissues by negative staining and electron microscopy (15). Remarkably, all the media contained large numbers of SLS-like aggregates with globular propeptide terminal extensions, strikingly similar to the intracellular SLS aggregates (Figure 33). These observations, together with those of similar aggregates in homogenates of lathyritic tissues, suggest that procollagen SLS secretory packages are secreted intact into the extracellular compartment. Subsequent biochemical analysis has shown that the aggregates are intact even after complete removal of the N-propeptide and extensive cleavage of the C-propeptide (83). Thus the character of assembly appears to be determined by the presence of the propeptides. Procollagen with its associated intermediate forms may well play an active role in directing fibril formation.

6. Lateral packing of molecules in fibrils

6.1. Introduction

Much of the data on which current concepts of collagen fibril structure are based come from X-ray diffraction. Since periodic structural features give rise to sharp X-ray diffraction spots, whilst less ordered features give a more diffuse intensity distribution, workers in this field have tended to concentrate attention on those aspects of the data that indicate three-dimensional crystallinity. The first evidence

for such crystallinity in collagen fibrils was obtained by North and co-workers (84) who observed a series of sharp spots, so-called Bragg reflections, in the equatorial region of the X-ray diffraction pattern from moist, stretched, rat tail tendon. (It is, of course, this region, at right-angles to the fibre axis, that gives information about lateral packing arrangements). The spots indicate that, when the structure is seen in projection down the fibril axis, structural units are arranged in a crystalline array in at least part of the fibril. This array can be specified by a unit cell, sides a, b and angle γ and the points in the array can be connected to give a set of Bragg planes, each with a particular separation d (Figure 36). Since, to a good approximation, the perpendicular distance of each equatorial (or near-equatorial) reflection from the meridian (i.e. the central line parallel to the fibre axis) is inversely proportional to one particular Bragg plane separation, the parameters a, b and γ may be determined by optimising the agreement between the observed and calculated Bragg plane spacings. In this way North and co-workers deduced the unit cell a = 6.2 nm, b = 7.6 nm, γ = 125°. Improved X-ray diffraction patterns obtained by Miller and Wray (85) were later indexed on a tetragonal unit cell with base a = b = 7.6 nm, γ = 90°, although

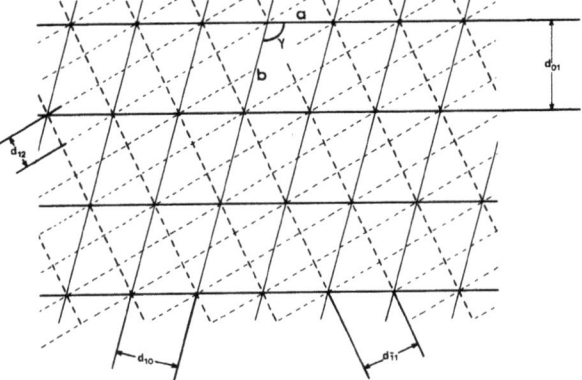

Figure 36. Bragg planes, spaced by d_{10}, d_{01}, etc, connecting points in a crystalline array.

Figure 33. Culture medium of chick embryo tendon cells 24 hr after isolation showing procollagen SLS aggregates (15). The line is the length of a collagen molecule without propeptide extensions, ~300 nm.

Figure 34. (a) Ultrathin transverse section of formaldehyde-fixed, PTA/UA-stained collagen fibril from rat tail tendon clamped under tension during embedding, showing concentrically oriented 4.5 nm periodicity (104). (b) The 4.5-nm periodicity is confirmed by optical diffractometry from the region of the fibril section near the edge indicated by the arrow. The marker = 50 nm.

Figure 35. Perspective view of two adjacent crystalline domains in a segment of a tentative model for the three-dimensional packing of molecules in a rat tail tendon collagen fibril (104). Open circles portray sections through individual molecules, and closed circles represent molecular ends; i.e. molecules indicated by open circles are staggered by nD (n = 1, 2, 3 or 4) with respect to the molecules indicated by closed circles. The two nearer surfaces are the exterior surfaces of the two domains. The ~4-nm spacing corresponds to the distance between rows of the molecular ends and these rows are oriented parallel to the nearby exterior surfaces. Molecules are tilted by about 5° in planes which make an angle of 30° with the nearby exterior surface.

most of the reflections could be accounted for by the smaller unit cell a = b = 3.8 nm, γ = 90° (86). More recently, even closer agreement between predicted and observed spacings, with fewer unobserved predicted reflections, has been obtained with the unit cell a = 2.7 nm, b = 3.9 nm, γ = 105° (87).

It is important to note that the positions of the X-ray reflections do not yield any information about the contents of the unit cell. This information is contained in the intensities of the reflections; because of the phase problem, however, it is not possible to deduce the unit cell contents directly from the diffraction pattern. In principle the contents could be determined by isomorphous replacement but this has not yet been achieved. Alternatively the validity of a proposed model may be tested by calculating the predicted intensity distribution and comparing this with the experimental data. Of all the models yet proposed, only quasihexagonal molecular packing (87) has been shown to give satisfactory agreement with both the positions and intensities of the observed Bragg reflections.

Before the Hodge-Petruska scheme was established, lateral packing models failed to take into account the relative axial positions of collagen molecules in the fibril. Any three-dimensional model must, of course, account for the axial D-periodicity of the fibril; in any D-repeating unit, therefore, there must be equal numbers of molecules in each of the five staggered positions (see Figure 2 and reference 62). The implications of the Hodge-Petruska scheme were first considered by Smith who pointed out that, for hexagonal packing, not more than two-thirds of the intermolecular contacts could involve 1D staggers; only in a molecular monolayer (e.g. the usual representation of the Hodge-Petruska scheme) can all intermolecular contacts be 1D-staggered and this led Smith to propose the five-stranded microfibril (88) in which a molecular monolayer five molecules wide is wrapped into a cylinder of indefinite length as the basic structural unit of the collagen fibril. Neighbouring microfibrils would then be staggered by any integral multiple of D and, since each microfibril is itself D-periodic, the overall D-periodicity of the fibril is maintained. The diameter of this proposed microfibril is approximately 4 nm. A model in which cross-linked tetramers of 1D-staggered molecules join to form a microfibril topologically identical to a Smith microfibril was independently proposed by Veis and co-workers (89).

Noting the close correspondence between the sizes of the five-stranded microfibril and the X-ray-derived tetragonal unit cell, Miller and Parry proposed a microfibril model based on the X-ray diffrac-

tion data (86). Furthermore, the observed propeller-shaped distribution (fanning) of the near-equatorial X-ray intensity indicated a tilting of the molecules with respect to the fibrils axis. This gave rise to the further suggestion that the microfibril was a rope-like supercoil. It should be noted, however, that agreement between predicted and observed X-ray diffraction intensities for this model has never been demonstrated.

The difficulties which arose when attempts were made to use the five-stranded microfibril model to explain the observed intensity distribution and to account for the measured density of fibrils led several workers to suggest alternative models (for review see reference 87), all based on the tetragonal unit cell (~3.8 nm. af093.8 nm or multiples thereof). These alternatives included two- and eight-stranded microfibrils as well as molecular crystals in which there were no microfibrillar subassemblies between molecule and fibril. Increasing levels of complexity, such as kinked molecules, paracrystalline distortion and different structures in the gap and overlap zones were introduced to account for the positions and intensities of the observed reflections, in particular the characteristic distribution of equatorial and near-equatorial reflections corresponding to lateral spacings of approximately 1.3 nm. There is a danger in increasing the complexity in this way. The increased number of parameters required to account for the data tends to undermine any supposed demonstration of agreement between predicted and observed diffraction patterns. Furthermore, since all these models were based on the ~3.8 nm 3.8 nm tetragonal unit cell, there were either five molecular cross-sections per unit cell and the density was too low, or ten molecular cross-sections per unit cell and the density was too high.

The quasihexagonal model for the three-dimensional crystalline packing of molecules in a fibril, proposed in 1979 (87), not only provided a satisfactory explanation for the positions and intensities of the observed Bragg reflections but was consistent with the observed density. This model marked a radical departure from others in that a new indexing scheme was proposed, based on a unit cell (a = 2.7 nm, b = 3.9 nm, γ = 105°) which gave closer agreement with observed spacings and fewer unobserved predicted reflections than earlier tetragonal cells. The three principal intermolecular Bragg plane spacings are unequal, each corresponding to one of the three most intense near-equatorial spacings in the X-ray diffraction pattern (i.e. 1.26 nm, 1.33 nm and 1.37 nm). A simple straight tilting of molecules accounts for the observed distribution of equatorial and off-equatorial intensity. Agreement between predicted and observed X-ray diffraction patterns has now been demonstrated (90).

The two distinguishing features of the quasihexagonal model are the unit cell and the extent and orientation of the straight molecular tilt. In fact a large number of quasihexagonal packing schemes fit these criteria. Only two of them, however, are generated by uniform 1D- or 2D-staggers in the three principal directions. An example of a non-uniform stagger pattern is the so-called 'compressed microfibril' model (91). Analysing the positions of Bragg reflections on the various 'row lines' parallel to the meridian provides a means for distinguishing between models. Recently, sufficiently detailed X-ray patterns have been obtained to permit such an analysis (92). The results show a strong preference for the axial packing arrangement of Figure 37, with uni-

forms staggers of 1D, 1D and 2D in the three principal directions. This model, in which molecules in 4D-staggered contact occur in extended (puckered) sheets across the structure , is also more consistent with cross-linking data (93, 94). Models with non-uniform stagger patterns can be eliminated as they predict up to five times the number of row line reflections, and these have not been observed. A recent modification (95) of the 'compressed microfibril' model removes this difficulty; the model does, however, require the gap zone to be disordered.

Collagen fibrils show crystalline order to varying degrees. The characteristic 3.8 nm row line has been detected in a number of tissues composed of type I collagen, particularly after formaldehyde fixation and staining in phosphotungstic acid (96). It has also been observed in reconstituted fibrils (97). Even in the most crystalline specimens, however, there is always an underlying diffuse equatorial intensity with a broad maximum corresponding to a spacing of 1.1 to 1.5 nm. Some specimens do not exhibit three-dimensional crystallinity at all (11) but only the diffuse equatorial scattering. The profile of the diffuse equatorial intensity indicates a liquid-like disorder in the lateral molecular packing (98). This combination of axial order (i.e. the D-periodicity) and lateral disorder is analogous to the structure of smectic A liquid crystals (99), an analogy which is particularly appropriate as a transition from laterally disordered packing to crystalline hexagonal packing of tilted molecules is characteristic of a smectic A→B transition (100).

6.2. Fibrils in cross-section

Some traces of the crystalline order in collagen fibrils ought to be detectable by electron microscopy. With suitable preparative methods, lateral periodicities of ~4 nm or microfibrils with diameters of this magnitude should be recognisable in the image.

When native or reconstituted collagen fibrils are examined by electron microscopy after negative staining, especially with phosphotungstic acid at neutral pH, a filamentous substructure is revealed. The filaments appear in both gap and overlap zones, though more clearly in the less tightly packed gap zone, and run approximately parallel to the fibril axis. Estimates of filament width range from 1.5 to 4 nm, and the filaments have been variously interpreted as single molecules or as microfibrils. It has to be accepted that dimensions of this magnitude cannot be accurately determined by negative staining. The random punctate appearance of collagen fibrils

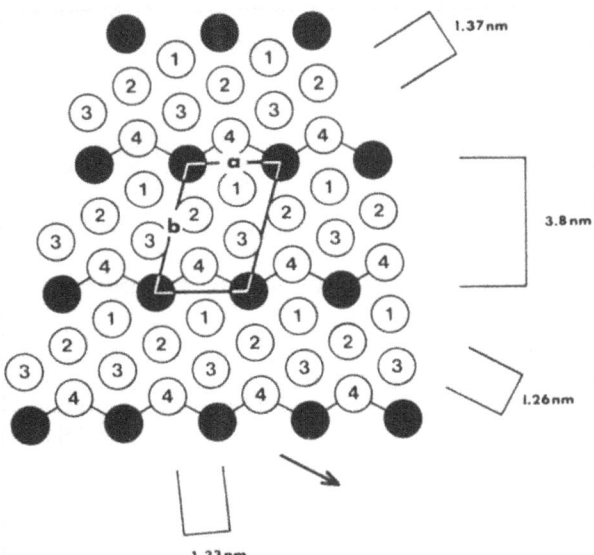

Figure 37. Tentative model for the packing of molecules in a crystalline domain of a rat tail tendon collagen fibril. The model is based on experimental data and interpretation provided by Fraser and MacRae (92); the axial packing arrangement shown here is the same as that in scheme 2A of their Figure 5. As in our Figure 35, the ends of molecules (at height 0) are portrayed as closed circles. Molecules represented by numbered open circles are staggered by 1D, 2D, 3D, 4D with respect to the molecules at height 0. The unit cell has a = 2.7 nm, b = 3.9 nm, γ = 105°. The drawing represents a projection down the fibril axis. The a-axis of the unit cell is in the plane of the diagram (i.e. normal to the fibril axis) but the b-axis projects below the plane and is therefore foreshortened slightly in projection. Bragg planes with spacings 1.26 nm, 1.33 nm, 1.37 nm, 3.8 nm are indicated. Molecules are straight-tilted with respect to the fibril axis by about 5° in a plane indicated by the arrow. The model is consistent with the formation of extended sheets of cross-linked, 4D-staggered molecules (93, 94); these cross-links are shown here by solid lines linking molecules at heights 0 and 4D.

in stained ultrathin transverse section has likewise been interpreted in terms of single molecules or microfibrils. These apparent variations in filament width, probably due to variations in the contribution of the stain, preclude discrimination between microfibrillar and molecular crystal structures on the basis of the filamentous appearance alone.

A more convincing demonstration of the existence of a microfibril would be the observation of a filament, of maximum lateral dimension 4 nm, showing a clearly defined D-periodicity. As noted in Section 4.3, D-periodic filaments of this width have been detected by optical diffraction analysis of obliquely banded collagen fibrils (62). Nevertheless, because of the unknown filament depth, these observations are insufficient to establish with certainty the existence of a microfibril. Filaments of width 3.5–4.0 nm, reconstituted from solution *in vitro*, have been purported to be D-periodic on the basis of X-ray diffraction observations of D-periodicity in oriented gels of the same material (101). Filament widths were measured on unstretched preparations whereas the gels were oriented by stretching for X-ray diffraction, so it is possible that further aggregation could have occurred during the stretching procedure, giving rise to the D-periodicity. No D-periodicity was apparent in electron micrographs of the filaments.

Periodic lateral interfilament separations of 5.5–10 nm have been reported, following convolution analyses of electron micrographs of collagen fibrils in longitudinal cryosections of rat tail tendon, and the data are claimed to support the existence of an 8-nm wide filamentous subunit (102). It is known, however, from the positions of the row lines in X-ray diffraction patterns, that the native 3.8-nm spacing can swell to as much as 17.5 nm when fibres are stained (103). Moreover, cryoprotectants (such as DMSO) can disrupt the structure of native fibrils (106) and fixatives (such as glutaraldehyde) may induce aggregation. These results must therefore be interpreted with caution.

It has been noted earlier that interpretation of the electron-optical image of a biological specimen is very much a matter of understanding the effects of preparation procedures. Procedures that can be demonstrated, by some independent means, to preserve the principal crystallographic features of the native collagen fibril should, therefore, offer the most promising approach to the study of those features by electron microscopy. This has been the basis of a recent study of ultrathin sectioned rat tail tendon fibrils, using X-ray diffraction to monitor all stages in preparation, from fixation, staining and dehydration to the infiltration and polymerisation of resin (104).

A procedure was found in which the characteristic ~4 nm lateral periodicity was preserved, with increased contrast, in the X-ray diffraction pattern of the embedded fibre. Using the same preparative procedure, this spacing could then be visualised in electron micrographs of ultrathin transverse sections of collagen fibrils (Figure 34). Optical diffraction analysis of the micrographs confirmed the periodicity and showed that it arose from concentrically oriented parallel lines of increased density in the outer regions of fibril cross sections. The micrographs provided no evidence of a two-dimensional unit cell, and in any one region of the fibril cross-section the 4-nm periodicity was restricted to one direction, with no indication of periodicities in other directions. The quasihexagonal model (87) predicts a ~2.5-nm periodicity at about 75° to the 4-nm periodicity, whilst the tetragonal packing schemes predict two orthogonal 4-nm periodicities. Thus neither packing scheme can be rejected on the basis of the electron microscopic data. Nevertheless, the quasihexagonal model, derived from X-ray diffraction, can be combined with the electron microscopic data to give the tentative lateral packing model shown in Figure 35 (104). In this model the molecules are tilted by ~4° with respect to the fibril axis. The tilted molecules occur in domains in the peripheral region of the fibril and therefore appear to follow a helical path around the fibril, consistent with the 'helical appearance' observed in collagen fibrils after treatment with acids or denaturing solvents or after freeze fracture (105–107).

It may be significant, with fibrillogenesis in mind, that molecules on the outer surface of the model (the presumed fibril surface) are in non-staggered axial register; so too are the molecules in the SLS-like aggregates which occur in the media of cultured fibroblasts immediately prior to fibrillogenesis (15). When it is seen in conjunction with the axial packing scheme of Figure 37, this stacking of non-staggered molecules in sheets parallel to the surface has another possible consequence; adjacent 4D-staggered molecules could form an extended array of covalently linked molecules (93) parallel to the surface and possibly circumferentially round the fibril.

Despite numerous observations of filaments in collagen fibrils, the question of whether or not these filaments correspond to D-periodic microfibrils is still unanswered. There is electron microscopic evidence from studies of fibril diameter distributions in a wide range of tissues that diameters are not uniformly distributed but, rather, that there exists a quantised 8-nm increment in the diameter of the smaller fibrils (14). Each 8-nm increment could correspond to the accretion of a single layer of 4-nm diameter microfibrils around the fibril periphery. Alternatively, it could correspond to the addition of a single layer of quasihexagonal packing, one unit cell thick (i.e. 4 nm). The 4-nm periodicity seen in the fibril cross-sections is consistent with the quasihexagonal model for three-dimensional molecular packing or with the presence of D-periodic microfibrillar subassemblies. The two schemes, microfibrillar and molecular crystalline, are not mutually exclusive. Fibril formation could occur via microfibrillar subassemblies which undergo lateral compression on assembly, giving rise to quasihexagonal packing as growth proceeds.

7. Concluding remarks

As the principal tensile component of the extracellular matrix, the collagen fibril plays a key structural role in connective tissues, providing strength and conferring form while allowing flexibility. With the aim of understanding how these mechanical properties can be related to the underlying molecular design, much of this chapter has been concerned with the molecular architecture of the fibril, as established by electron microscopy and other techniques. Yet collagen is only one component of the extracellular matrices of connective tissues. The composition and organisation of these matrices vary enormously from one site to another. Not only do the nature and relative abundance of the different extracellular components vary between tissues but they can also change with time in a single tissue. Specific interactions between collagen and other components appear to be all-important. In mature tendon, for example, where dermatan sulphate is the most abundant glycosaminoglycan, electron microscopy shows that the proteoglycan is in a D-periodic arrangement on the outside of the collagen fibril (108). We do not yet know, however, if these non-collagenous components are exercising control of collagen fibril growth or if, as seems more likely, the collagen is acting as a scaffolding on which the pressure-resisting elements of the extracellular matrix are being assembled. Much remains to be learnt about these collagen-proteoglycan relationships.

The last few years have seen renewed interest in the molecular arrangement in fibrils, and a clearer picture is now emerging. An important step, resulting from the elucidation of the complete amino acid sequence of the type I collagen molecule, has been the recognition that the origins of the axial intermolecular stagger lie in that sequence, particularly in the distribution of charged and hydrophobic residues. The basic mechanism which governs the formation of the molecular arrangement is manifestly one of self-assembly, similar in principle to that occurring in other systems in which protein subunits come together spontaneously to form multiunit aggregates. Nevertheless, it is now evident that the assembly process in the collagen system is much more complex than was originally envisaged and that many unanswered questions remain. Crystallinity is not an invariable feature of all collagen fibrils and has to be reconciled with the existence of liquid-like disorder in molecular packing (98) and with NMR-based evidence that the contact regions between molecules in the fibril are fluid (109). What, then, are the constructional principles which govern fibril formation and give rise to a high-tensile strength fibril which is apparently partly crystalline and partly disordered? These constructional principles must also explain how lateral accretion becomes energetically less favourable with increase in fibril diameter, so that fibrils (rather than tactoids with a much greater range of diameters) form, even when fibrils are reconstituted in vitro. Size limitation is still an ill-understood topic in molecular biology. Noting that certain connective tissue disorders are characterised by abnormal collagen fibril diameter distributions, an answer to this question could have a broader significance in biology and in medicine.

Acknowledgements

Grateful acknowledgement is due to the Medical Research Council and the Science Research Council for the award of research grants which enabled some of the research studies described here to be undertaken. The authors thank Mrs. A. Lucas for assistance with the preparation of the figures and Mrs. S. Moyser for typing the manuscript. Thanks are also due to Mrs. P.M. Armitage, Miss F.M.M. Butler, Dr. M.J. Capaldi, Dr. D.F. Holmes, Mr. K.E. Kadler, Dr. K.M. Meek and Dr. M. Tzaphlidou for the use of electron micrographs and other illustrations.

References

1. Bornstein P, Traub W: The chemistry and biology of collagen. In: The Proteins, third ed., vol 4. Neurath H, Hill RL (eds), London and New York, Academic Press, 1979, pp 411–632.
2. Ramachandran GN, Ramakrishnan C: Molecular structure. In: Biochemistry of collagen. Ramachandran GN, Reddi AH (eds), New York and London, Plenum Press, 1976, pp 45–84.

3. Thomas JC, Fletcher GC: Dynamic light scattering from collagen solutions. II. Photon correlation study of the depolarized light. Biopolymers 18: 1333–1352, 1979.

4. Gelman RA, Piez KA: Collagen fibril formation in vitro. A quasielastic light scattering study of early stages. J Biol Chem 255: 8098–9102, 1980.

5. Fietzek PP, Kühn K: The primary structure of collagen. Int Rev Connect Tissue Res 7: 1–60, 1976.

6. Tomlin SG: The structure of collagen fibres. In: Proceedings of the International Wool Textile Research Conference Australia, 1955, vol B. Crewther WG (ed), Melbourne, CSIRO, 1956, pp 178–192 (see Fig 2).

7. Hodge AJ, Schmitt FO: The charge profile of the tropocollagen macromolecule and the packing arrangement in native-type collagen fibrils. Proc Natl Acad Sci USA 46: 186–197, 1960.

8. Kühn K, Zimmer E: Eigenschaften des Tropokollagen-Moleküls und deren Bedeutung für die Fibrillenbildung. Z Naturforsch 16b: 648–658, 1961.

9. Hodge AJ, Petruska JA: Recent studies with the electron microscope on ordered aggregates of the tropocollagen marcromolecule. In: Aspects of Protein Structure. Ramachandran GN (ed), London and New York, Academic Press, 1963, pp 289–300.

10. Miller A: Molecular packing in collagen fibrils. In: Biochemistry of Collagen. Ramachandran GN, Reddi AH (eds), New York and London, Plenum Press, 1976, pp 85–136.

11. Brodsky B, Eikenberry EF: Characterization of fibrous forms of collagen. Methods Enzymol 82: 127–174, 1982.

12. Brodsky B, Eikenberry EF, Cassidy K: An unusual collagen periodicity in skin. Biochim Biophys Acta 621: 162–166, 1980.

13. Stinson RH, Sweeny PR: Skin collagen has an unusual D-spacing Biochim Biophys Acta 621: 158–161, 1980.

14. Parry DAD, Craig AS: Growth and development of collagen fibrils in connective tissue. In: Ultrastructure of the Connective Tissue Matrix. Ruggeri A, Motta PM (eds), The Hague, Martinus Nijhoff, 1984, pp 34–64.

15. Bruns RR, Hulmes DJS, Therrien SF, Gross J: Procollagen segment-long-spacing crystalities. Their role in collagen fibrillogenesis. Proc Natl Acad Sci USA 76: 313–317, 1979

16. Miyahara M, Njieha FK, Prockop DJ: Formation of collagen fibrils in vitro by cleavage of procollagen with procollagen proteinases. J biol Chem 257: 8442-8448, 1982.

17. Wood GC, Keech MK: The formation of fibrils from collagen solutions. I The effect of experimental conditions: kinetic and electron microscope studies. Biochem J 75: 588–598, 1960.

18. Hall CE: Introduction to Electron Microscopy, First ed. New York and London, McGraw-Hill, 1953 (see Fig 12.12c on p. 410).

19. Tromans WJ. HorneRW, Gresham GA, Bailey AJ: Electron microscope studies on the structure of collagen fibrils by negative staining. Z Zellforsch 58: 798–802, 1963.

20. Olsen BR: Electron microscope studies on collagen. I. Native collagen fibrils. Z Zellforsch 59: 184–198, 1963.

21. Chapman JA: The staining pattern of collagen fibrils. I. An analysis of electron micrographs. Connect Tissue Res 2: 137–150, 1974.

22. Bruns RR, Gross J: High-resolution analysis of the modified quarter-stagger model of the collagen fibril. Biopolymers 13: 931–941, 1974.

23. Kühn K, Grassmann W, Hofmann U: Über den Aufbau den Kollagenfibrille aus Tropokolllagenmolekülen. Naturwissenschaften 47: 258–259, 1960.

24. Unwin PNT: Beef liver catalase structure. Interpretation of electron micrographs. J Mol Biol 98: 235–242, 1975

25. Chapman JA, Hardcastle RA: The staining pattern of collagen fibrils. II. A comparison with patterns computer-generated from the amino acid sequence. Connect Tissue Res 2: 151–159, 1974.

26. Meek KM, Chapman JA,Hardcastle RA: The staining pattern of collagen fibrils. Improved correlation with sequence data. J Biol Chem 254: 10710–10714, 1979.

27. Hulmes DJS, Miller A, Parry DAD, Piez KA, Woodhead-Galloway J: Analysis of the primary structure of collagen for the origins of molecular packing. J Mol Biol 79: 137–148, 1973.

28. Chapman JA, Holmes DF, Meek KM, Rattew CJ: Electron-optical studies of collagen fibril assembly. In: Structural Aspects of Recognition and Assembly in Biological Macromolecules. Balaban M, Sussman JL, Traub W, Yonath A (eds), Rehovot and Philadelphia, International Science Services, 1981, 1, pp 387–401.

29. Tzaphlidou M, Chapman JA, Meek KM: A study of positive staining for electron microscopy using collagen as a model system. I. Staining by phosphotungstate and tungstate ions. Micron 13: 119–131, 1982.

30. Tzaphlidou M, Chapman JA, Al-Samman MH: A study of positive staining for electron microscopy using collagen as a model system. II. Staining by uranyl ions. Micron 13: 133–145, 1982.

31. Katayama E, Nonomura Y: Quantitative analysis of the mechanism of negative staining with native collagen fibrils and polar tropomyosin paracrystals. J Biochem 86: 1495–1509, 1979.

32. Butler FMM:Electron Microscope Studies of Negatively Stained Collagen. MSc Thesis, University of Manchester, 1982.

33. Leibovich SJ; Weiss JB: Electron microscope studies of the effects of endo- and exopeptidase digestion on tropocollagen. A novel concept of the role of terminal regions in fibrillogenesis. Biochim Biophys Acta 214: 445–454, 1970.

34. Ghosh SK, Mitra HP: Oblique banding pattern in collagen fibrils reconstituted in vitro after trypsin treatment. Biochim Biophys Acta 405: 340–346, 1975.

35. Weiss JB: Enzymic degradation of collagen. Int Rev Connect Tissue Res 7: 101–157, 1976.

36. Comper WD, Veis A: The mechanism of nucleation for in vitro collagen fibril formation. Biopolymers 16: 2113–2131, 1977.

37. Comper WD, Veis A: Characterization of nuclei in in vitro collagen fibril formation. Biopolymers 16: 2133–2142, 1977.

38. Gelman RA, Poppke DC, Piez KA: Collagen fibril formation in vitro. The role of the non-helical terminal regions. J Biol Chem 254: 11741–11745, 1979.

39. Helseth DL, Veis A: Collagen self-assembly in vitro. Differentiating specific telopeptide-dependent interactions using selective enzyme modification and the addition of free amino telopeptide. J Biol Chem 256: 7118–7128, 1981.

40. Bensusan HB, Scanu A: Fiber formation from solutions of collagen. II. The role of tyrosyl residues. J Am Chem Soc 82: 4990–4995, 1960.

41. Haworth RA, Chapman JA: A study of the growth of normal and iodinated collagen fibrils in vitro using electron microscope autoradiography. Biopolymers 16: 1895–1906, 1977.

42. Capaldi MJ, Chapman JA: The C-terminal extra-helical peptide of type I collagen and its role in fibrillogenesis in vitro. Biopolymers 21: 2291–2313, 1982.

43. Bruns RR, Gross J: Band pattern of the segment-long-spacing form of collagen. Its use in the analysis of primary structure. Biochemistry 12: 808–815, 1973.

44. Hulmes DJS, Miller A, White SW, Doyle BB: Interpretation of the meridional X-ray diffraction pattern from collagen fibres in terms of the known amino acid sequence. J Mol Biol 110: 643–666, 1977.

45. Helseth DL, Lechner JH, Veis A: Role of the amino-terminal extra helical region of type I collagen in directing the 4D overlap in fibrillogenesis. Biopolymers 18: 3005–3014, 1979.

46. Weiss JB: Preparation of segment-long-spacing collagen. In: The Methodology of Connective Tissue Research. Hall DA (ed), Oxford, Joynson-Bruvvers, 1976, pp 73–80.

47. Kühn K: Segment-long-spacing crystallites, a powerful tool in collagen research. Collagen Rel Res 2: 61–80, 1982.

48. Chapman JA, Armitage PM: An analysis of fibrous long spacing forms of collagen. Connect Tissue Res 1: 31–37, 1972.

49. Bowden JK, Chapman JA: The precipitation of segmented-long-spacing collagen by inorganic triphospate and perdisulphate ions. Connect Tissue Res 1: 109–112, 1972.

50. Doyle BB, Hukins DWL, Hulmes DJS, Miller A, Woodhead-Galloway J: Collagen polymorphism. Its origins in the amino acid sequence. J Mol Biol 91: 79–99 1975.

51. Beer M, Wiggins JW, Tukel D, Stoeckert CJ: Biological structure determination through atomic microscopy. Chemica Scripta 14: 263–266, 1978/79.

52. Fernandez-Madrid F, Noonan S, Riddle J, Karvonen R, Sasaki D: Intracellular processing of procollagen induced by the action of colchicine. J Anat 130: 229–241, 1980.

53. Pérez-Tamayo R: The occurrence and significance of SLS crystallites in vivo. Connect Tissue Res 1: 55–60, 1972.

54. Imura S. Tanaka S, Takase B: Intracytoplasmic segment long spacing fibrils in chondrosarcoma. J Electron Microsc (Tokyo) 24: 87–95, 1974.

55. Weinstock M,Leblond CP: Synthesis, migration, and release of precursor collagen by ondotoblasts as visualized by radioautography

after [³H]-proline administration. J Cell Biol 60:92–127, 1974.

56. Chapman JA: Preparation of fibrous long spacing collagen. In: The Methodology of Connective Tissue Research. Hall DA (ed), Oxford, Joynson-Bruvvers, 1976, pp 63–72.

57. Kajikawa K, Nakanishi I, Yamamura T: The effect of collagenase on the formation of fibrous long spacing collagen aggregates. Lab Invest 43: 410–417, 1980.

58. Bard JBL, Chapman JA: Polymorphism in collagen fibrils precipitated at low pH. Nature (Lond) 219: 1279–1280, 1968.

59. Bruns RR:Supramolecular structure of polymorphic collagen fibrils. J Cell Biol 68: 521–538, 1976.

60. Williams BR, Gelman RA, Poppke DC, Piez KA: Collagen fibril formation. Optimal in vitro conditions and preliminary kinetic results. J Biol Chem 253: 6578–6585, 1978.

61. Bruns RR, Trelstad RL, Gross J: Cartilage collagen. A staggered substructure in reconstituted fibrils. Science 181: 269–271, 1973.

62. Doyle BB, Hulmes DJS, Miller A; Parry DAD, Piez KA, Woodhead-Galloway J. A D-periodic narrow filament in collagen. Proc R Soc Lond B186: 67–74, 1974.

63. Wood GC: The precipitation of collagen fibrils from solution. Int Rev Connect Tissue Res 2: 1–31, 1964.

64. Wood GC: The formation of fibrils from collagen solutions. 2. A mechanism of collagen fibril formation. Bioch J 75: 598–605, 1960.

65. Cassel JM, Mandelkern L, Roberts DE: The kinetics of the heat precipitation of collagen. J Am Leather Chem Assoc 57: 556–575, 1962.

66. Gelman RA, Williams BR, Piez KA: Collagen fibril formation. Evidence for a multi-step process. J Biol Chem 254: 180–186, 1979.

67. Brennan M, Davison PF: Role of aldehydes in collagen fibrillogenesis in vitro. Biopolymers 19: 1861–1873, 1980.

68. Silver FH, Trelstad RL: Type I collagen in solution. Structure and properties of fibril fragments. J Biol Chem 255: 9427–9433, 1980.

69. Silver FH: Type I collagen fibrillogenesis in vitro. Additional evidence for the assembly mechanism. J Biol Chem 256: 4973–4977, 1981.

70. Trelstad RL, Hayashi K, Gross J: Collagen fibrillogenesis. Intermediate aggregates and suprafibrillar order. Proc Natl Acad Sci USA 73: 4027–4031, 1976.

71. Bard JBL, Chapman JA: Diameters of collagen fibrils grown in vitro. Nature (Lond) 246: 83–84, 1973.

72. Holmes DF, Chapman JA: Axial mass distributions of collagen fibrils grown in vitro. Results for the end regions of early fibrils. Biochem Biophys Res Commun 87: 993–999, 1979.

73. Capaldi MJ, Holmes DF, Chapman JA: Collagen fibrillogenesis in vitro. Effects of precipitating conditions (in preparation).

74. Fitton Jackson S, Smith RH: Studies on the biosynthesis of collagen. I. The growth of fowl osteoblasts and the formation of collagen in tissue culture. J Biophys Biochem Cytol 3: 897–912, 1957.

75. Goldberg B, Green H: An analysis of collagen secretion by established mouse fibroblast lines. J Cell Biol 22: 227–258, 1964.

76. Prockop DJ, Kivirikko KI, Tuderman L Guzman NA: The biosynthesis of collagen and its disorders. N Engl Med 301: 13–23, 77–85, 1979.

77. Lapière CM, Nusgens B, Pierard G, Hermanns JF: The involvement of procollagen in spatially orientated fibrillogenesis. In: Dynamics of Connective Tissue Macromolecules. Burleigh PMC, Poole AR (eds), Amsterdam North Holland, 1975, pp 33–50.

78. Wick G, Olsen BR, Timpl R: Immunohistologic analysis of fetal and dermatosparactic calf and sheep skin with antisera to procollagen type I. Lab Invest 39: 151–156, 1978.

79. Fleischmajer R, Timpl R, Tuderman L, Raisher L, Weistner M, Perlish JS, Graves PN: Ultrastructural demonstration of extension aminopropeptides of type I and III collagens in human skin. Proc Natl Acad Sci USA 78: 7360–7364, 1981.

80. Lapière CM, Nusgens B: Polymerisation of procollagen in vitro Biochim Biophys Acta 342: 237–246, 1974.

81. Cho M-I, Garant PR: Sequential events in the formation of collagen secretion granules with special reference to the development of segment-long-spacing aggregates. Anat Rec 199: 309–320, 1981.

82. Wright GM, Leblond CP: Immunohistochemical localisation of procollagen. III. Type I procollagen antigenicity in osteoblasts and prebone (osteoid). J Histochem Cytochem 29: 791–804, 1981.

83. Hulmes DJS, Bruns RR, Gross J: On the state of aggregation of newly secreted procollagen. Proc Natl Acad Sci USA 80: 388–392, 1983.

84. North ACT, Cowan PM, Randall JT: Structural units in collagen fibrils. Nature (Lond) 174: 1142–1143, 1954.

85. Miller A, Wray JS: Molecular packing in collagen. Nature (Lond) 230: 437–439, 1971.

86. Miller A, Parry DAD: Structure and packing of microfibrils in collagen. J Mol Biol 75: 441–447, 1973.

87. Hulmes DJS, Miller A: Quasi-hexagonal molecular packing in collagen fibrils. Nature (Lond) 282: 878–880, 1979.

88. Smith JW: Molecular patterns in native collagen. Nature (Lond) 219: 157–158, 1968.

89. Veis A, Anesey J, Mussell S: A limiting microfibril model for the three-dimensional arrangement within collagen fibres. Nature (Lond) 215: 931–934, 1967.

90. Miller A, Tocchetti D: Calculated X-ray diffraction pattern from a quasi-hexagonal model for the molecular arrangement in collagen. Int J Biol Macromol 3: 9–18, 1981.

91. Trus BL, Piez KA: Compressed microfibril models of the native collagen fibril. Nature (Lond) 286: 300–301, 1980.

92. Fraser RDB, MacRae TP: Unit cell and molecular connectivity in tendon collagen. Int J Biol Macromol 3: 193–200, 1981.

93. Bailey AJ, Light ND, Atkins EDT: Chemical cross-linking restrictions on the models for the molecular organization of the collagen fibre. Nature (Lond) 288: 408–410, 1980.

94. Hulmes DJS, Miller A: Molecular packing in collagen. Nature (Lond) 293: 239–240, 1981.

95. Piez KA, Trus BL: A new model for packing of type-I collagen molecules in the native fibril. Biosci Rep 1: 801–810, 1981.

96. Jesior J-C, Miller A, Berthet-Colominas C: Crystalline three-dimensional packing is a general characteristic of type I collagen fibrils. FEBS Lett 113: 238–240, 1980.

97. Eikenberry EF, Brodsky B: X-ray diffraction of reconstituted collagen fibers. J Mol Biol 144: 397–404, 1980.

98. Woodhead-Galloway J, Machin PA: Modern theories of liquids and the diffuse equatorial X-ray scattering from collagen. Acta Cryst A32: 368–372, 1976.

99. Hukins DWL, Woodhead-Galloway J: Collagen fibrils as examples of smectic-A biological fibres. Mol Cryst Liq Cryst 41: 33–39, 1977.

100. Wendorff JH: Scattering in liquid crystalline polymer systems. In: Liquid Crystalline Order in Polymers. Blumstein A (ed), London and New York, Academic Press, 1978, pp 1–41.

101. Veis A, Miller A, Leibovich SJ, Traub W: The limiting collagen microfibril. The minimum structure demonstrating native axial periodicity. Biochim Biophys Acta 576: 88–98, 1979.

102. Squire JM, Freundlich A: Direct observation of a transverse periodicity in collagen fibrils. Nature (Lond) 288: 410–413, 1980.

103. Brodsky (Doyle) B, Hukins DWL, Hulmes DJS, Miller A, White S, Woodhead-Galloway J: Low angle X-ray diffraction studies on stained rat tail tendons. Biochim Biophys Acta 535: 25–32, 1978.

104. Hulmes DJS, Jesior J-C, Miller A, Berthet-Colominas C, Wolff C: Electron microscopy shows periodic structure in collagen fibril cross-sections. Proc Natl Acad Sci USA 78: 3567–3571, 1981.

105. Rayns DG: Collagen from frozen factured glycerinated beef heart. J Ultrastruct Res 48: 59–66, 1974.

106. Stolinski C, Breathnach AS: Freeze-facture replication and surface sublimation of frozen collagen fibrils. J Cell Sci 23: 325–334, 1977.

107. Ruggeri A, Benazzo F, Reale E, Collagen fibrils with straight and helicoidal microfibrils: a freeze-fracture and thin-section study. J Ultrastruct Res 68: 101–108, 1979.

108. Scott JE, Orford CR: Dermatan sulphate-rich proteoglycan associates with rat tail tendon collagen at the d band in the gap region. Biochem J 197: 213–216, 1981.

109. Jelinski LW, Torchia DA: Investigation of labelled amino acid side-chain motion in collagen using ¹³C nuclear magnetic resonance. J Mol Biol. 138: 255–272, 1980.

Authors' address:
Department of Medical Biophysics
University of Manchester Medical School
Manchester M13 9PT, U.K.

CHAPTER 2

Growth and development of collagen fibrils in connective tissue

DAVID A.D. PARRY and ALAN S. CRAIG

1. Introduction

The form and integrity of the animal body is largely dependent upon the composition and spatial arrangement of the structural composite known as connective tissue. Each member of this complex family of tissues contains the same building blocks – collagen, glycosaminoglycans, glycoproteins, elastic fibres, cellular constituents, water and minerals – but they are assembled in a multitude of different ways. Ultimately, the interactions of these components with one another bestow the appropriate mechanical attributes upon the tissue.

Connective tissues play very diverse roles in the vertebrates. For instance, the visco-elastic properties of the vitreous humour are important in helping to maintain the shape of the eye. The randomly orientated collagen fibrils in this tisssue are reported to have very small diameters (~10 nm). Consider also the transparent cornea, which forms the anterior projecting portion of the eye and which acts as its primary light-collecting element. Here the collagen fibrils have diameters ~17 or 24 nm and are arranged parallel to one another in lamellae with each subsequent lamella being rotated by about 90° relative to its neighbours. In contrast yet again, cartilage has evolved as a pressure-bearing tissue. The most superficial of the collagen fibrils in articular cartilage lie approximately parallel to the surface but elsewhere show little obvious organisation until in close proximity to the underlying calcified material, where they become orientated almost perpendicular to the surface. The collagen fibrils in cartilage typically have diameters in the range 30–65 nm. Tendon and ligament exemplify those connective tissues which have high tensile strength and which are designed to transmit force efficiently; they have high collagen contents and consist of axially orientated fibrils ranging in diameter up to 550 nm. Another connective tissue of major biological importance is skin, which plays a relatively passive role as a bound-

ary between an animal and its environment and in specialised cases (such as fish skin) may also have an exotendinous role. Bone and dentine, which have the requirements of rigidity and hardness, are examples of connective tissues which become calcified as they mature. Clearly, the content and composition of the chemical components present in a tissue together with the size and organisation of the collagen fibrils have a profound effect on the mechanical properties of the tissue.

At least six genetically distinct collagen molecules have been characterised or partially characterised. With a few minor exceptions, collagen molecules of all types contain three similarly directed α-chains (for a detailed account of molecular structure and regularities therein, see (1)). Each α-chain has much of its sequence in the form $(Glycine-X-Y)_n$ where X and Y represent some amino or imino acid residue. The three-stranded, coiled-coil conformation adopted by such regions in the collagen molecule is thought to differ little between the collagen species and a molecular model for this structure has now been refined using a linked-atom, least-squares analysis in conjunction with the X-ray diffraction data from highly stretched tendon (2). Important interspecific structural differences, however, do exist in that (a) the number of triplets in types I, III and IV collagen molecules are 338, 341 and ~430 respectively (b) the chain composition of types I, II and III collagen molecules are $[\alpha(I)]_2\alpha2$, $[\alpha1(II)]_3$ and $[\alpha1(III)]_3$ respectively (c) the carbohydrate contents of most type I, II, III and IV collagens are ~0.4, 4, 0.4 and 12% respectively. Note, however, that the carbohydrate content of tendon and cornea, which are both predominantly type I tissues, are quite different (~0.4 and 6% respectively) (d) type IV collagen molecules, in contrast to other genetic types, do not appear to aggregate into a fibrillar form *in vivo* (e) with the exception of the carboxy-terminal telopeptide in the α2(I) chain, the amino- and carboxy-telopeptide regions of the α-chains in types I,

II and III collagen molecules each contain a lysine residue capable of forming a covalent cross link.

The ratios of the genetic types of collagen vary from one tissue to another and commonly change with age. For example young skin contains a high content of type III collagen which is gradually replaced by type I collagen as the tissue ages. In cornea, type II collagen is initially laid down by the epithelium but this ultimately becomes a minor constituent as type I collagen synthesis predominates with increasing age. Further, the major molecular species of collagen is type I in tendon, ligament and bone, type II in cartilage and vitreous humour and type IV in basement membrane whilst type V collagen forms a minor constituent of the total collagen in bone. These few examples emphasise that different tissues contain collagen molecules of different length, chain composition, amino acid sequence, degree of hydroxylation and glycosylation, and cross-linking capability, Further, it is rare to find a tissue which contains but a single molecular species of collagen – in almost every case different molecular species co-exist.

Many fundamental questions regarding fibrillogenesis have been posed but few have yet been answered. For instance, what effects (if any) do the genetically distinct collagens have on the mode of aggregation of molecules into fibrils? Can genetically distinct collagen molecules co-exist within a single fibril? What factors control fibril size distribution and rate of growth? Is the growth of collagen fibrils controlled by the degree of molecular glycosylation or the type and content of the glycosaminoglycans in the matrix? What mechanism causes the collagen fibrils of most tissues to decrease in size after the onset of maturity? How is a tissue able to be remodelled when the mechanical conditions which it normally experiences are suddenly altered? These and many other important problems are now being tackled in laboratories around the world from both biological and medical standpoints. Ultimately such studies will lead to an understanding at the molecular level of remodelling and wound healing, and also the effects of ageing on the mechanical and physical properties of the many and varied connective tissues.

Untangling the inter-relationships between the components in connective tissue is a daunting task; but as this review will attempt to show, significant progress is now being made. In particular, some understanding is being gained of the ways in which collagen molecules pack together and aggregate into fibrils with different functional properties.

2. Methods of ultrastructural research

Collagen fibrils in connective tissue are composed of molecules with diameters ~1.4 nm. Thus if the growth and development of such fibrils is to be observed, techniques must be employed which have a resolving power close to 1 nm. As the light microscope does not meet this requirement, electron microscopy must be the method of choice. Transmission electron microscopes commonly have a resolving power ~0.5 nm, but unfortunately structural details at a molecular level are often masked or destroyed by the preparative procedures currently employed for the 'preservation' of hydrated biological specimens. The technique, nonetheless, provides important ultrastructural information on the development of collagen fibrils.

Collagen fibrils with diameters <100 nm can be placed directly on to support grids and visualised after contrasting by heavy metal shadowing or staining. Such an approach has been used with reconstituted collagen fibrils and with those from tissues which have been mechanically or chemically dispersed. Whole mount specimens, when negatively stained, show collagen fibrils to be composed longitudinally of alternating light and dark bands with an axial periodicity of D (67 nm). As the stain preferentially occupies the gap region of the D-periodic structure, the resulting stain intensity reflects the mass distribution of the fibril. Alternatively, positively stained whole mount specimens typically produce up to 13 striations within the D-period. These striations are produced by the affinity of the metal ions for the charged residues, resulting in a stain intensity reflecting the charge distribution along the fibril(3). Finally, whole mount specimens shadowed with heavy metals provide information on the surface topography of the fibrils.

Most of the electron microscope studies on collagenous tissues have been done using thin sections of epoxy-embedded material. Interpretation of the high resolution images is often difficult due to artefacts inherent to the severe preparative procedures employed. However, low resolution microscopy has allowed the spatial arrangements of fibrils, elastic fibres and cells to be studied (Figure 1). Thin sectioning has also enabled the fibrils to be observed in transverse section, and from such studies quantitative measurements can be made which lead to an understanding of the mechanisms and patterns of fibril growth that are discussed in this review. Material prepared for thin sectioning can be either positively or negatively post-stained or stained *en bloc* during fixation or dehydration. Further, thin sec-

36

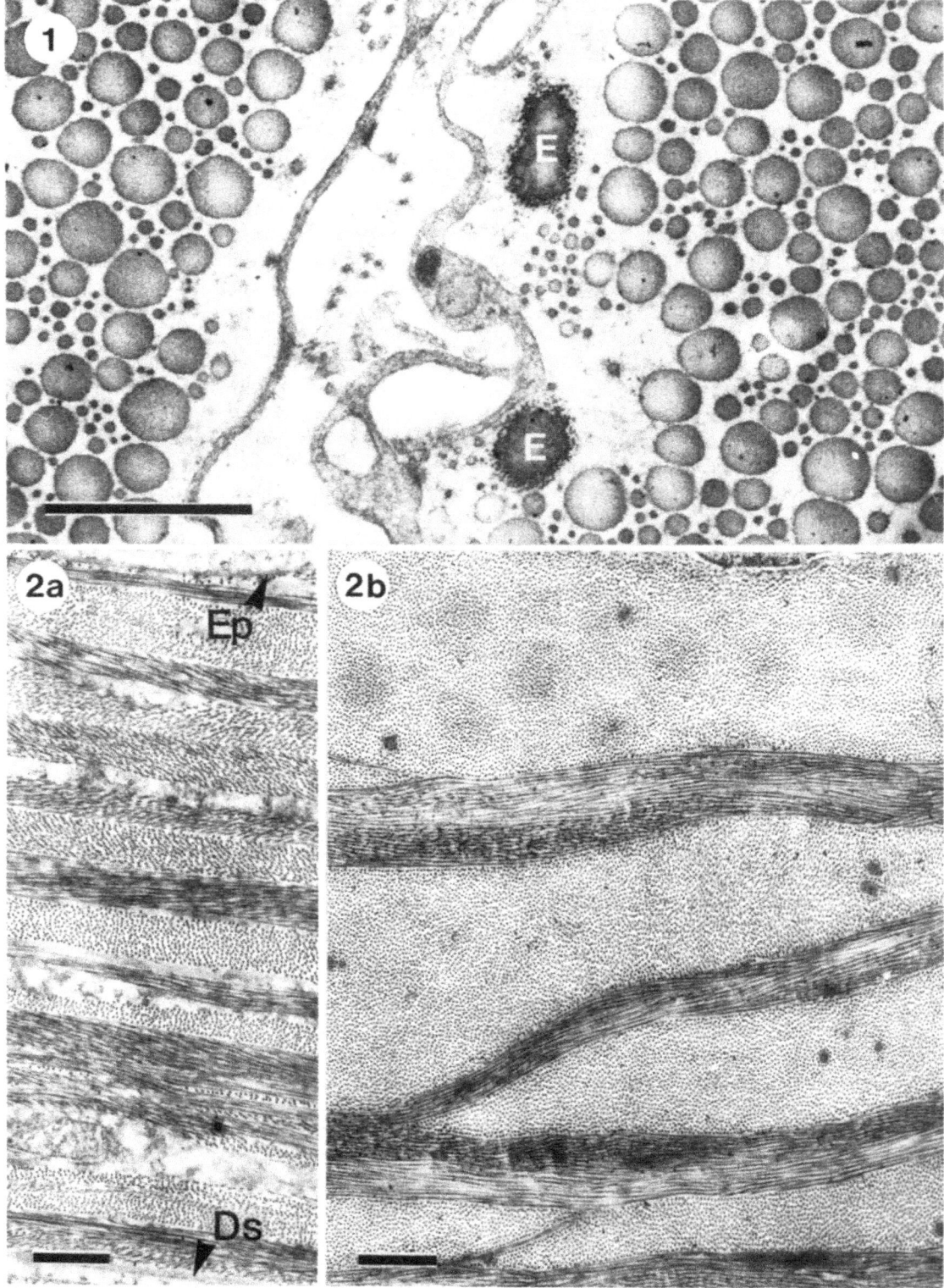

tioning techniques permit the use of histochemical staining methods specific for connective tissue components. For instance, the proteoglycans of the matrix may be visualised using Thorotrast, Ruthenium Red or Alcian Blue.

Collagenous tissues which have not been embedded may also be sectioned if frozen. Ultracryotomy avoids the rigours of fixation, dehydration and embedding and allows imaging of fibrils sectioned in a hydrated and close to an *in vivo* state. Alternatively, rapidly frozen specimens may be fractured, and replicas made of their surfaces after varying degrees of etching. These techniques (freeze-fracturing and freeze-etching) also allow visualisation of the internal structure of the fibril.

Indirect ways of studying the ultrastructure of connective tissues include fibre diffraction techniques. These methods involve irradiating an orientated specimen with a well-collimated beam of X-rays or neutrons and recording the diffraction pattern so formed on film or with position-sensitive detectors. The positions of the diffraction maxima, together with their intensities, provide information on the molecular structure and packing. Unfortunately, however, an unambiguous model of molecular conformation and organisation can rarely be determined from such data.

Diffraction techniques also suffer from the limitation that the large scale features of a structure (>100 nm) cannot readily be determined, since the diffracted beam containing this information lies very close to the undeviated main beam and is usually imperfectly resolved from it. With electron microscopy, such large scale features are those which are most easily seen. However, diffraction techniques allow hydrated tissues to be studied and data can thus be recorded from specimens which more closely approximate the *in vivo* state than do those prepared for electron microscopy. A coordinated approach involving both X-ray diffraction and electron microscopy is thus needed if a comprehensive study of connective tissue ultrastructure is to be achieved.

3. Developmental studies of collagen fibrils.

Although many have investigated the growth and development of collagen fibrils *in vitro* (see chapter 1), the results of those studies will not be reported here; instead this review will limit itself to a discussion of the *in vivo* results as determined primarily from electron microscopy. It is appropriate here to emphasise that, until recently, very little reliable data on the collagen fibril diameter distributions in connective tissues have been available. In the past, many investigators have failed to report the ages of the specimens studied or have failed to take sufficient quantitative measurements to render their fibril diameter distributions meaningful. Further, care has not always been taken to calibrate the magnification of the electron microscope for each set of micrographs taken (using, say, a grating replica) and in yet other instances, only a mean or range of fibril diameters has been reported without any reference to the form of the distribution from which it was derived.

Before the collagen fibril diameter distributions for various tissues at different ages are reported here, it is important to establish several gross structural features of the collagen fibril which will form the basis of future interpretation:
(a) Collagen fibrils are polar entities; they are composed of similarly directed collagen molecules.
(b) There is no correlation between the polarities of neighbouring fibrils in orientated connective tissues; fibrils randomly point 'up' and 'down' (as in tendon; 4, 5).
(c) There is no convincing evidence that anastomosing fibrils occur *in vivo* in any connective tissue in spite of occasional reports to the contrary.
(d) Collagen fibrils are circular in cross-section, a result recently confirmed by low-angle X-ray diffraction of foetal chick metatarsal tendon (6, 7).
(e) The axial ratio of a collagen fibril is always very high in mature connective tissue and values in excess of 10^4 are probably not uncommon. In-

Figure 1. Transverse section of an eight-week-old rat tail tendon showing large numbers of collagen fibrils of circular cross-section embedded in an amorphous matrix containing elastic fibres (E). The field is traversed by two cytoplasmic processes from the cellular constituent of tendon, the fibrocytes (reprinted from Parry and Craig (4) with permission of John Wiley and Sons). Bar = 1 μm.
Figure 2. Sections through the corneal stroma showing the organisation of the collagen fibrils into lamellae: (a) Section through the entire stromal thickness of the thin cornea of snake. The stroma is bounded anteriorly by the epithelium (Ep) and posteriorly by Descemet's membrane (Ds). The lamellae are typically ~0.3 to 0.8 μm thick and are frequently seen with their constituent collagen fibrils in alternating transverse and longitudinal section. Bar = 1 μm. (b) Section through a central portion of the thick cornea of magpie. The lamellae are usually 0.5–3 μm in thickness, and the collagen fibrils have a similar diameter (~25 nm) to that seen in the snake (both panels are reprinted from Craig and Parry (8) with permission of Academic Press). Bar = 1 μm.

deed, the ends of a collagen fibril are rarely, if ever, seen in the electron microscope unless the connective tissue is either pathological or has been subjected to excessive stresses. Craig and Parry (unpublished data) have followed the courses of 1,000 fibrils in tendon over 14 serial sections (~1 μm total thickness) without observing either a fibril terminating or originating. It is thus probable that collagen fibrils in all connective tissues are several millimetres long and, in all probability, several centimetres long in some tissues.

From these considerations it is reasonable, though not unequivocal, to interpret the collagen fibril diameter distributions described in the following text as arising from 'infinitely' long cylindrical fibrils of various diameters rather than arising from a sampling of diameters in sections through a homogeneous population of tapering fibrils. It must also be appreciated that the preparative procedures employed for electron microscopy may have modified the fibril dimensions and thus the absolute values reported here may be in error. However, since all specimens studied by the authors were treated in a similar way, and these constitute the bulk of the data reported, it is likely that a comparison of fibril sizes at different ages in the same tissue will be meaningful. Further, there is always the potential problem of sampling errors in quantitative electron microscopy. In the examples described by the authors, the only criteria for taking micrographs were that the fibrils appeared reasonably circular in section and they were sufficiently well stained to enable quantitative measurements to be made. Within these limitations, the data in the following sections can be described.

3.1. Cornea

The structure of the vertebrate cornea has evolved as a result of the need to function in two distinctive ways. Firstly, the cornea must be transparent and have a refractive index greater than that of air or water in order that it may act as the primary light-collecting element of the eye in both terrestrial and aqueous environments. Secondly, it must have sufficient structural integrity to maintain its shape and allow it to play its important protective role, being the most anterior tissue of the eye. The collagenous nature of the corneal stroma has simultaneously met both demands.

The periphery of the cornea is continuous with the sclera and is limited anteriorly by a stratified squamous epithelium. Posterior to this lies the stroma, consisting of parallel arrays of small diameter col-

lagen fibrils arranged in layers (lamellae). These lamellae lie parallel to the surface of the cornea and each is rotated ~90° with respect to its neighbour. The number of these lamellae, and their widths, vary considerably between species (Figure 2a, b) and give rise to corneal stromata ranging in thickness from ~10 μm in the snake to ~550 μm in man (8). As in other connective tissues (Table 1) the fibrils are embedded in a matrix of hydrated proteoglycans consisting primarily, in this case, of chondroitin sulphate and corneal keratan sulphate. The cellular component of the stroma consists of long slender fibrocytes (keratocytes) which are circumferentially disposed.

The stroma is limited posteriorly by Descemet's membrane in which a fibrillar network can be observed by electron microscopy. The membrane, which has been shown by chemical analyses to contain collagen as a major component, is separated from the aqueous humour of the anterior chamber of the eye by a typical squamous endothelium.

In the developing cornea, type II collagen is synthesised by the epithelium and incorporated into fibrils within the stroma. In later development type I collagen is synthesised by the increasing number of keratocytes; this results in the adult corneal stroma being composed almost entirely of this molecular species.

The collagen fibrils of the corneal stroma have a uniformity of diameter which is almost unique among adult connective tissues. Although a wide range of mean fibril diameters have been reported (10–45 nm), it has recently been shown (8) that the collagen fibril diameters of the corneal stroma are ~24 nm in cartilaginous fish, amphibians, reptiles, birds and mammals (figure 3) and ~17 nm in the bony fish. The only exceptions to this are the 17 nm collagen fibrils found in the corneal stroma of an adult sealion and the anomalous ~20 nm fibrils demonstrated in stromata of three neonatal animals; capuchin monkey, squirrel monkey and hippopotamus (8).

An early report (9) had suggested that the diameters of the corneal collagen fibrils in the rat increase from ~19 nm in the most anterior regions of the stroma to ~34 nm in the most posterior regions, adjacent to Descemet's membrane. Further, others (10) had measured the collagen fibril diameters in the anterior, central and posterior portions of the rabbit stroma and obtained values of 18.6, 19.8 and 17.8 nm respectively. These results indicated that there was no change in fibril diameter with location. In order to resolve these conflicting results, Craig and Parry (8) studied the corneas of snake, magpie, chick, guinea pig, rat and man at each of six equi-

Table 1. Percentage wet weights of connective tissue components in mature animals[a]

Tissue	Collagen	Glycosamino-glycans	Water	Elastin
Tendon	30%	0.03–0.3%	65%	1.5%
Skin	30%	0.03–0.35%	60–72%	0.2%
Bone[b]	5–20%	0.4%	30–50%	0.1%
Fibrocartilage	20%	0.6%	75%	0.1–0.2%
Elastic cartilage	16%	3–4%	70%	5–6%
Hyaline cartilage	5–18%	5–11%	75%	<0.1%
Cornea	12–15%	0.2–1%	80%	–
Aorta	5–15%	0.2-2.5%	70–75%	7–15%
Whartons Jelly	12%	0.3%	88%	–
Ligamentum nuchae	9%	–	55%	35%
Vitreous humour	0.25%	0.02%	99%	–

[a] The total percentage wet weight does not always add up to 100% in this table due to lack of inclusion of cellular components and non-collagenous proteins. Where no value has been listed, this signifies that no quantitative estimate is available. The values quoted are approximations only; they have been derived from a range of values quoted in the literature.

[b] Bone contains ~45% minerals.

spaced positions across the width of the corneal stroma. Their observations indicate that the diameter of the corneal collagen fibrils remains unchanged across the stroma except for some species where the fibrils are smaller in the lamella adjacent to Descemet's membrane (see Table 3 in reference 8). In the magpie these fibrils had diameters ~20 nm whilst in the rat and chick the mean diameters lay close to 17 nm, a value similar to that seen in the foetal corneas studied.

To date, collagen fibril diameters in foetal corneas have been measured in but two vertebrate species, man and rat, and in each case the mean diameters lay close to 17 nm (8). The corneas from these animals together with that of guinea pig were also studied at various ages between birth and senescence, and it was demonstrated that there was no significant change in mean fibril diameters (~ 24 nm) throughout these stages of life (8). Thus cornea is unusual amongst connective tissues in that it has uniform diameter fibrils which remain unchanged in diameter throughout post-natal life.

In summary, the following can be said of the collagen fibrils of the vertebrate corneal stroma:
(a) They grow during foetal development and by birth have attained the diameters which exist in adult life.
(b) They are uniform in diameter in adult animals, being ~17 nm in bony fish and ~24 nm in all other classes of vertebrates.
(c) They maintain their adult diameters throughout remaining life.
These conclusions imply that a very precise diameter-regulating-mechanism operates on the collagen fibrils in the vertebrate corneal stroma. Whilst the strict uniformity of fibril diameter is unlikely to lend itself to any unique *mechanical* properties, it could be important in the ultrastructural organisation of collagen fibrils within the lamellae, which in turn may endow the tissue with its unique light-scattering properties.

3.2. Tendon and ligament

Tendons and ligaments are dense regular connective tissues; they are comprised of parallel collagen fibrils close-packed in an hydrated proteoglycan-rich matrix containing fine elastic networks. However, in some specialised 'elastic ligaments' such as the *ligamentum nuchae* of grazing mammals, the elastic fibres are the dominant fibrillar elements (Table 1). The fibroblasts responsible for the synthesis of collagen, elastin and the proteoglycans are the only cellular elements present in tendon. These elongate cells are arranged parallel to the bundles of collagen fibrils and have long cytoplasmic processes ramifying throughout the tissue. These features can all be seen in Figure 1. Such axially orientated collagenous tissues, being flexible and having high tensile strength, are well suited for their mechanical role – the efficient transmission of force.

Many connective tissues, including tendon, have a macroscopic crimp with a periodicity which is typically ~100 μm. The crimp, which may be removed by stretching the tissue by ~3–4%, is believed to act as a compliance mechanism. For instance, it is considered that a small increase in the length of a tendon, as a result of straightening the crimped fibre bundles, will prevent damage to the tendon during

40

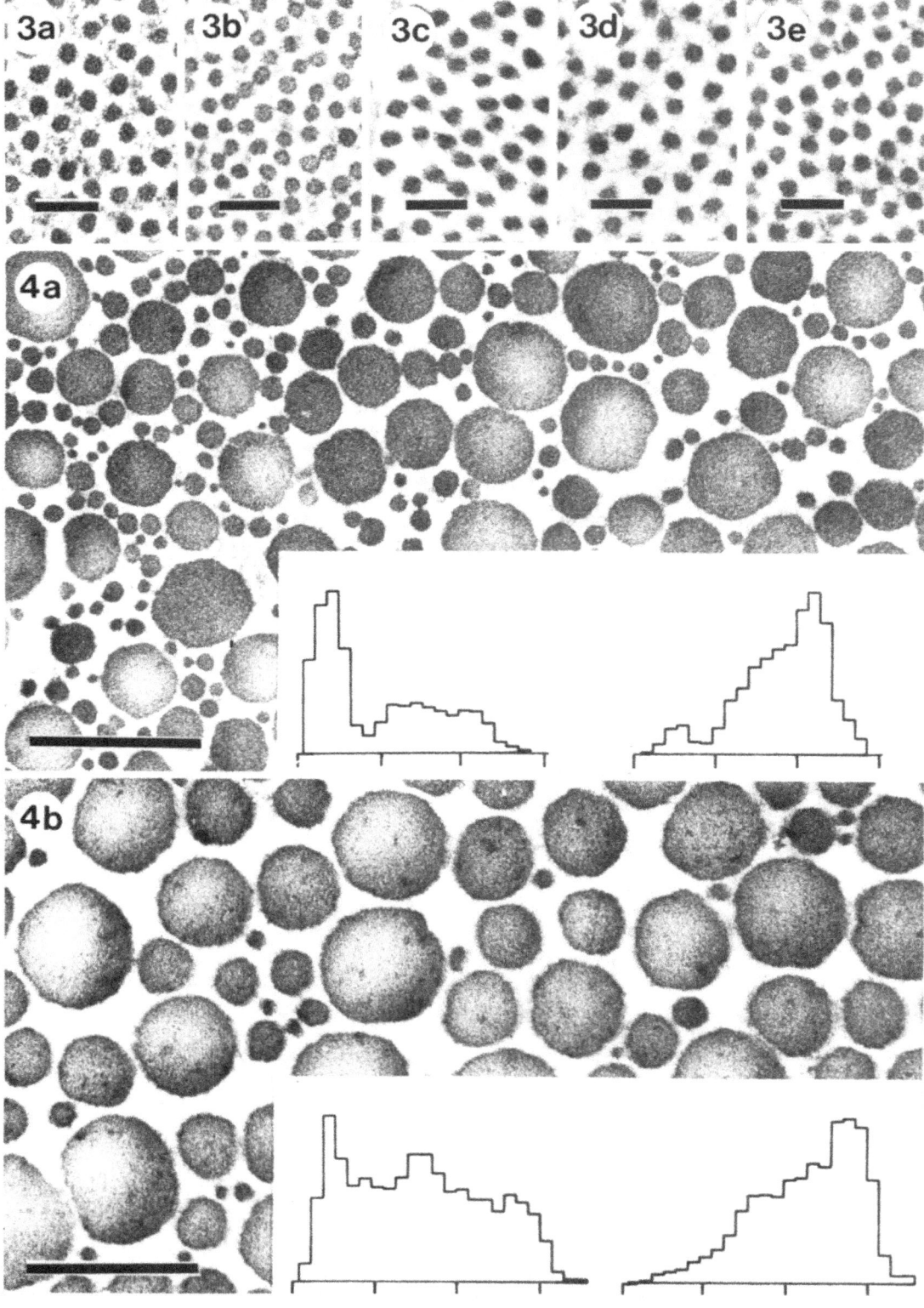

the initial rapid contraction of its parent muscle.

Collagen fibril diameter distributions have been measured as a function of age for a number of tendons and ligaments from human, rat, horse, chicken, sheep and guinea pig. The results show many interesting trends. In the case of horse, two tendons (common digital extensor and superficial flexor) and one ligament (suspensory) were studied from the age of 2 months prior to birth right through to senescence (5). Before birth, the unimodal distributions of fibril diameter were broad, similar in form and with mean values close to 100 nm. However, at maturity the superficial flexor tendon and the suspensory ligament both had bimodal fibril diameter distributions whereas that for the extensor tendon had remained unimodal (Figure 4). In all three cases, the mass-average diameters of the fibrils had increased (Table 2). At senescence, the distributions once again became very similar to one another and each contained a high percentage of small diameter fibrils, with a consequent reduction in the mass-average diameter from the maximum values recorded at maturity. It is not known whether the largest fibrils present at maturity breakdown mechanically through a 'fatigue' mechanism of some sort or whether the turnover of collagen results in the newly synthesised collagen being unable to grow into large diameter fibrils (see Section 7).

An extensor tendon, a flexor tendon and a diaphragmatic ligament from guinea pig have been studied between birth and maturity (Table 2). In each case, the collagen fibril diameter distributions at birth were fairly broad and unimodal. The mass-average diameter of the collagen fibrils increased with age though, once again, the form of the distributions at maturity may remain unimodal (diaphragmatic ligament) or become bimodal (extensor and flexor tendons).

Avian metatarsal tendon has been extensively investigated during the foetal stages of development (Figure 5; 6, 7, 11). In most cases, the collagen fibril diameter distributions were unimodal and sharp (standard deviations of the distributions < 3.5 nm) and had mean diameters lying close to a multiple of 8

nm (Table 2). The widths of the diameter distributions increased rapidly at ages beyond ~ 17 days foetal. One particular specimen of chick metatarsal tendon (18d foetal) was shown to have a broad 'unimodal' collagen fibril diameter distribution, though careful measurement revealed that it could be resolved into a number of populations each with a mean value lying close to a multiple of 8 nm (Figure 6; 7). At maturity, the same tendon had a bimodal distribution of collagen fibril diameters.

The ultrastructure of human tendons has not been studied in much detail though Dyer and Enna (12) have reported some data on *plantaris, extensor indicis, flexor digitorum superficialis* and *extensor digitorum longus* from adults. Although the distributions of fibril diameters were not published in that work, it appears from the text that these distributions were bimodal. Schwarz (13) has reported age-related data on human Achilles tendon where the distribution of fibril diameters was unimodal and sharp for a 35-cm foetus but broad and bimodal for a 42-year-old adult. The data recorded by Schwarz, although limited in quantity, does confirm that the fibrils increase in diameter between birth and maturity.

The collagen fibril diameter distributions in six tendons or ligaments from rat have also been measured as a function of age (Table 2; 4, 14, 15, 16). The results from these studies may be summarised as follows. Firstly, the collagen fibril diameter distributions were unimodal and sharp (standard deviation <3.5 nm) during the foetal stages of development and at birth. The mean diameters of the fibrils lay close to a multiple of 8 nm. Secondly, after birth the mass-average diameter of the fibrils increased rapidly until maturity. The form of the collagen fibril diameter distribution at this age may be either unimodal (tibial collateral ligament) or bimodal (tail tendon, *flexor digitorum longus* and Achilles tendon). As with the chick metatarsal tendon, some distributions of collagen fibril diameter were resolved into constituent populations which had mean values close to multiples of 8 nm (16). Finally, beyond maturity the mean and mass-average diameter of the collagen

←

Figure 3. Transverse sections through the corneal stromata of five orders of vertebrates (a) cartilaginous fish (dogfish); (b) amphibian (salamander); (c) reptile (snake); (d) bird (magpie); (e) mammal (rabbit). In each case, the collagen fibrils have a mean diameter ~25 nm (reprinted from Craig and Parry (8) with permission of Academic Press). Bar = 0.1 μm.

Figure 4. Transverse sections of a ligament and a tendon from a five year old horse: (a) suspensory ligament showing a bimodal distribution of fibril diameters; (b) common digital extensor tendon showing an unimodal distribution of fibril diameters. Insets show the number and mass distributions of the collagen fibrils plotted as a function of diameter. The graduation marks on the horizontal (diameter) axes are 100 nm apart. It can be seen that the mass associated with the larger diameter fibrils accounts for the bulk of the collagen present in the tissue irrespective of the precise form of the number distribution (reprinted from Parry, Craig and Barnes (5) with permission of the Royal Society of London). Bar = 0.5 μm.

Table 2. The form, mean and mass-average diameters of the collagen fibril diameter distributions in connective tissues as a function of age

Tissue	Source	Foetal	Birth	Maturity	Senescence	Age	Mean diameter (nm)	Mass–average diameter (nm)
Tail tendon	Rat	*	*	* *	* *	18d F	24	24
						0d	30 & 49	31 & 49
						2–3w	97	115
						5w	133	210
						8w	185	320
						13–14w	204	340
						13mo	139	333
						2y	146	319
Tibial collateral ligament	Rat	(*)	*	*	–	5d	47	48
						8w	105	140
						16w	128	173
Flexor digitorum longus tendon	Rat	(*)	*	* *	–	5d	41	41
						8w	126	175
						16w	133	204
Achilles tendon	Rat	(*)	*	* *	–	5d	51	52
						8w	107	160
						16w	128	214
Forelimb flexor tendon	Rat	*	*	–	–	18d F	24	24
						0d	31	31
						2d	33	35
						5d	43	45
Hindlimb flexor tendon	Rat	*	*	–	–	18d F	18	18
						0d	32	32
						2d	35	37
						5d	33	33
Common digital extensor tendon	Horse	*	*	*	* *	9mo F	105	141
						1y	170	198
						1.5y	185	252
						3y	165	211
						5y	162	238
						19y	37	157
Superficial flexor tendon	Horse	*	*	* *	* *	9mo F	100	120
						1y	96	177
						3y	51	143
						5y	53	187
						19y	35	139
Suspensory ligament	Horse	*	*	* *	* *	9mo F	110	147
						1y	111	197
						5y	103	191
						19y	43	152
Metatarsal tendon	Chicken	*	*	* *	–	11d F	32	33
						12d F	39	39
						13d F	42	43
						14d F	40	41
						17d F	41	41
						18d F	51	55
						A	137	240
Diaphragmatic ligament	Guinea pig	(*)	*	*	–	0d	45	52
						7d	38	47
						17d	59	74
						66d	68	87
						A	88	132
Extensor tendon	Guinea pig	(*)	*	* *	–	0d	72	91
						7d	79	102
						17d	80	106
						66d	128	166
						A	161	221

Table 2. (Continued).

Tissue	Source	Foetal	Birth	Maturity	Senescence	Age	Mean diameter (nm)	Mass–average diameter (nm)
Flexor tendon	Guinea pig	(*)	*	* *	–	0d	70	86
						7d	82	115
						17d	66	90
						66d	61	98
						A	71	121
Extensor tendon	Sheep	*	*	–	–	60d F	39	39
						72d F	41 & 49	41 & 50
						120d F	68	94
						5w	87	111
Flexor tendon	Sheep	*	*	–	–	60d F	25 & 33	26 & 34
						72d F	33 & 41	34 & 41
						120d F	69	81
						5w	91	113
Ligament	Sheep	*	*	–	–	70d F	25 & 32	26 & 33
						72d F	40	41
						120d F	65	80
						5w	90	112
Achilles tendon	Human	*	*	* *	–	–	–	–
Plantaris tendon	Human	–	–	* *	–	–	–	–
Extensor indicis tendon	Human	–	–	* *	–	–	–	–
Flexor digitorum superficialis tendon	Human	–	–	* *	–	–	–	–
Extensor digitorum longus tendon	Human	–	–	* *	–	–.	–	–
Flexor digitorum profundus tendon	Rabbit	–	–	* *	–	A (compressive side)	71	150
						A (tensional side)	125	204
Cornea	>30 different animals	*	*	*	*	Bony fish past birth.	17	17
						All other animals past birth.	24	25
Skin	Rat	*	*	*	*	18d F	24	25
						0d	32	33
						2d	38	38
						5d	31 & 38	31 & 38
						8w	113	121
						16w	130	137
						20mo	91	97
Skin	Sheep	*	*	*	–	60d F	23	23
						60 & 72d F	30	31
						120d F	83	87
						0d	81	88
						A	64	67
Skin	Guinea pig	(*)	*	*	–	0d	75	84
						7d	85	88
						17d	71	77
						66d	86	90
						A	68	72
Skin	Human	*	*	*	*	14w F	25	25
						24w F	60	62
						5d	67	70
						20y	89	93
						70y	75	77

Table 2. (Continued).

Tissue	Source	Foetal	Birth	Maturity	Senescence	Age	Mean diameter (nm)	Mass–average diameter (nm)
Skin	Chicken	(*)	*	*	–	0d	50	52
						A	92	94
Skin	Greyhound	(*)	(*)	*	–	A	82	89
Skin	Lamprey	(*)	(*)	*	–	A	55	55
Skin	Trout†	–	–	**	–	18mo	61	119
							129	192
Skin	Rat tail	–	–	**	–	5mo	108	154
						1y	115	156
Foot–pad skin	Guinea pig	(*)	*	*	–	0d	67	70
						66d	54	61
						A	53	61
Sclera	Human	*	–	–	–	–	–	–
Fibrocartilage	Rat	–	–	**(?)	–	8w	35	57
Fibrocartilage	Human	–	–	**(?)	–	4y	63	–
						7y	65	–
						16y	58	–
						35y	43	–
						67y	48	–
Hyaline (articular) cartilage	Human	(*)	*	**(?)	**(?)	maturing	26	28
						mature	53	66
						old	54	82
Nucleus pulposus	Human	–	–	**(?)	–	4y	27	–
						7y	30	–
						16y	26	–
						35y	34	–
						67y	37	–
Aortic valve	Human	(*)	(*)	*	–	–	–	–
Lung	Human	(*)	(*)	*	–	–	–	–
Arachnoid membrane	Human	(*)	(*)	*	–	–	–	–
Sural nerve endoneurium	Human	(*)	(*)	*	–	–	–	–
Endoneurium	Rat	(*)	*	*	–	5d	31	32
						8w	44	45
						16w	48	48
Paratenon	Rat	(*)	*	*	*	0w	40	40
						2–3w	42	44
						5w	50	51
						8w	50	51
						13–14w	50	52
						13mo	45	46
						2y	38	40

*, ** Unimodal and bimodal distribution of collagen fibril diameters respectively.

(*) Unimodal fibril diameter distribution has not been measured, but its form has been deduced from data collected at a later stage of development.

† The considerable local variation in fibril diameter distribution in different layers of the skin.

F, A Foetal and adult respectively.

fibrils in rat tail tendon decreased from the maximum value noted at maturity. The largest fibrils frequently became mis-shapen and had the appearance of fibrils which were breaking down.

Developmental studies on sheep ligament and extensor and flexor tendons (16) have indicated a pattern of growth similar to that seen in tendons or ligaments from the horse and the guinea pig. In the latter half of the foetal development of sheep (gestation period ~145d) the collagen fibrils in these tissues develop rapidly. At about 60–72d foetal, the fibrils have a uniform diameter with a mean value which is close to a multiple of 8 nm (Figure 7). Occasionally the fibril diameter distributions can be

45

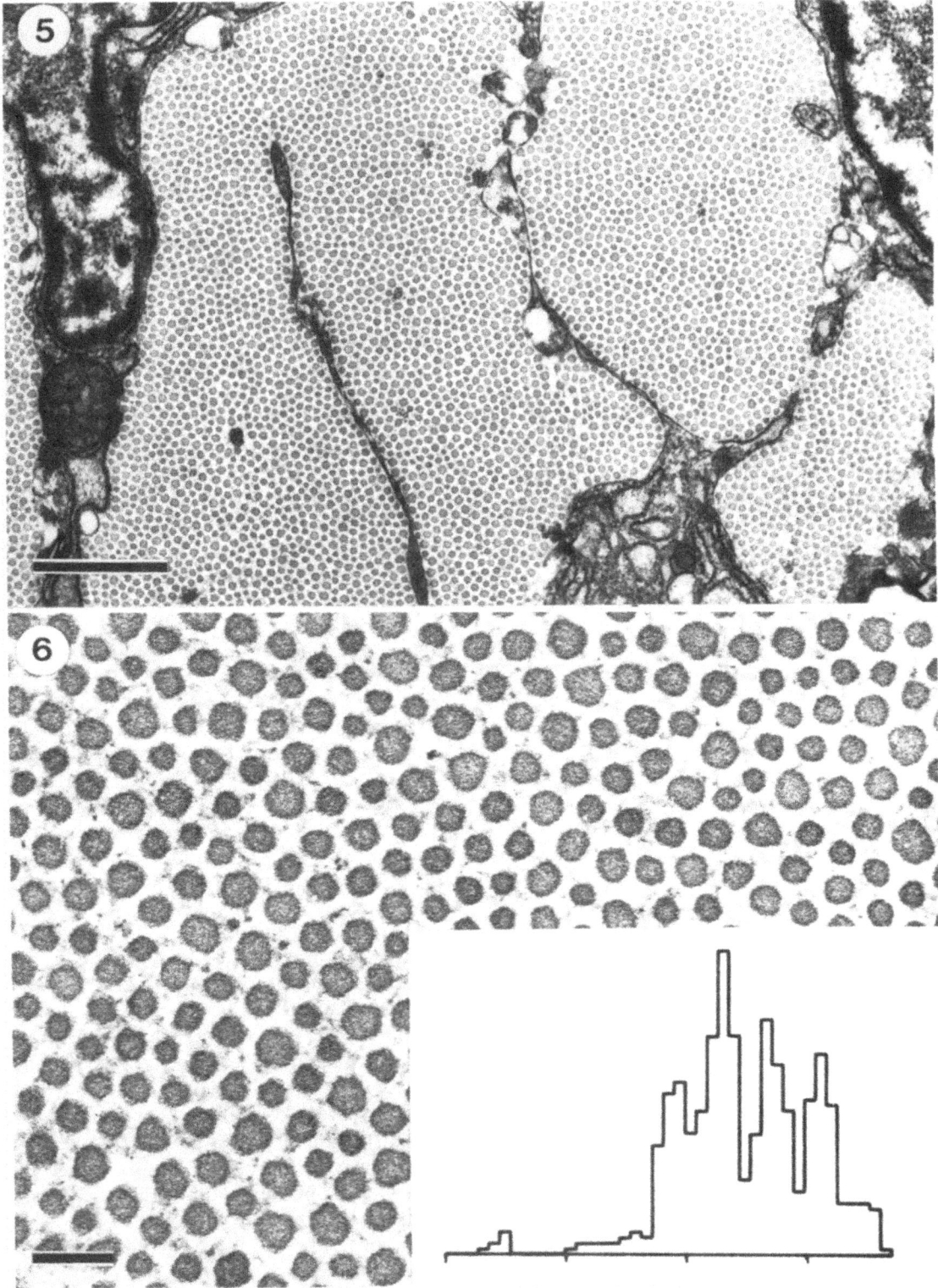

Figure 5. Transverse section of 18-day foetal chick metatarsal tendon. Bundles of collagen fibrils may be seen interleaved between the cellular processes of the fibrocytes. Bar = 1 μm.

Figure 6. Transverse section of 18-day foetal chick metatarsal tendon. The number distribution of collagen fibril diameters (inset) shows that the fibrils appear to have preferred diameters which are close to multiples of 8 nm. The graduation marks on the horizontal (diameter) axis are 20 nm apart. Bar = 0.2 μm.

partially resolved into components similar to that already noted for the chicken and rat. By 120d foetal, the collagen fibril diameter distributions for all three tissues were broad and unimodal and had mass-average diameters ≃80–90 nm. At the age of 5w after birth, the mass-average diameters increased to ~110 nm. Thus the tissues had collagen fibril diameter distributions which were similar at each of the ages studied.

3.3. Skin

Skin is the tissue which forms the physical limit between an animal and its environment. In most vertebrates its outermost layer is composed of a keratinising stratified squamous epithelium. The basal (germinal) layer of the epidermis is commonly undulating and is separated from the underlying dermis by a basement membrane. The most superficial part of the dermis is thrown into papillae which interdigitate with the undulations of the epidermis.

The collagen fibrils of the papillary layer of the dermis are usually smaller in diameter than those deeper in the dermis, and are rather randomly orientated though many of the fibrils within the papillary core lie at right angles to the epidermal surface. The remainder of the dermis (the reticular layer) forms the greater part of the skin and consists of elongate fibrocytes, bundles of collagen fibrils and the associated hydrated proteoglycan matrix (Table 1).

In developing skin type III collagen is synthesised by the fibrocytes but, with increasing age, type I collagen synthesis takes over and becomes the dominant genetic type in the mature tissue. Tajima and Nagai (17) have shown that the sizes of the collagen fibrils, the ratio of type I to type III collagen and the content and composition of the glycosaminoglycans all vary in a systematic way with distance below the epidermis. Thus skin is not a macroscopically homogeneous connective tissue; significant biochemical and ultrastructural variations do occur with depth. Further, gross histological variations in the skin are to be found at different sites on the body of a given animal. With these limitations in mind, electron microscope studies of the transverse sections of a number of developing skins have still been able to reveal common trends in growth and development.

Skins may be divided into two functional groupings. Firstly, there are those skins which respond to environmental stimuli but which nonetheless have an essentially passive mechanical role. In such cases, the elastic, shear- and pierce-resistant skin must limit the exchange of water and other materials between the animal and its environment and also act as an energy dispersive device. The latter facility will minimise the effects of accidental contact with an object during locomotion and attack either by predators or animals of the same species (18, 19). Thus, vulnerable underlying tissues are protected against extensive and debilitating damage. Secondly, there are those skins which have an active role in addition to the passive one previously noted. For example, shark skin is as strongly attached to the underlying musculature as the latter is to the backbone; consequently the skin acts like an external tendon (20). Since tendon has the ability to store energy and return about 60% of it usefully to the animal when the stress is removed (19), the exotendinous action of some skins clearly play an important locomotory role.

The 'passive' skins, such as those from human, rat and sheep, generally comprise collagen fibrils with a relatively small spread of diameters. Anastomosing bundles of these collagen fibrils in the skin form a complex network with little obvious three-dimensional organisation though it is not uncommon for the fibrils to lie within planes which are approximately parallel to the surface of the skin. The fibril bundles show varying degrees of crimp and this is related to the orientation of the Langer lines.

The 'passive' vertebrate skins from a selection of altricious animals (human, rat and greyhound) have been studied. In the early stages of development, rat skin has a very sharp unimodal distribution of collagen fibril diameters (24 ± 2 nm at 18d foetal) but between birth and maturity the collagen fibril grow rapidly in size (131 ± 20 nm at 16 w). Beyond maturity, however, the mean fibril diameter decreases (91 ± 17 nm at 20 months). Similarly, human skin shows the same trend (Table 2) as does chick skin over the age range studied (birth to maturity). The maximum values of the mass-average diameters of the collagen fibrils in human, chick, greyhound and rat skin are 93, 94, 89 and 137 nm respectively and were all attained at a time close to the onset of maturity.

The 'passive' skins from precocious vertebrates such as sheep and guinea pig develop in a slightly different way. The collagen fibrils in sheep skin increase rapidly in mean diameter during foetal development and reach a maximum value at a time close to birth (Table 2). For mature sheep skin, the mean collagen fibril diameter decreases from 81 ± 17 nm at birth to about 64 ± 10 nm. Similarly, in guinea pig skin the collagen fibrils have a relatively constant diameter during the period from birth to 66d (Table 2), but at adulthood the mean fibril diam-

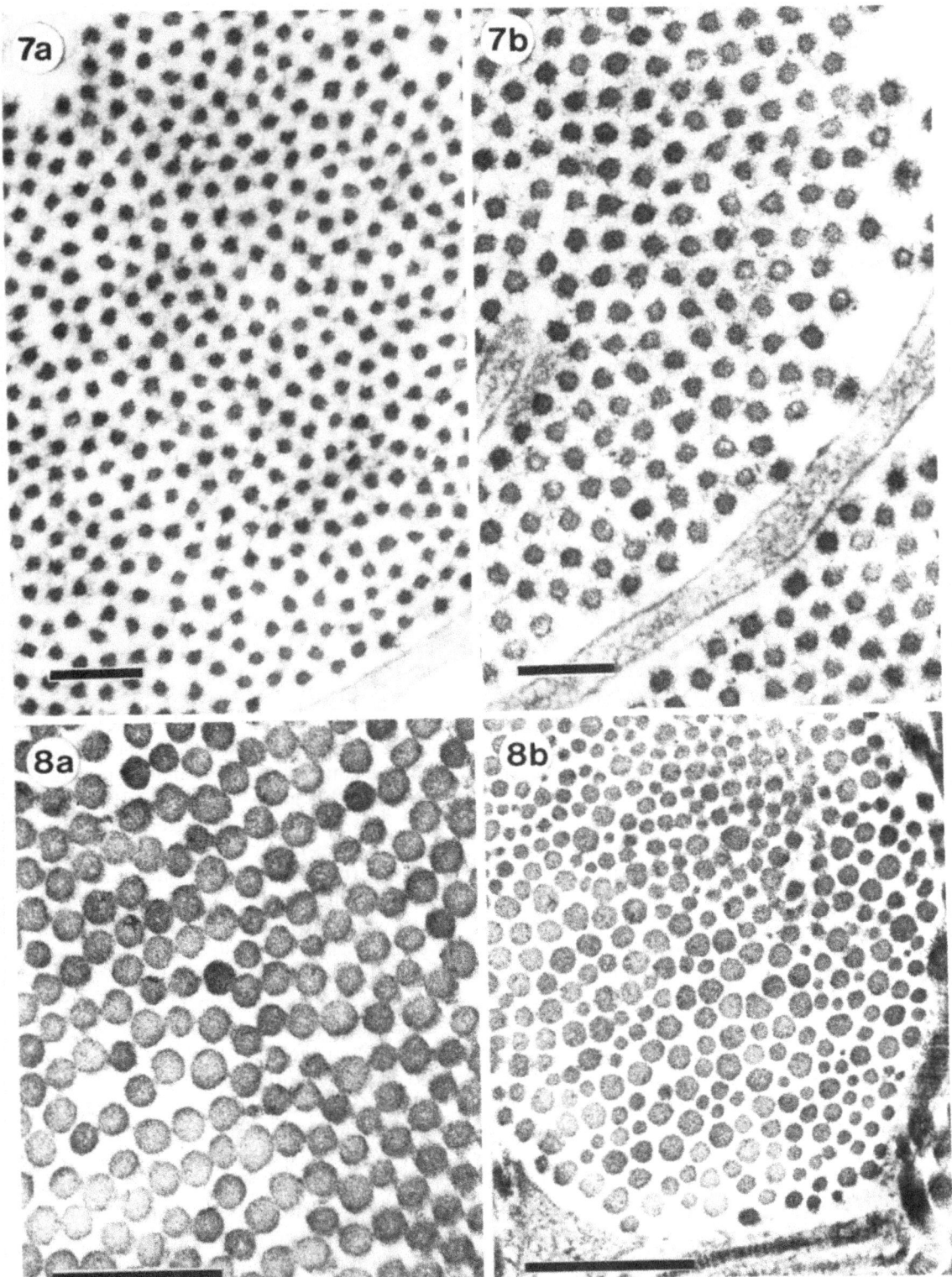

Figure 7. Transverse section of superficial flexor tendon from a foetal sheep. The micrographs show uniform and sharp distributions of fibril diameter (standard deviation of the distributions <3.5 nm) with means ~24 (a) and 40 nm (b) respectively. Bars = 0.2 μm.
Figure 8. Transverse section of (a) adult greyhound skin (mass average diameter ~89 nm) and (b) 66-day guinea pig footpad skin (mass average diameter ~61 nm). Bars = 0.5 μm.

eter decreases. Thus in both sheep and guinea pig, the collagen fibrils have their largest diameter at birth (or at an age close to birth), and this reflects the high degree of development of these animals when born. Interestingly, the maximum mass-average diameters of the collagen fibrils in sheep and guinea pig skins are attained at birth and are 88 and 90 nm respectively; these values are very close to the maxima, attained at maturity, for human, chick and greyhound skins (93, 94 and 89 nm respectively). It can thus be seen that the passive skins studied to date all have a maximum mass-average diameters ~90 nm (Figure 8a), the only exception is the skin of the rat. The higher value obtained for this skin (~137 nm) may reflect the unusual growth pattern of the rat, an animal that never attains true maturity in that in continues to grow throughout its adult life. On these grounds, it can be criticised as a laboratory model in that it is atypical of other animals.

The skin of the footpad of guinea pig has also been studied (Figure 8b; Table 2). In this case, the collagen fibrils have diameters which are lower than those reported for back skin at corresponding ages. Footpad skin is predominantly a pressure-bearing tissue and consequently is not expected to contain large diameter fibrils (21). Indeed it is expected that a low tensile strength but good creep-resistant tissue should be composed predominantly of smaller diameter collagen fibrils as observed. In each of the examples so far listed, the collagen fibril diameter distributions have had a unimodal form with a relatively narrow range of sizes.

In contrast, the 'active' skins from shark (20) and trout (22) contain layers of well-orientated collagen fibrils rather than the more random organisation found in most 'passive' skins. These layers are arranged helicoidally and in an alternating sense around the body of the fish in such a way that fibrils in adjacent layers commonly make an angle ~60° with one another. Thus collagen fibrils in such skins are cyclically stressed and relaxed as the fish swims. 'Active' skins, however, are not confined to the fish but are found also in amphibians (tadpole) and mammals (whale and rat tail skin). Tadpole and whale skins also exhibit a prominent layered structure. Two active skins studied in some detail were those from the trout and the tail of the rat (22). In trout skin, the distribution of fibril diameters differed significantly from one location to another; the broad distributions were either bimodal (mean diameter = 129 nm, mass-average diameter = 192 nm) or right skewed (mean diameter = 61 nm, mass-average diameter = 119 nm, see figure 9). Rat tail skin had a collagen fibril diameter distribution which

was markedly bimodal at the age of 5 months but which became right skewed by 12 months. At these ages, however, the mean and mass-average diameters remained virtually unchanged at ~110 and ~155 nm respectively. Fibrils with diameters ~300 nm were contained within all of these distributions; such large diameter fibrils do not occur in the passive skins. Thus the collagen fibril diameter distributions for active skins are intermediate between that for a passive skin and a tendon.

It is apparent from these results that the ultrastructural organisation of the collagen fibrils in skin varies substantially between animals and even between different parts of the same animal. Further, the form of the collagen fibril diameter distribution ranges from unimodal and sharp, right through to bimodal and broad, and, as will be shown in Section 4, is a direct consequence of the particular mechanical role played by the skin.

3.4. Cartilage

Cartilage is a resilient connective tissue which has evolved through its ability to withstand high compressive stresses. In addition to its cellular component (the chondrocytes) and the water-rich proteoglycan matrix, cartilage consists of a network of small diameter collagen fibrils (Table 1); these endow the tissue with its limited tensile attributes. The fibrils are composed predominantly (though not entirely) of highly glycosylated type II collagen molecules. The proteoglycan aggregate of the matrix, which provides much of the all-important compressive properties, consists of chondroitin sulphate and keratan sulphate chains covalently linked to protein 'cores' which are in turn attached via link proteins to a hyaluronic acid molecule. Cartilage, like cornea, is essentially an avascular tissue and thus the colloidal properties of its matrix, apart from contributing to its resilience, are very important for the nutrition of its cells. Cartilage is classically recognised as having three distinctive forms; hyaline cartilage, elastic cartilage and fibrocartilage. It must be remembered, however, that the term hyaline cartilage is used rather loosely to specify any of a variety of cartilage types which cannot be more precisely defined as either elastic cartilage or fibrocartilage.

Hyaline cartilage is that encountered in the embryo as a template for most bones of the axial and appendicular skeletons. It persists in adult animals in long bone epiphyseal plates, on joint surfaces of bones (articular cartilages), on the ventral ends of ribs and as structural elements of the trachea and larynx. Hyaline cartilage has chondrocytes in lacunae which are elliptical in section immediately beneath the perichondrium or under the free surface of articular cartilage. In these superficial positions the long axes of the lacunae lie parallel to the surface of the cartilage since their chondrocytes arise from the circumferential chondrogenic layer of the perichondrium by a mechanism known as 'appositional growth'. Deeper within the tissue the lacunae become larger and more angular as they become swollen by 'interstitial growth' which gives rise to isogenous groups of chondrocytes.

Few sources of hyaline cartilage have been studied ultrastructurally though some have suggested that the collagen fibrils have diameters which typically lie in the range 35–120 nm (23). One hyaline cartilage that has been investigated is human articular cartilage. In this tissue the most superficial fibrils lie parallel to the surface and provide the tangential forces required to withstand swelling pressure, whereas the deepest fibrils, which act as connecting elements between cartilage and the underlying bone, are approximately perpendicular to the surface (24). The mean and mass-average diameter of the collagen fibrils in maturing, mature and old tissue show a steady increase (Table 2; 15, 25). In the old cartilage studied, the fibrils ranged in diameter from 5 to 160 nm and had a right skewed distribution of sizes that approximated to a bimodal form (25). Maroudas and Muir (26) have noted that the collagen content of articular cartilage decreases and the glycosaminoglycan content increases from the superficial to the deeper aspects of the tissue. Further, it is well established that the tensile and compressive properties of the cartilage also vary significantly with site and depth. Considerable care must therefore be taken in interpreting the electron microscope observations.

In the *nucleus pulposus* of human invertebral disc the collagen fibrils have diameters ~30 nm (Table 2), and it has been established that fibril diameter increases slowly with age (27). Dahmen (28) has also studied this tissue and has suggested that the fibril diameter distribution may be bimodal. Others have noted that the collagen fibrils in young rabbit *nucleus pulposus* have diameters close to 30 nm.

Elastic cartilage is characterised by the occurrence of chondrocytes scattered single or in isogenous groups throughout the proteoglycan matrix, an abundance of elastic fibres and a relatively small complement of collagen fibrils. At the present time, however, there are no collagen fibril diameter distributions published for elastic cartilage though it is known that the fibrils have a mean diameter which is not grossly different from that noted for other forms of cartilage. In the mammalian body, elastic cartilage is found in the pinna of the ear, the walls of the auditory and eustachian tubes, the epiglottis and in parts of the corniculate and cuneiform cartilages.

Fibrocartilage is that encountered in some specialised areas of dense connective tissue of the body. It often has poorly defined limits but is characterised by the presence of typical chondrocytes in accompanying pockets of matrix, lying among a dense network of orientated collagenous elements. Fibrocartilage occurs in some articular cartilages, the sites of insertion of some tendons into bones and in the *annulus fibrosus* of the intervertebral discs. The *annulus fibrosus* forms a peripheral link between the adjacent vertebrae and is constructed from 10 to 15 lamellae arranged in concentric cylindrical sheets. Polarising microscopy has shown that the collagen fibres are orientated at 40–70° to the vertebral axis with opposite directions of tilt occurring in adjacent lamellae (29). Type I collagen predominates in the outermost lamellae of the *annulus fibrosus* whereas in the inner lamellae it is type II collagen which is dominant. It has also been noted that the type II collagen-rich inner regions are usually mechanically weaker than the type I collagen-rich outermost regions and also that the inner part of the *annulus fibrosus* is more liable to clefting. It thus seems probable that the mechanical properties of this tissue will differ significantly from one location to another.

The collagen fibril diameters in *annulus fibrosus* have been measured by several groups though the actual form of the distributions have never been published. However, it has been reported (15, 27, 28) that the collagen fibril diameter distributions were often bimodal or right skewed (Figure 10). Mean diameters (or the range of diameters) have been noted for rat (15), human (27) and pig (30). The mean values lie generally in the range 35–65 nm (Table 2) whereas the mass-average diameters are probably ~20 nm higher (i.e. 70 ± 20 nm). The age-related study of human *annulus fibrosus* (27) has shown that the mean diameter of the collagen fibrils decreases with increasing age, though it must be emphasised that this does not necessarily imply that the mass-average diameter will show the same trend. It is quite possible for the mean value of the distribution to decrease whilst the mass-average diameter increases. As will be discussed in detail in Section 4, these two values are related to the tensile strength of a tissue and its creep-resistant properties.

From the limited data reported here, it is apparent that there is little definitive information on the collagen fibril diameter distributions in cartilage of any

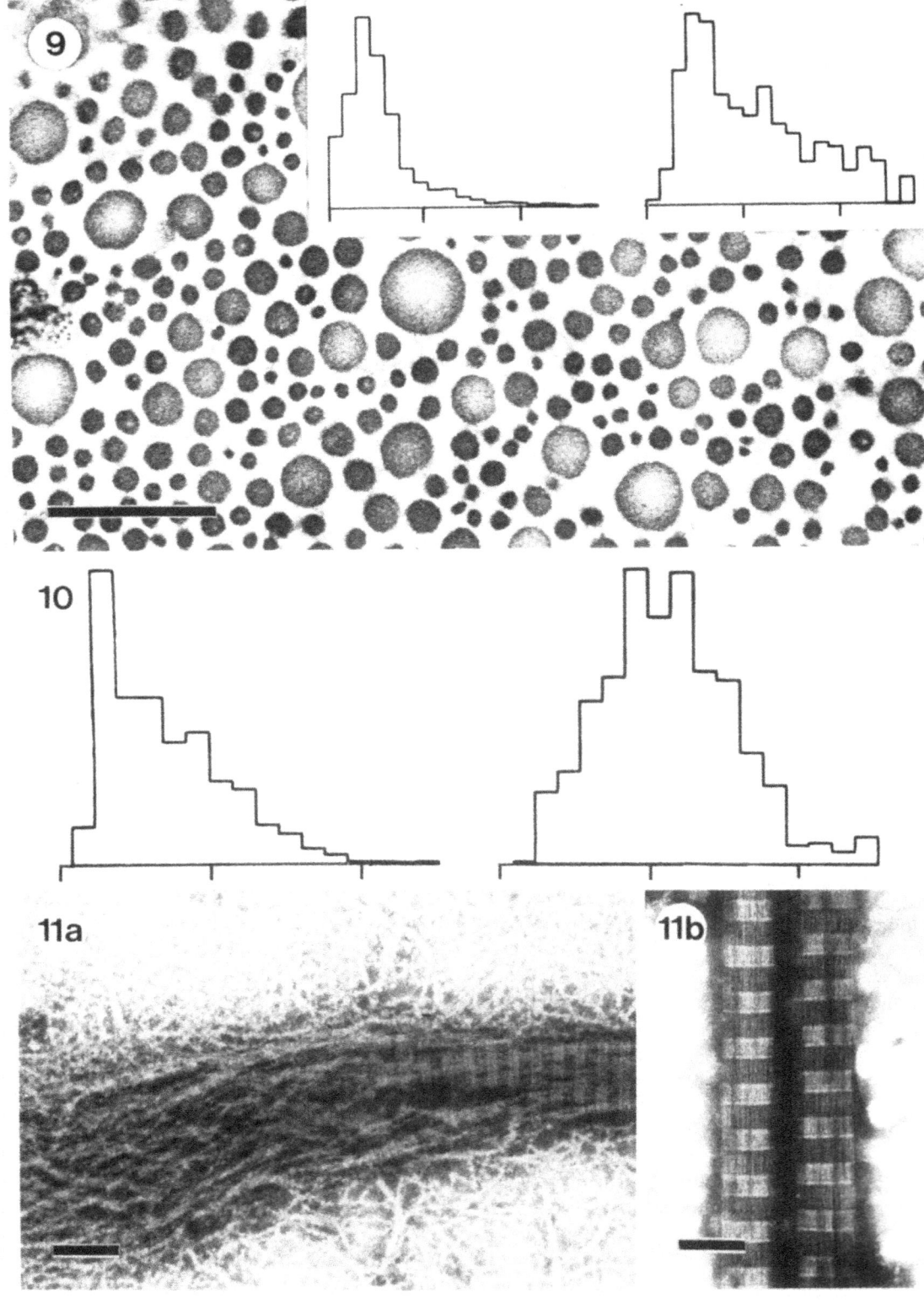

type. Indeed, even in the work that has been undertaken, it can be argued that most of the results are not easily interpreted; they cannot be related to one another, as the location of the specimen taken from the cartilage has not been specified precisely. Clearly, relatively little is known at present about the mechanism of growth and development of collagen fibrils in cartilage.

3.5. Bone and dentine

Bone and dentine differ significantly from other connective tissues in that the component collagen fibrils act as sites of nucleation for the deposition of mineral salts. The important physical properties of bone are that it is a lightweight, elastic material having both high compressive (~150 MPa) and tensile (50–250 MPa) strengths. Bones have both metabolic and mechanical roles; they not only act as a readily available source of calcium for the homeostatic regulation of this element in the blood and body fluids, but they also act as structural components providing internal support in the vertebrate body. Bones provide the sites for attachment and insertion of the muscles and tendons which are essential for locomotory and respiratory movement. In addition, bones form protective cages for the sanction of the vital organs of the thoracic and cranial cavities. At the macroscopic level it can be seen that bone has an outer layer, which is composed of cortical or compact bone, and an inner region composed of trabecular or cancellous bone. There is no sharp boundary between these two types, rather one blends naturally into the other. The outer surface of most bones is covered by an osteogenic layer known as the periosteum. The cellular component of bone is very small in terms of the total mass of the tissue and in compact bone consists largely of osteocytes lying in lacunae. The osteocytes, lacunae and their surrounding matrix are organised into lamellae which lie either parallel to the surface near the periphery of bones, remain ordered in small pockets of interstitial bone or are disposed circumferentially about haversian canals to form haversian systems (osteons).

The mechanical properties of this tissue are largely attributable to the interstitial substance, which has two major components – the organic matrix and the inorganic salts. The organic matrix is largely collagenous with a small amount of glycosaminoglycans (Table 1). The collagen fibrils of the matrix are composed largely of type I collagen molecules although some type V collagen is also known to be present. In lamellar bone, the collagen fibrils are arranged in layers which are concentrically wound around the haversian canals. This arrangement of fibrils gives rise to alternating dark and light bands when the haversian systems are viewed in polarised light. The organic matter of bone consists of hydroxyapatite crystallites and amorphous calcium phosphate. Further, some citrate, carbonate, fluoride, sodium and magnesium ions are found complexed to or in association with the mineral. The microcrystals of mineral are intimately associated with the collagen fibrils but are not randomly arranged along them, rather they appear to be D-periodically disposed. It is thought that the gap regions of the fibrils may act as the nucleation sites for mineralisation (31, 32). The inorganic constituents are responsible for the hardness of bone whereas the organic components of the matrix (and the collagen fibrils in particular) are responsible for its toughness and resilience. These features taken together make bone an ideal skeletal material ideally adapted for its chemical and mechanical functions.

Dentine, although having a chemical composition which is very similar to that of bone, has a greater hardness and a structure which is considerably different. Macroscopically, it appears radially striated due to the presence of innumerable canals, known as the dentinal tubules, which radiate from the pulp cavity. These tubules are occupied by cytoplasmic

←

Figure 9. Transverse section of the 'active' skin from trout. In this tissue, the collagen fibrils have a right skewed distribution of diameters (left inset) similar to that seen for many tendons. Such a distribution of diameters is consistent with trout skin having an exotendinous role. The mass of collagen in this skin is divided equally between the smaller and the larger diameter fibrils as can be seen in the mass distributions of diameters plotted at right inset. The graduation marks on the horizontal (diameter) axis are 100 nm apart. Bar = 0.5 μm.

Figure 10. Number distribution (left) and mass distribution (right) of collagen fibril diameters in *annulus fibrosus* from an eight-week-old rat. The graduation marks on the horizontal (diameter) axis are 50 nm apart.

Figure 11. Negatively stained preparation of collagen fibrils: (a) Rat tail tendon after treatment with acetic acid. The D-periodic fibril can be seen to fray out into what appear to be subfibrillar elements. The diameter of such elements has been reported to lie in the range 3–12 nm. Bar = 0.2 μm. (b) Whole mount preparation of stretched cuvierian tubules from the sea cucumber *Holothuria forskali*. The collagen fibrils are longitudinally striated and this may be seen most clearly in the darker staining 'gap' regions of the fibril. Bar = 0.1 μm.

processes of odontoblasts which are located on the pulp cavity extremities of the dentine. Embedded in the organic matrix between the dentinal tubules are the collagen fibrils which appear to be preferentially orientated parallel to the long axis of the tooth and perpendicular to the dentinal tubules. As in bone the collagen fibrils are predominantly composed of type I collagen molecules.

Ultrastructurally, bone and dentine have not proved easy tissues to study since their mineral content has rendered them opaque in the transmission electron microscope. However, this problem can partially be overcome by studying either demineralised bone or immature bone prior to mineralisation. Very limited quantitative information on the growth and development of collagen fibrils in bone have been published though Robinson and Watson (33) have reported that the collagen fibrils in human bone increase in diameter with age from a minimum of about 15–40 nm to a maximum of about 150 nm. Jackson (34) has also reported that in the fibroblastic layer of the periosteum of avian embryonic bone the collagen fibrils have diameters of 10 and 25 nm at 9 and 16 days respectively, whereas in the osteoblastic layer, the diameters of newly formed fibrils are typically 40 nm. Jackson summarised her results by stating that 'individual fibrils of growing bone enlarge, which agrees with the general experience that intercellular fibrils continue to increase in diameter with increasing age'.

Some workers have stated that mature bone contains collagen fibrils with diameters in the range 50–70 nm whereas others have suggested that the distribution of diameters appear to be unimodal with a mean value ~80 nm. The authors have studied phalangeal bone from neonatal guinea pigs and showed that the collagen fibril diameters ranged from about 20 to 140 nm with a mean value close to 80 nm. Thus the present data are consistent with the growth and development of collagen fibrils in bone tissue being similar to that observed for the non-mineralising connective tissues. However, as little comparative data on collagen fibril diameters in developing bone is available, it is possible that particular bones from various animals will exhibit different distributions of fibril diameters.

Information of collagen fibril diameters in predentine and dentine is also rather sparse. Collagen fibrils in the proximal region of rat predentine have small diameters which are typically 20–30 nm though in the distal predentine the mean fibril diameter is known to increase (35). Further, Watson and Avery (36) have noted that in the proximal predentine of the hamster incisor the collagen fibrils are about 10–30 nm in diameter but that the diameters increase to about 60 nm in the most distal predentine i.e. at the mineralisation front. However, in the dentinal region of the rat incisor, the collagen fibrils range in diameter from about 20 to 60 nm with little evidence of any specific orientation being present. In addition, some very large diameter fibrils (~100–200 nm), known as the von Korff fibrils, are occasionally seen extending from the pulp to the dentinal matrix and passing between the odontoblasts (37).

3.6. Other connective tissues

Although many connective tissues, other than those already described in the earlier parts of this section, have been studied by electron microscopy, rarely have their constituent collagen fibrils been studied as a function of age. A little information, however, is available and some of these results will now be presented. The collagen fibrils in a few other tissues which have some unusual characteristics are also described.

Paratenon is that connective tissue which lies peripherally about a tendon. The collagen fibrils are generally small in diameter and are embedded in a matrix rich in hyaluronic acid. It has been suggested that hyaluronic acid may facilitate a smooth sliding motion between the tendon and the connective tissue fascia through which it passes, though this point is not universally accepted. All of the collagen fibril diameter distributions measured for the paratenon of rat tail tendon were shown to be unimodal in form; the mass-average diameters increased between birth and maturity before decreasing as the rat approached senescence (Table 2; 22). This pattern of growth is typical of the collagen fibrils in the connective tissues of altricious animals (see Section 4) and is further supported by data obtained from the diameter distributions measured for the collagen fibrils constituting the endoneurium of rat sciatic nerve (Table 2; 15).

Two other specialised connective tissues, namely arachnoid membrane and sclera, were studied by Schwarz (13). The former is a loose connective tissue which covers the brain and spinal cord and, in man, was shown to contain collagen fibrils with mean diameters ~40 nm at 30y and ~70 nm at 70y. Sclera, which is a dense collagenous membrane protecting the posterior chamber of the eye, is continuous with the cornea. The collagen fibrils in human sclera are much larger than in cornea and also exhibit a greater range of diameters. The mean collagen fibril diameters in sclera (Table 2) increase steadily from the foetal stage of development right through to adulthood (13).

The *tapetum fibrosum* from the sheep's eye consists of several hundred layers of well-orientated collagen fibrils with diameters ~122 ± 21 nm (38). Even though the thicknesses and orientations of the lamellae are somewhat variable, the thicker lamellae, and hence the majority of the collagen fibrils, have a similar orientation in the *tapetum*. The role of the *tapetum* is to act as a light-reflecting device; it is located immediately behind the retina of the animal, and it functions in such a way that light, which has been transmitted through the retina, is reflected or scattered back for further processing. It is interesting to contrast the structural features of the transparent cornea, which contains small diameter collagen fibrils (~24 nm) ordered in lamellae rotated by ~90° with respect to their neighbours, and the reflecting *tapetum*, which is also layered but which contains large diameter collagen fibrils (~122 nm), many of which have a similar orientation. Unfortunately, no studies on the size of the collagen fibrils as a function of age have been carried out on this fascinating tissue.

Two connective tissues which contain very small diameter collagen fibrils, but which have not been studied as a function of age, are vitreous humour and spongin. Vitreous humour contains very little collagen but a great deal of water (Table 1). Filaments in the vitreous humour have been identified as collagenous on the basis of chemical analysis and the presence of a D-period in low-angle X-ray diffraction patterns and in electron micrographs. Also, X-ray diffraction patterns of orientated specimens of vitreous filaments (39) have shown a 0.29-nm meridional reflection characteristic of collagen and a low-angle equatorial reflection at a spacing of 9.3 nm, a value which probably represents the diameter of the constituent collagen fibrils *in vivo*. Electron microscope measurements of fibril diameters of whole mount specimens have ranged from 10 to 15 nm, and it seems possible that these higher values have arisen from flattening of the fibrils on the support grids and/ or metal build-up on the shadowed preparations.

Similar size fibrils have also been observed in tissues from invertebrates. In particular, two forms of spongin fibres, known as spongin A and spongin B, have been studied ultrastructurally; spongin A fibres contain long collagen fibrils with diameters ~20 nm, and spongin B is comprised of bundles of fibrils with diameters < 10 nm (40). Since these measurements were also made on whole mount preparations, the values quoted may again prove to be an over-estimate of the true fibril diameter if specimen flattening has occurred.

4. Mechanical properties of developing connective tissue

Since connective tissues play a predominantly mechanical role, it is important to understand how the different chemical constituents endow the tissue with the mechanical properties that it requires. It may be considered that the collagen fibrils in connective tissue provide much of the tensile strength and the hydrated glycosaminoglycan components of the matrix account for the compressive properties. Such a fibril-matrix composite is optimally designed to resist fracture and crack propagation in the tissue. Further, a crimp in the fibre bundles present in some connective tissues, provides a compliance mechanism which allows for the rapid initial extension of the tissue (~3–4%) without stretching the constituent collagen fibrils.

Before discussing how the mechanical properties of a tissue may be related to collagen content, fibril size and fibril orientation, it is necessary to understand something of the way in which the collagen fibril diameter distributions in a diversity of connective tissues vary as a function of age. Parry, Barnes and Craig (21) undertook such an analysis and, as a result, were able to put forward eight general observations. The data on which those conclusions were based have been increased substantially over the last few years (Table 2), and most of the conclusions have received further confirmation. The eight points may be briefly summarised and updated as follows:

(a) During foetal development and at birth, the distribution of collagen fibril diameters is unimodal. The standard deviations of the distributions vary considerably between the different tissues.

(b) The form of the collagen fibril diameter distributions at birth is sharp for altricious animals (i.e. those which do not actively forage for themselves for a reasonable period after birth) but broad for precocious animals (i.e. those which are capable of active locomotion from a time shortly after birth).

(c) The mass-average diameter of collagen fibrils increases between birth and maturity. The only exceptions known at present are the fibrillar collagen from the corneal stromata of all animals, and from the skins of the two precocious mammals studied.

(d) Orientated type I tissues which are subjected to long-term, high-stress levels have a bimodal distribution of collagen fibril diameters at maturity.

(e) As tissues, other than cornea, age beyond maturity, the mean and the mass-average diameters of the collagen fibrils decrease. There is some evidence that cartilage may prove to be an exception to this general observation.

(f) The ultimate tensile strengths of connective tissues are positively correlated with the mass-average diameter of the collagen fibrils.

(g) The collagen fibril diameter distribution present in a connective tissue is closely related to the magnitude and duration of stress levels encountered by that tissue.

(h) The collagen fibril diameter distribution and the mechanical properties of a tissue are strongly correlated.

This latter point was elaborated in some detail by the authors (15,21) who pointed out that the overall form of the collagen fibril diameter distributions was dependent on two different factors, namely its tensile requirements and its creep-resistant properties. Since a connective tissue with a high tensile strength contains collagen fibrils with large diameters (>150 nm), it was postulated that the density of stable covalent cross links occurring laterally between molecules increases with fibril diameter and hence that large diameter fibrils have a greater tensile strength than do small ones. This is a consequence of the substantial depletion of cross links about the periphery of small diameter fibrils since here the periphery relative to the internal area becomes significant. However, the density of the axial cross linking between linear arrays of collagen molecules, staggered by 4D with respect to one another within the fibril, is not related to the diameter of that fibril. Some connective tissues, such as Achilles tendon in human, require well-defined, creep-resistant properties if they are to function in the necessary manner. It was proposed that this would be achieved if the connective tissue contains a high proportion of small diameter fibrils, since the surface area per unit volume of collagen is greater for small fibrils than it is for large diameter ones. For example, a single collagen fibril of diameter D has a surface area per unit length of πD. If this same fibril could be divided into n smaller fibrils each with diameter d, then $n \pi (d/2)^2 = \pi (D/2)^2$ and the total surface area per unit length of the small fibrils would be $n\pi d = \sqrt{n} \pi D$, i.e. the surface area per unit length of n collagen fibrils would be $\sqrt{n}$ times larger than for a single fibril of the same volume. Consequently the interface between fibrils and matrix is greatly enhanced if the connective tissue has a high proportion of small fibrils. It is not immediately clear how the interactions between the matrix and the collagen fibrils, or between fibrils themselves, will endow the tissue with its creep-resistant properties, but it would be presumed that interactions between the collagen and the proteoglycans play an important role. It is known that one of the glycosaminoglycans (dermatan sulphate) interacts strongly with collagen and may in fact penetrate the collagen fibrils thus linking them together via the proteoglycans without resort to covalent interactions. The potential number of such interactions is proportional to the area of the interface between fibrils and matrix.

Thus the requirement that the tissue be strong demands the inclusion of large diameter collagen fibrils in the tissue, whereas the requirement for creep-resistance necessitates the incorporation of a high percentage of small diameter fibrils. As the authors have pointed out (21) 'in many cases, a tissue containing a unimodal collagen fibril diameter distribution has sufficient fibrils of the correct size to inhibit creep and maintain the necessary tensile strength. In other cases, the requirements of creep inhibition and appropriate tensile strength can only be achieved by a distribution of fibril diameters which appears obviously bimodal'.

Reliable mechanical data for connective tissues have generally proved difficult to obtain, and the data which have been collected may have seriously underestimated, possibly 10-fold, the in vivo functional strength of connective tissues (19, 41). In spite of these difficulties, the trend of tensile properties with age can be assessed and in recent years data have been collected for rat skin (42, 43, 44), tail tendon (43, 45), aorta (43) and bone (46). In each case, the tensile moduli were shown to increase between birth and maturity but to decrease beyond this point. These results closely parallel both the measured changes in mass-average diameter of the collagen fibrils in rat skin and tail tendon and the qualitative changes in fibril size noted for aorta and bone. Thus the mechanical data currently available provide support for the authors' hypothesis that the tensile properties of a connective tissue are closely correlated to the mass-average diameter of the constituent collagen fibrils. Additional support for this idea comes from data obtained on the ultimate tensile strength of different connective tissues. For instance, the measured ultimate tensile strength and Young modulus of mature tendon (60 MPa, 400–1200 MPa), 'passive' skin (12 MPa, 20–100 MPa) and cartilage (1.2 MPa, ~16 MPa) follow the same trend as the mass-average diameter of the collagen fibrils present in the tissue namely 160–240 nm, 90–130 nm and 70 nm respectively. Although no mechanical data are available for 'active' skins, the form of their collagen fibril diameter distributions would make it seem probable that the tensile properties of 'active' skin will be intermediate between those of 'passive' skin and tendon.

The proposed correlation between collagen fibril mass-average diameter and tensile properties is necessarily a simplification, since components other than collagen in connective tissue will contribute to the mechanical properties. Nonetheless, this correlation suggests that the collagen fibrils do play the dominant role. Fibril orientation, which is another factor of importance in determining the overall mechanical attributes of a tissue, differ both within a particular tissue and from one tissue to another (c.f. cornea 3.1, tendon 3.2, skin 3.3 and cartilage 3.4). Further, those tissues with the greatest tensile strength are those with the highest collagen contents. Although many factors influence the mechanical properties of a particular tissue *in vivo*, it has now been clearly established that the collagen fibril diameter distribution is of paramount importance.

A tissue is continually remodelled between birth and maturity in order to adjust to its changing mechanical requirements. In addition, mature tissue must have a homeostatic regulatory mechanism if the collagen fibril diameter distribution is to remain close to that required. It is believed that the duration and level of stress is communicated to the cells via a feed-back mechanism (21, 47), which incorporates a control over the synthesis of the glycosaminoglycans . This implies that differing relative amounts of the various glycosaminoglycans have a direct influence on the ultimate size distribution of the fibrils and consequently, if the mechanical requisites of a tissue change, the collagen fibril diameter distribution may be modified accordingly.

5. Subfibrillar structure

So far in this review, the observed forms of the collagen fibril diameter distributions for a variety of connective tissues have been described as a function of age. These results have been interpreted in terms of the changing mechanical requirements of the tissue with increasing age (Section 4) and also in terms of the different functional attributes, such as tensile strength and creep-resistance, required of the tissue *in vivo*. Thus the reasons for the growth and development of collagen fibrils are now beginning to become apparent, though clearly the details are not understood in other than a very general way. In this section, an attempt will be made to understand the actual mode of growth of the collagen fibrils; electron microscope plus X-ray diffraction results which relate to the substructure of the collagen fibril will be presented. Part of this analysis has been reported previously (48), but the salient points will be re-iterated here for the sake of completeness.

5.1. Electron microscope observations

Many electron microscopists have reported either that collagen fibrils have a lateral periodicity greater in magnitude than the diameter of a single molecule or that the fibrils exhibit a filamentous appearance under certain conditions of preparation and observation. While the obvious conclusion is that these data imply that collagen molecules aggregate into some sort of discrete filamentous structure, the evidence will be objectively considered and the likelihood of possible alternatives assessed. The questions will be posed as to whether the electron microscope results are capable of interpretation in terms of the aggregation of single collagen molecules into a non-filamentous fibril.

The breakdown of fibrils after chemical treatment often leads to subfibrillar elements being seen in the electron microscope (Figure 11a). The measured diameters of the filaments have been reported to lie in the range 3.0–3.5 nm (49), 4–5 nm (50) and 4.5–12.0 nm (51), and in many cases the filaments were shown to spiral about the long axes of the collagen fibrils. There is little doubt that the observations do relate to the aggregation of a number of collagen molecules; but if a filamentous model is to be a viable one, then the molecular aggregation must be specific (i.e. non-random). It must also be shown that all of the filaments have the same diameter and that this diameter is unchanged along their length. At present, technical difficulties preclude such quantitative measurements of these features.

There is no electron microscope evidence of a regular two-dimensional lattice in transverse section of collagen fibrils from any source, with the possible exception of dogfish egg capsule (52). However, a 4.6-nm tangentially orientated period has been seen in ultrathin transverse sections of the large fibrils from rat tail tendon (53) (see also Chapter 1, Section 6.2). Also Squire and Freundlich (54) have observed a 5.5–10.0-nm period in longitudinal cryosections from this same tissue. Further, negatively stained isolated collagen fibrils invariably show a longitudinal striated appearance in the gap region of the fibril (Figure 11b). The periodicity of these striations is greater than the diameter of a single molecule. These observations imply *either* that molecules have aggregated laterally into a discrete filamentous structure *or* that single molecules have packed together in a lattice capable of explaining the specific long-range periodicities observed. Which of these alternatives is correct cannot be determined at present.

Freeze-fracture techniques have revealed a filamentous appearance in collagen fibrils from a wide variety of connective tissues (see also Chapter 4), (50, 55, 56). In predominantly Type I- or Type II-containing tissues (with the exception of cornea), these filaments lie approximately parallel to the long axis of the fibril, whereas in Type III-containing tissues the filaments apparently spiral, about the axis of the fibrils, in a right-handed manner with a pitch angle ~18°. The diameters of these filaments have been reported to be about 4–5 nm (50) and 7–8 nm (55). It has been argued that such observations arise from artefacts introduced by the use of cryoprotectants (57) or that they represent the surface deposition of glycosaminoglycans on the fibril (58, 59). The freeze-fracture results can therefore be explained in terms of a filamentous structure if the observations are taken at face value, but the possibility that such filaments are produced by artefact must not be overlooked.

Parry and Craig (60, 61) have studied many foetal and immature connective tissues and have shown that the collagen fibrils frequently have a very sharp distribution of diameters (standard deviation < 3.5 nm). In such cases, they have shown that the mean fibril diameters lie close to multiples of 8 nm. This observation, together with the fact that collagen fibrils in cornea remain at a specific diameter throughout active life (8), suggest that the fibrils are indeed composed of units larger in lateral dimensions than that of a single molecule.

Recently Eikenberry and colleagues (6, 7) have shown that their electron microscope and X-ray diffraction results from foetal chick metatarsal tendon are most easily explained if the collagen fibril diameter distributions have a narrow but finite width. If this is so, then the spread of diameters measured by electron microscopy is real and is not entirely a consequence of the difficulty of measurement. The implication of this result is that, whilst the fibril appear to have preferred diameters which are multiples of 8 nm, fibrils with intermediate diameters do indeed exist; this would indicate that growth does not occur through the assembly of filaments but conversely through rapid molecular addition between stable groupings (i.e. those which are multiples of 8 nm in diameter).

Since collagen aggregates in a variety of ways *in vitro*, it has been possible to study the polymorphic forms and hence determine the manner in which

molecules interact with one another (62). One such form is obliquely striated and has D-periodic filaments of lateral dimensions 3.7–4.0 nm, which are shifted axially ~9.5 nm with respect to their neighbours (63). While it is possible that an *in vitro* observation may not reflect the *in vivo* situation, the implication of these results is that a filamentous substructure in the collagen fibril may exist.

Much of the electron microscope data, which has been obtained using a variety of techniques, are consistent with the existence of a filamentous structure. However, it is important to be aware that electron microscope studies are always open to the criticism that the preparative procedures employed may induce artefacts in the tissue and thus hide the true structure of the *in vivo* state. Thus, while the evidence points to a filamentous structure for the fibril and provides little direct support for a 'molecular' fibril, the latter possibility cannot be excluded since one interpretation of the X-ray diffraction results (Section 5.2), obtained using native hydrated tissue, provides support for this concept.

5.2. X-ray diffraction results

X-ray diffraction patterns from tendon show discrete Bragg reflections which indicate that a crystalline lattice exists within the collagen fibrils (64–68). There is also diffuse scatter in the equatorial and near-equatorial region of the diffraction pattern which may be interpreted as arising from less ordered portions of the fibril. The positions of the Bragg reflections have been measured (65, 69, 70) and interpreted in terms of two different unit cells (a) a tetragonal unit cell with a = b = 2 af093.847 nm, c = 4 af0967 nm (65, 66) or a closely related triclinic unit cell with a = 7.56 nm, b = 7.55 nm, c = 4 af0966.8 nm, $\beta = 93°$ and $\gamma = 90°$ (69) or (b) a triclinic ('quasi-hexagonal') unit cell with a = 3.96 nm, b = 2.72 nm, c = 67.5 nm, $\alpha = 88.6$–89.7°, $\beta = 94.2$–94.9° and $\gamma = 104.7°$ (70, 71). The first of these unit cells was postulated to contain four five-stranded Smith-type microfibrils (72), each of diameter ~3.8 nm and related to one another by a 4_3 screw axis, whereas it was proposed that the quasi-hexagonal unit cell accommodated individual molecules, which were straight and tilted ~5° with respect to the fibril axis (70, 71). As both the quality and quantity of X-ray data have increased, it has become apparent that the smaller of the two unit cells i.e. the quasi-hexagonal cell, can explain all of the data in a satisfactory manner. Also, Fourier transform calculations of the quasi-hexagonal model have shown quantitative agreement with the observed intensity distribution in the equatorial and near-equatorial region of the diffraction patterns (73). Such agreement has not been obtained for the Smith-type microfibril model, though Trus and Piez (74) have argued that a compressed microfibril model may be compatible with the X-ray data. In such a model, the five-stranded microfibril is compressed to fit on a quasi-hexagonal lattice; the strands may be either straight or supercoiled. Piez and Trus (75) have recently developed their model further by suggesting that the portions of the collagen molecules in the overlap region lie on a quasi-hexagonal lattice whereas those portions of the molecules in the gap region are disordered. The Fourier transform of such a model has been calculated for the equatorial and near-equatorial region and has been shown to be in excellent agreement with the observed data.

A problematical aspect of the quasi-hexagonal model is the angle through which the molecules are tilted. This is estimated to be ~5° for tendon and implies that the minimum fibril diameter containing such molecules is ~290 sin 5° (25 to 30 nm). Thus fibrils with smaller diameters, such as occur in foetal tissue, cannot be accounted for by a simple quasi-hexagonal model. In addition, the fanning angle in the near-equatorial region of the X-ray patterns is much larger for skin than for tendon and this observation, together with the fact that the D-period in skin is smaller than in tendon (65.2 versus 67.0 nm; 76), strongly indicates that collagen molecules are tilted by a larger angle in skin (~12°) than in tendon (~5°). If this interpretation is correct, then the projected length of a collagen molecule in skin (i.e. the minimum fibril diameter) is ~290 sin 12° (~60 nm). Thus the idealised quasi-hexagonal scheme cannot explain the structure of collagen fibrils in skin that have a diameter less than 60 nm. Clearly, whilst much of the X-ray data support the quasi-hexagonal model proposed by Hulmes and Miller (71), there are aspects of it which do not satisfy some of the other data currently available.

5.3. Correlation of electron microscope and X-ray data

In Section 5.1 it was shown that the electron microscope evidence taken *as a whole* favoured the idea that the collagen fibrils were composed of filaments. However, it was recognised that much of the data could be criticised on the basis that they may be manifestations of artefacts introduced by the preparative procedures. None of the electron microscope evidence provides direct support for the quasi-hexagonal mode though, equally, none of the evidence can be shown to exclude such a model.

In Section 5.2, it was shown that the X-ray diffraction patterns from 'in vivo' tendons could be explained, both as regards the positions of the reflections and their intensities, in terms of *either* a quasi-hexagonal lattice containing straight collagen molecules tilted ~5° with respect to the fibril axis or a compressed microfibril model with an ordered overlap and a disordered gap structure. The quasi-hexagonal model as postulated does not readily offer an explanation for the structure of collagen fibrils in tendon or skin with diameters of less than 30 or 60 nm respectively. Further, since the largest structural unit in the quasi-hexagonal model is that of an individual molecule ~1.4 nm in diameter, the model does not account for either the filamentous appearance of collagen fibrils or the 5–10-nm lateral periods observed in the electron microscope. Equally it can be argued that the X-ray evidence does not preclude a definite pattern of covalent lateral cross-links between molecules which could define a minimum structural unit. Squire and Freundlich (77) have noted that 'even if molecular clumping in collagen fibrils occurs as a result of the preparative procedures, periodic clumping, however caused, is most likely to be a reflection of a periodic variation in the intermolecular interactions in collagen fibrils.

Recently Craig and Parry (48) have made some attempt to reconcile the quasi-hexagonal model with the electron miscroscope observations. It was considered that the fibril may indeed have a filamentous substructure with discrete numbers of collagen molecules cross linked laterally to one another and that the filaments could be constructed from straight tilted molecules packing on a quasi-hexagonal lattice (48, 74). In addition, it has been suggested (48, 78) that crystal dislocations (i.e. local regions in which the molecules are distorted either in a smooth continuous manner or in a sharp discontinuous way) could allow axially ordered arrays of the collagen molecules to follow a super-helical path around the axis of the fibril. This will endow the collagen fibril with the cylindrical symmetry observed in the electron microscope and deduced from X-ray diffraction studies (6, 7). This modification to the quasi-hexagonal scheme, which leads to the concept of crystalline domains within the fibril, would allow fibrils of *all* sizes to have the same fundamental structure.

Although the Smith-type microfibril model has been supported by the electron microscope evidence, it has been unable to account satisfactorily for

58

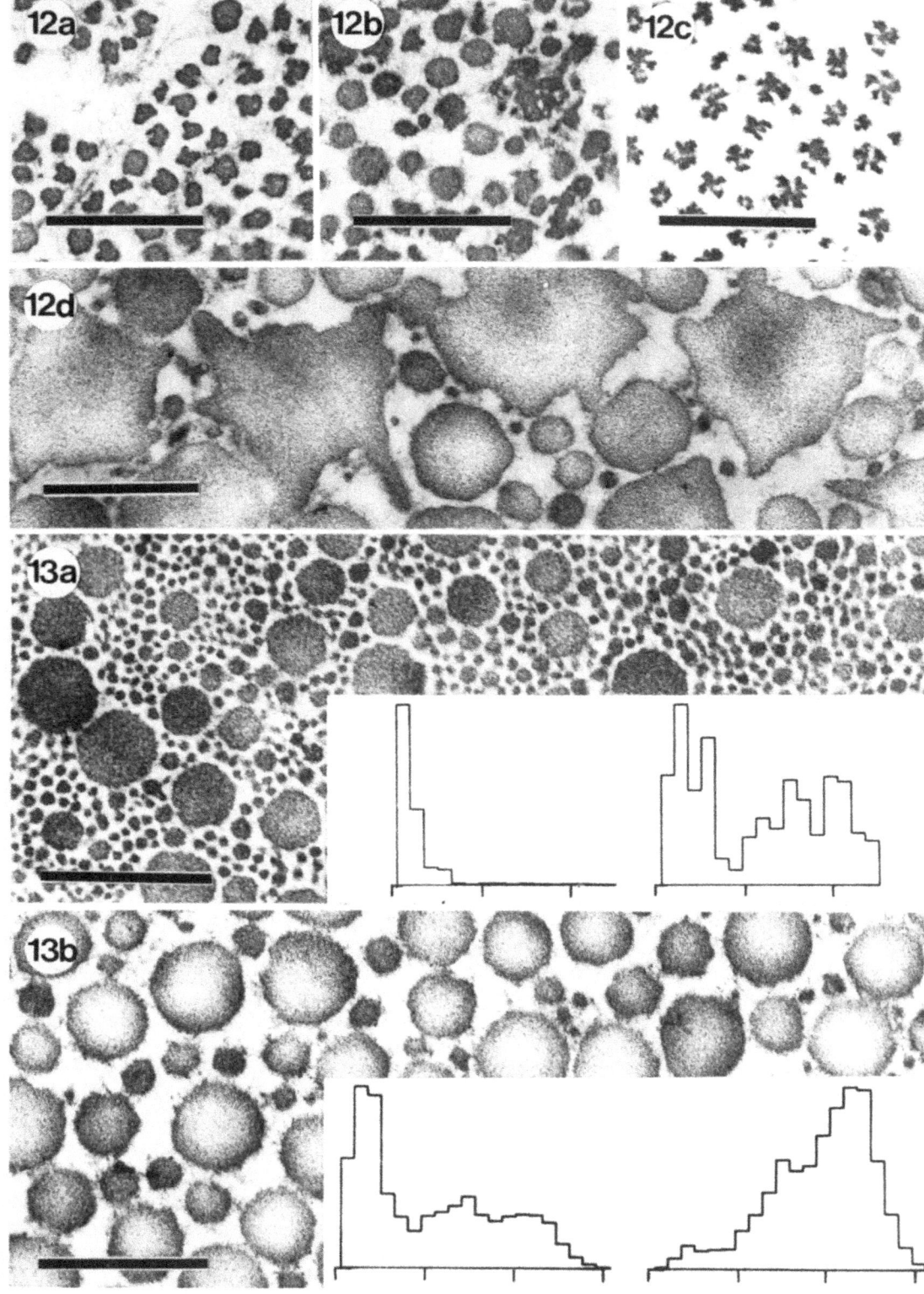

the intensity distribution in the X-ray diffraction patterns. However, the most recent microfibril model of Piez and Trus (75) appears to be compatible with all of the electron microscope and X-ray data. Their model is based on the quasi-hexagonal lattice but retains many of the features of a Smith-type microfibril. For instance, in the compressed five-stranded microfibril, those portions of the collagen molecules in the overlap regions are tilted ~5° and fall on a quasi-hexagonal lattice. Such a model appears to be a viable and attractive alternative to the dislocated quasi-hexagonal model previously discussed.

6. Collagen assembly disorders

A diversity of hereditable and acquired diseases arise from or lead to defects in the mechanism of collagen fibrillogenesis. The likely aetiology of some of these disorders have been listed by Light and Bailey (44) and range widely from cistron-translation in the Marfan syndrome, the presence of aldehyde blocking agents in homocystinuria, low levels or activities of enzymes such as lysyl hydroxylase in Ehlers-Danlos Syndrome type VI (EDS-V1) and *Osteogenesis imperfecta* and lysyl oxidase in EDS-V, Menkes kinky hair syndrome and lathyrism, and unusual metabolic regulation in Progeria and the Werner syndrome. Other disorders include the usually rare and mostly hereditable diseases such as *Cutis laxa*, scurvy, rickets, diabetes, dermatosparaxis, osteopetrosis and rheumatoid arthritis (44).

The scope of this review, however, demands that discussion be confined to a brief account of examples of a few of these disorders studied by electron microscopy, which present a ready diagnosis through either apparent fibrillar malformations or through a changed collagen fibril diameter distribution. For more wide-ranging accounts of collagen assembly disorders, the reader must refer to the medically orientated literature, though it should be added that virtually all of the work described therein has been descriptive rather than quantitative.

6.1. Fibrillar malformations

A most extreme degree of collagen fibrillar malformation is displayed in the hereditable disorder known as dermatosparaxis, which has been shown to occur in Norwegian sheep, Belgian and Texan cattle and a Himalayan cat. The disorder is caused by abnormally low levels of procollagen aminopeptidase in the tissue and a subsequent lack of cleavage of the amino-terminal extension from the procollagen molecules. The persistence of the terminal peptide leads to a cross linking deficiency which results in a gross disorganisation of the fibrils, which have a stellate or pinwheel appearance in transverse section (Figure 12a; 79).

Other connective tissue dysplasias show similarly malformed fibrils, though usually to a lesser extent than seen for the dermatosparactic tissues. These include the dysplastic greyhound dermis (Figure 12b; 80), human emphysematous lung (81), 'tight-skin' mutant mouse dermis (82), dermal lesions of humans suffering from Buschke-Ollendorff syndrome (83) and the dermis of man in EDS-I (84).

A non-pathological, but striking example of fibrillar malformation, has been shown to occur in the cuvierian tubules of the sea cucumber *Holothuria forskali* (85). In the native tissue, layers of collagen fibrils are arranged in mutually rotated layers similar to that seen in cholesteric liquid crystals. When the tubules are ruptured, the collagen fibrils are ejected and become axially and irreversibly aligned into a fibre which is highly extensible and which has little tensile strength. Bailey and colleagues (86) showed that although the axial head-to-tail cross links between collagen molecules were present in this tissue,

Figure 12. Transverse sections of collagen fibrils from tissues which display gross fibrillar malformations: (a) Tendon of a newborn dermatosparactic lamb showing irregular collagen fibrils. (b) Skin from a greyhound suffering from a dermal dysplasia which manifests itself by a highly fragile and hyperextensible skin. Some of the larger fibrils appear to be breaking down and smaller fibrils of irregular section can also be seen. (c) Collagen fibrils from the highly hyperextensible cuvierian tubules of the sea cucumber (*Holothuria forskali*). The fibrils are known to be lacking in the 'stable' covalent cross-links. (d) Collagen fibrils from the tail tendon of a senescent (2 y) rat. The larger fibrils have bizarre forms and appear to be breaking down. This is consistent with the reduction in mass-average diameter of collagen fibrils noted in all senescent tissues studied. Bars = 0.5 μm.

Figure 13. Transverse sections of collagen fibrils from the suspensory ligament of the horse. (a) 3-year-old horse suffering from laminitis and resulting in a disuse atrophy of the ligament and a preponderance of small diameter fibrils. (b) 1-year-old normal horse showing the expected bimodal distribution of collagen fibril diameters. Insets show the number and mass distributions of the collagen fibrils plotted as a function of diameter. The graduation marks on the horizontal (diameter) axes are 100 nm apart. It can be seen that the mass associated with the larger diameter fibrils accounts for the bulk of the collagen present in the normal ligament, while the mass is evenly divided between the large and small fibrils in the pathological specimen. Bars = 0.5 μm.

the 'stable cross links' linking molecules together transversely were absent. Thus the fibrils formed are not laterally stabilised and gross changes in fibril morphology may readily be seen especially after the tissue has been stressed (Figure 12c). These data imply that the tensile properties of connective tissues are closely dependent upon the number and density of transverse stable cross links within the collagen fibrils.

While it seems probable that malformed collagen fibrils in pathological connective tissues will not have the same mechanical properties as their counterparts in normal tissue, there is little quantitative mechanical data to substantiate this statement. Some mechanical data are available from dermatosparactic tissue, but much more data are required from other tissues containing malformed fibrils if a correlation between fibril morphology and mechanical properties is to be established beyond doubt. It must be remembered that malformed fibrils are also seen in cuvierian tubules and in senescent tissues such as in 2-year-old rat tail tendon (Fig. 12d). It follows that fibril malformation may either be a consequence of an hereditable or an acquired disorder, which leads to errors in fibrillar assembly, or alternatively may arise after normal fibrillogenesis and occur as a result of mechanical forces experienced *in vivo*.

6.2. Anomalous fibril diameter distributions

Some dysplasias of connective tissues are characterised by distributions of collagen fibril diameters which are markedly changed from normal. Examples of such changed distributions have been shown in cornea (leucoma in man, 13; stromal dystrophy in man, 87; scar tissue in rabbit, 88), in skin (EDS-I, II, IV and V in man, 89, 90) and in tendon (Dupuytren's contracture of the palmar aponeurosis of man, unpublished data of Craig, Flint and Parry; superficial flexor tendon from horse suffering from contracted tendon, 5). In the latter example, the collagen fibril diameter distribution measured was unimodal, in contrast to the bimodal distribution expected from a tendon of this type (5). The total collagen content of the diseased and normal specimens appeared to be similar when calculated as the area covered by the collagen fibrils in transverse section. In longitudinal section, however, it was seen that many fibrils were ruptured and exhibited D-periods (40–90 nm) differing significantly from the commonly accepted value (~67 nm). The tendon also seemed largely devoid of elastic fibres.

Appreciable changes have also been shown to occur in the collagen fibril diameter distribution, and in the collagen and elastic tissue contents, of the suspensory ligament from a horse suffering from laminitis (5). This painful disorder of the lamina of the hoof causes the horse to become lame; it attempts to avoid the use of the affected limb and a resulting 'disuse atrophy' is reflected in the ultrastructure of the suspensory ligament from that limb. As can be seen in Figure 13a, although the dysplastic tissue has a bimodal collagen fibril diameter distribution similar to that seen for normal tendon, the relative number of fibrils in the two peaks is changed dramatically from about 50:50 to about 95:5 (Figure 13b; 5). Also the mean fibril diameters for each of the modes in the diameter distribution for the pathological tissue are slightly less than the corresponding values for the normal tissue. The diseased tissue appeared largely devoid of the elastin component in the constituent elastic fibres and also had an apparent collagen content of about two-thirds that of normal tissue. These changes are all consistent with, and could be predicted from, a reduction of the mechanical requirements of this tissue when removed from its normal long-term, high-stress role.

As pointed out earlier in this section, there is little mechanical data on tissues with collagen fibril diameter distributions that differ significantly from the norm. Thus, while it has not yet been proved that the mechanical properties of pathological tissue with an anomalous fibril size distribution will be different from that for the corresponding native tissue, it seems very likely that such a correlation must exist.

7. Concluding remarks

This review has described the gross morphology of a diversity of connective tissues and presented ultrastructural data on the growth and development of their constituent collagen fibrils. Such fibrils, which have been shown to vary widely both in mean diameter and in range of sizes in different connective tissues, are often found preferentially orientated in layers or lamellae such as occur in cornea, tapetum and in some types of skin, cartilage and bone. Not all connective tissues have such a structure. For instance, the collagen fibrils in tendon are axially orientated and in some other tissues, such as vitreous humour, have only random organisation. Thus while the gross features of most connective tissues are now reasonably well established, there is a continuing need for detailed age-related studies of the constituent collagen fibrils.

One of the major conclusions deduced from the

data presented is that the collagen fibril diameter distributions are closely related to the mechanical properties of the tissue. In particular, collagen fibrils provide most of the tensile strength and the matrix provides the ability of the tissue to withstand compressive stresses. The smaller diameter collagen fibrils contribute to the creep-resistant properties of the tissue and thus enhance its resilience. Such an attribute is a consequence of the increased surface area per unit volume of collagen in small diameter fibrils and the concomitant increase in the potential number of interactions between fibril and matrix or between fibril and fibril, the latter probably being mediated via an orthogonal arrangement of proteoglycans. The larger diameter collagen fibrils, by virtue of their greater intermolecular cross link density, are likely to provide much of the tensile properties of the tissue. Fibrils of all sizes embedded in an hydrated proteoglycan gel will contribute to the ability of the tissue to resist fracture and crack propagation. Even though something of the fundamental relationship between collagen fibril diameter distributions and mechanical properties is now understood, there is still a great need for reliable data, both mechanical and ultrastructural, to confirm the relationships postulated. For practical reasons, however, mechanical studies are usually performed *in vitro* and the data obtained rarely equate to the conditions experienced *in vivo*. Nonetheless, the trends exhibited by the mechanical data as a function of age are likely to be genuine. The trend that the mass-average diameter of the collagen fibrils increases during foetal development but decreases during senescence, is observed in almost all of the connective tissues studied and parallels the changes in the mechanical moduli measured. However, a few specialised tissues, such as cornea, have proved to be exceptions. The maximum mass-average diameter of the collagen fibrils occurs either at birth or at maturity and is a characteristic of both the animal and the tissue.

Three major questions arise from this work:

(i) How do the proposed subfibrillar models of collagen relate to the mechanism of fibril growth and development?

(ii) What is the cause and mechanism of the decrease in mean collagen fibril diameter in the connective tissues of an animal as it approaches old age?

(iii) How may a tissue be remodelled when the stress and strains to which it is normally subjected become changed?

Although there are no definitive answers to any of these questions at the present time, there are a number of points which may prove to be relevant in a discussion of these topics.

In the case of the first question posed, one interpretation of the quantitative analyses of high quality X-ray diffraction patterns of tendon has indicated that the collagen fibrils are composed of a quasi-hexagonal arrangement of straight molecules tilted ~5° with respect to the fibril axis. This model does not imply a structure intermediate in size between that of the molecule and the fibril and cannot account for the structure of small diameter fibrils (Section 5). Electron microscopy, however, has indicated that groupings of collagen molecules may form a subfibrillar element. As has been suggested earlier in this review, it does seem possible that these apparently contradictory points of view may be combined into a single model. Covalent cross links could occur laterally between molecules and define a minimum structural unit whilst the molecular organisation could still be basically that of straight tilted molecules. Periodically occuring discontinuities of crystal dislocations could then allow the axially cross linked molecules to follow a super-helical path around the axis of the fibril. Such an arrangement would not only endow the fibril with the cylindrical symmetry shown by electron microscopy and recently confirmed by X-ray diffraction studies, but would account for the structure of small fibrils. Alternatively, the quantitative X-ray diffraction data were shown to be compatible with a compressed microfibril model. In this structure, it was proposed that those portions of the collagen molecules in the overlap regions of the fibril were ordered and arranged on a quasi-hexagonal lattice whilst the portions in the gap regions were laterally disordered. This scheme is also consistent with the electron microscope data in that it accounts for the observed filamentous appearance of the fibrils and allows for the existence of small diameter fibrils. Collagen fibril growth may occur by molecular addition if the quasi-hexagonal model is correct, but alternatively may arise from the addition of five-stranded microfibrils. If fibril growth is by the addition of single molecules and the 8 nm incremental growth pattern of collagen fibrils observed by electron microscopy is correct, then molecular accretion must occur in a discontinuous fashion. Irrespective of the sub-fibrillar structure, at least three mechanisms of collagen fibril growth seem feasible. Firstly, fibrils present in the early stages of development may act as nuclei for further growth, which could occur by the simple peripheral accretion of molecules or microfibrils. Secondly, fibrils composed entirely of newly synthesised collagen molecules may appear at their definitive sizes.

Such fibrils would have diameters dependent upon the stage of development of the tissue. Finally, it is possible that fusion of similarly directed fibrils could occur, although there is no direct evidence to support this idea.

The second question posed – 'What is the cause and mechanism of the decrease in mean collagen fibril diameter in the connective tissues of an animal as it approaches old age?' – is equally difficult to answer, though three mechanisms are worthy of consideration. Firstly, molecular or microfibrillar depletion at the periphery of existing fibrils would lead to a reduction in their diameters and give rise to a population of smaller diameter fibrils. Secondly, it is possible that the large numbers of small fibrils seen in some senescent tissues (particularly tendon) are the products of assembly of newly synthesised collagen molecules. Finally, it is possible that high stresses imposed on a tissue for long periods of time may induce 'fatigue' and subsequent mechanical failure in the tissue; a situation analogous to that which occurs in metals in long-term, high-stress situations. The 'fatigue' mechanism would presumably involve a rupturing of some of the covalent cross links between molecules and could give rise to the irregular shaped fibrils seen in some very old tissue. Such fibrils would then become more susceptible to enzymatic attack and could ultimately breakdown into a population of small diameter fibrils.

Three mechanisms have been discussed for fibril growth and for fibril breakdown. In either instance, such mechanisms are not mutually exclusive and some combination of them is not unlikely. Further, the mechanisms for fibril growth and development must ultimately take into account data on the turnover rates of collagen in different connective tissues. Such data are sparse, though values for periodontal ligament (2d), cornea (7d) and tendon (300d) have been reported.

Although the third question – 'How may a tissue be remodelled when the stresses and strains to which it is normally subjected become changed?' – is concerned with material not described in this review, it is relevant to point out that substantial progress is now being made in this area. Evidence is being accumulated by the authors and their colleagues (Drs M.H. Flint and G.C. Gillard) to suggest that a close correlation exists between the proteoglycans present in a tissue and the collagen fibril size distribution. It has not yet been possible to establish whether the proteoglycans determine or limit the collagen fibril size in some way or whether the converse is true. Equally it is possible that both change concomitantly in response to some other stimulus. Current data, including the observation that the turnover rate of the glycosaminoglycans is much faster than it is for collagen, suggests that the former possibility is more likely i.e. that it is the proteoglycans that control fibril growth and development possibly through changing charge profiles in the micro-environment. The mechanical stresses experienced by fibroblastic cells may switch-on the synthesis of those glycosaminoglycans which are appropriate to modify the collagen fibril size distribution in line with the mechanical requirements of the tissue (91).

Those collagen fibril diameter distributions which have been measured as a function of age are now providing the beginnings of a data base for normal connective tissues. Consequently, differences between normal and pathological tissues may be more easily defined. Pathological connective tissues often contain either malformed fibrils or fibrils with a markedly different distribution of diameters and in both cases, the mechanical properties of the tissue will be altered.

When more is known of the mechanisms of fibril growth and development, the research worker will be in a better position to formulate methods of treatment for some connective tissue disorders. Certainly, this idea is not as far-fetched as it would have seemed just five or ten years ago. Collagen fibrillogenesis is at the forefront of current research in connective tissue and no one can doubt that significant advances will occur during the next decade.

Note added in proof

After the preparation of this chapter a number of significant papers have appeared in the literature which relate to the material covered in this review, and we present here a selection of those contributions. Henkel and Glanville (92) showed that type I and type III collagen molecules in both human leiomyoma and calf aorta are covalently linked, thus suggesting the presence of both molecular species within a single fibril. Fraser et al. (93), in a definitive piece of work, refined the packing of collagen molecules in tendon and showed that the overlap region of the fibrils contained a regular array of molecular segments tilted by about 4.2° and that the gap regions contained portions of molecules which were disordered. Timpl et al. (94), using a rotary shadowing technique, have presented a model for the packing of type IV collagen molecules in basement membrane. X-ray diffraction patterns of type II collagen in lamprey notochord sheath have indicated the presence of a unit cell which differs from that found in type I-containing tendons (95). New data on collagen fibril size in cartilage from rat has been presented by Zambrano et al. (96); mean diameters ($\pm$S.D.) are 21.5 ± 3.6 nm for ear elastic cartilage, 72.5 ± 19.7 nm for femoral articular cartilage, 28.6 ± 7.2 nm for tracheal hyaline cartilage and 57.8 ± 20.4 nm for knee-joint meniscus fibrocartilage. Low angle X-ray diffraction patterns of various cornea have indicated that the fibrils in vivo have diameters of about 38 nm (97, our interpretation of 98). The smaller values observed by electron microscopy (~25 nm) indicate that considerable shrinkage may occur when conventional electron microscope preparative techniques are used. These effects may be more severe when dealing with foetal tissues or with tissues containing small diameter fibrils.

References

1. Parry DAD, Craig AS: The molecular structure of collagen. In: The Chemistry and Biology of Mineralized Connective Tissue. Veis A (ed), Amsterdam, Elsevier/North Holland, 1981, pp 63–67.
2. Fraser RDB, MacRae TP, Suzuki E: Chain conformation in the collagen molecule. J Mol Biol 129: 463–481, 1979.
3. Chapman JA, Hulmes DJS: Electron microscopy of the collagen fibril. In: Ultrastructure of the Connective Tissue Matrix. Ruggeri A, Motta PM (eds), The Hague, Martinus Nijhoff, 1984, pp. 1–33.
4. Parry DAD, Craig AS: Quantitative electron microscope observations of the collagen fibrils in rat-tail tendon. Biopolymers 16:1015–1031, 1977.
5. Parry DAD, Craig AS, Barnes GRG: Tendon and ligament from the horse: an ultrastructural study of collagen fibrils and elastic fibres as a function of age. Proc R Soc Lond [Biol] 203: 293–303, 1978.
6. Eikenberry EF, Brodský B, Parry DAD: Collagen fibril morphology in developing chick metatarsal tendon. I/1 X-ray diffraction studies. Int J Biol Macromol 4: 322–328, 1982.
7. Eikenberry EF, Brodsky B, Craig AS, Parry DAD: Collagen fibril morphology in developing chick metatarsal tendon. II/2 Electron microscope studies. Int J Biol Macromol 4: 393–398, 1982.
8. Craig AS, Parry DAD: Collagen fibrils of the vertebrate corneal stroma. J Ultrastruct Res 74: 232–239, 1981.
9. Jakus MA: The fine structure of human cornea. In: The Structure of the Eye. Smelser GK (ed), New York, Academic Press, 1961, pp 343–366.
10. Cox JL, Farrell RA, Hart RW, Langham ME: The transparency of the mammalian cornea: J Physiol 210: 601–616, 1970.
11. Jackson SF: The morphogenesis of avian tendon. Proc R Soc Lond [Biol] 144: 556–572, 1956.
12. Dyer RF, Enna CD: Ultrastructural features of adult human tendon. Cell Tissue Res 168: 247–259, 1976.
13. Schwarz W: Morphology and differentiation of the connective tissue fibres. In: Connective Tissue. Tunbridge RE, Keech M, Delafresnaye JF, Wood GC (eds), Oxford, Blackwell, 1957, pp 144–155.
14. Parry DAD, Craig AS: Collagen fibrils and elastic fibres in rat-tail tendon: an electron microscopic investigation. Biopolymers 17: 843–855, 1978.
15. Parry DAD, Craig AS, Barnes GRG: Fibrillar collagen in connective tissue. In: Fibrous Proteins: Scientific, Industrial and Medical Aspects. Parry DAD, Creamer LK (eds), London, Academic Press, 1980, 2, pp 77–88.
16. Craig AS, Parry DAD: Growth and development of collagen fibrils in immature tissues from rat and sheep. Proc R Soc Lond [Biol] 212: 85–92, 1981.
17. Tajima S, Nagai Y: Distribution of macromolecular components in calf dermal connective tissue. Connect Tissue Res 7: 65–71, 1980.
18. Wainwright SA, Biggs WD, Currey JD, Gosline JM: Mechanical Design in Organisms. London, Edward Arnold, 1976.
19. Harkness RD: Mechanical properties of connective tisssue in relation to function. In : Fibrous Proteins: Scientific, Industrial and Medical Aspects. Parry DAD, Creamer LK (eds), London, Academic Press, 1979, 1, pp 207–230.
20. Wainwright SA, Vosbrugh F, Hebrank JH. Shark skin: function in locomotion. Science 202: 747–749, 1978.
21. Parry DAD, Barnes GRG, Craig AS: A comparison of the size distribution of collagen fibrils in connective tissues as a function of age and a possible relationship between fibril size distribution and mechanical properties. Proc R Soc. Lond [Biol] 203: 305–321, 1978.
22. Flint MH, Craig AS, Reilly HC, Gillard GC, Parry DAD: Collagen fibril diameters and glycosaminoglycan content of skins – indices of tissue maturity and function. Connect Tissue Res (in press).
23. Muir H, Bullough P, Maroudas A: The distribution of collagen in human articular cartilage with some of its physiological implications. J Bone Joint Surg 52: 554–563, 1970.
24. Aspden RM, Hukins DWL: Collagen organization in articular cartilage, determined by X-ray diffraction, and its relationship to tissue function. Proc R Soc Lond [Biol] 212: 299–304, 1981.
25. Dahmen G: Polarizing and electronmicroscopic investigations of maturing and old human articular cartilage. In: Connective Tissue and Ageing. Vogel HG (ed), Amsterdam, Excerpta Medica, 1973, pp 109–113.
26. Maroudas A, Muir H: The distribution of collagen and glycosaminoglycans in human articular cartilage and the influence on hydraulic permeability. In: Chemistry and Molecular Biology of the Intercellular Matrix. Balazs EA (ed), London, Academic Press, 1970, 3, pp 1381–1401.
27. Happey F, Naylor A, Palframan J, Pearson CH, Render RM, Turner RL: Variations in the diameter of collagen fibrils, bound hexose and associated glycoproteins in the intervertebral disc. In: Connective Tissues, Biochemistry and Pathophysiology. Fricke R, Hartmann F (eds), New York, Springer-Verlag, 1974, pp 67–70.
28. Dahmen G: Submikroskopische Untersuchungen an Wirbelbandscheiben. Z Rheumaforsch 22: 192–213, 1963.
29. Berthet C, Hulmes DJS, Miller A, Timmins PA: Structure of collagen in cartilage of invertebral disk. Science 199: 547–549, 1978.
30. Szirmai JA: Structure of the invertebral disc. In: Chemistry and Molecular Biology of the Intercellular Matrix. Balazs EA (ed), London, Academic Press, 1970, 3, pp 1279–1308.
31. Glimcher M, Krane SM: The organization and structure of bone and the mechanism of calcification. In: Treatise on Collagen, Vol 2, Pt B. Gould BS (ed), London, Academic Press, 1968, pp 67–251.
32. White SW, Hulmes DJS, Miller A, Timmins PA: Collagen–mineral axial relationship in calcified turkey leg tendon by X-ray and neutron diffraction. Nature (Lond) 266: 421–425, 1977.
33. Robinson RA, Watson ML: Crystal–collagen relationships in bone as observed in the electron microscope. III. Crystal and collagen morphology as a function of age. Ann NY Acad Sci 60:596–628, 1955.
34. Jackson SF: The fine structure of developing bone in the embryonic fowl. Proc R Soc Lond [Biol] 146: 270–280, 1957.
35. Weinstock M, Leblond CP: Synthesis, migration, and release of precursor collagen by odontoblasts as visualized by radioautography after [³H] proline administration. J Cell Biol 60: 92–127, 1974.
36. Watson ML, Leblond CP: The development of the hamster lower incisor as observed by electron microscopy. Am J Anat 95: 109–161, 1954.
37. Ross R: The connective tissue fiber forming cell. In: Treatise on Collagen, Vol 2, Pt A. Gould BS (ed), London, Academic Press, 1968, pp 1–82.
38. Bellairs R, Harkness MLR, Harkness RD: The structure of the tapetum of the eye of the sheep. Cell Tissue Res 157: 73–91, 1975.
39. Gross J, Matoltsy AG, Cohen C: Vitrosin: a member of the collagen class. J Biophys Biochem Cytol 1: 215–220, 1955.
40. Gross J, Sokal Z, Rougvie M: Structural and chemical studies on the connective tissue of marine sponges. J Histochem Cytochem 4: 227–246, 1956.
41. Cusack S, Miller A: Determination of the elastic constants of collagen by Brillouin light scattering. J Mol Biol 135: 39–51, 1979.
42 Vogel HG: Strain of rat skin at constant load (creep experiments): influence of age and desmotropic agents. Gerontology 23: 77–86, 1976.
43. Vogel HG: Influence of maturation and age on mechanical and biochemical parameters of connective tissue of various organs in the rat. Connect Tissue Res 6: 161–166, 1978.
44. Light ND, Bailey AJ: Covalent crosslinks in collagen. Characterization and relationships to connective tissue disorders. In: Fibrous Proteins: Scientific, Industrial and Medical Aspects. Parry DAD, Creamer LK (eds), London, Academic Press, 1979, 1, pp 151–177.
45. Torp S, Arridge RGC, Armeniades CD, Baer E: Structure–property relationships in tendon as a function of age. In: Structure of Fibrous Biopolymers, Colston Papers No. 26. Atkins EDT, Keller A (eds), London, Butterworths, 1975, pp 197–221.
46. Vogel HG: Influence of maturation and aging on mechanical and biochemical parameters of rat bone. Gerontology 25: 16–23, 1979.
47. Flint MH, Gillard GC, Merrilees MJ: The effects of local physical environmental factors on connective tissue organization and glycosaminoglycan synthesis. In: Fibrous Proteins: Scientific, Industrial and Medical Aspects. Parry DAD, Creamer LK (eds), London, Academic Press, 1980, 2, pp 107–119.
48. Craig AS, Parry DAD: The sub-structure of collagen fibrils. In: The Chemistry and Biology of Mineralized Connective Tissue. Veis A (ed), Amsterdam, Elsevier/North Holland, 1981, pp 107–112.
49. Lillie JH, MacCallum DK, Scaletta LJ, Occhino JC: Collagen structure: evidence for a helical organization of the collagen fibril. J Ultrastruct Res 58: 134–143, 1977.
50. Ruggeri A, Benazzo F, Reale E: Collagen fibrils with straight and helicoidal microfibrils: a freeze-fracture and thin-section study. J Ultrastruct Res 68: 101–108, 1979.

64

51. Bouteille M, Pease DC: The tridimensional structure of native collagenous fibrils, their proteinaceous filaments. J Ultrastruct Res 35: 314–338, 1971.
52. Knight DP, Hunt S: Fibril structure of collagen in egg capsule of dogfish. Nature (Lond) 249: 379–380, 1974.
53. Hulmes DJS, Jesior JC, Miller A, Berthet-Colominas C, Wolff C: Electron microscopy shows periodic structure in collagen fibril cross sections. Proc Natl Acad Sci USA 78: 3567–3571, 1981.
54. Squire JM, Freundlich A: Direct observation of a transverse periodicity in collagen fibrils. Nature (Lond) 288: 410–413, 1980.
55. Rayns DG: Collagen from frozen fractured glycerinated beef heart. J Ultrastruct Res 48: 59–66, 1974.
56. Reale E, Benazzo F, Ruggeri A: Differences in the microfibrillar arrangement of collagen fibrils. Distribution and possible significance. J. Submicro Cytol 13: 135–143, 1981.
57. Stolinski C, Breathnach AS: Freeze-fracture replication and surface sublimation of frozen collagen fibrils. J Cell Sci 23: 325–334, 1977.
58. Reed R: Freeze-etched connective tissue. Int Rev Connect Tissue Res 6: 257–305, 1973.
59. Breathnach AS: Application of the freeze-fracture technique to investigative dermatology. Br J Dermatol 88: 563–574, 1973.
60. Parry DAD, Craig AS: Electron microscope evidence for an 80Å unit in collagen fibrils. Nature (Lond) 282: 213–215, 1979.
61. Parry DAD, Craig AS: An 80Å unit in collagen fibrils. In: Structural Aspects of Recognition and Assembly in Biological Macromolecules. Balaban M, Sussman JL, Traub W, Yonath A (eds), Boston, International Science Services, 1981, 1, pp 377–386.
62. Doyle BB, Hukins DWL, Hulmes DJS, Miller A, Woodhead-Galloway J: Collagen polymorphism: its origins in the amino acid sequence. J Mol Biol 91: 79–99, 1975.
63. Doyle BB, Hulmes DJS, Miller A, Parry DAD, Piez KA, Woodhead-Galloway J: A D-period narrow filament in collagen. Proc R Soc Lond [Biol] 186: 67–74, 1974.
64. Miller A, Wray JS: Molecular packing in collagen. Nature (Lond) 230: 437–439, 1971.
65. Miller A, Parry DAD: Structure and packing of microfibrils in collagen. J Mol Biol 75: 441–447, 1973.
66. Fraser RDB, Miller A, Parry DAD: Packing of microfibrils in collagen. J Mol Biol 83: 281–283, 1974.
67. Miller A: Molecular packing in collagen fibrils. In: Biochemistry of collagen. Ramachandran GN, Reddi AH (eds), New York, Plenum Press, 1976, pp 85–136.
68. Brodsky B, Eikenberry EF: Characterization of fibrous forms of collagen. In: Methods in Enzymology 'Structural and Contractile Proteins,' Vol 82. Cunningham LW, Frediriksen DW (eds), New York, Academic Press, 1982, pp 127–174.
69. Fraser RDB, MacRae TP: The crystalline structure of collagen fibrils in tendon. J Mol Biol 127: 129–133, 1979.
70. Fraser RDB, MacRae TP: Unit cell and molecular connectivity in tendon collagen. Int J Biol Macromol 3: 193–200, 1981.
71. Hulmes DJS, Miller A: Quasi-hexagonal molecular packing in collagen fibrils. Nature (Lond) 282: 878–880, 1979.
72. Smith JW: Molecular pattern in native collagen. Nature (Lond) 219: 157–158, 1968.
73. Miller A, Tocchetti D: Calculated X-ray diffraction pattern from a quasi-hexagonal model for the molecular arrangement in collagen. Int J Biol Macromol 3: 9–18, 1981.
74. Trus BL, Piez KA: Compressed microfibril models of the native collagen fibril. Nature (Lond) 286: 300–301, 1980.
75. Piez KA, Trus BL: A new model for packing of type I collagen molecules in the native fibril. Biosci Rep 1: 801–810, 1981.
76. Brodsky B, Eikenberry EF, Cassidy K: An unusual collagen periodicity in skin. Biochim Biophys Acta 621: 162–166, 1980.
77. Squire JM, Freundlich A: Molecular packing in collagen. Nature (Lond) 293: 240, 1981.
78. Fraser RDB, MacRae TP, Suzuki E, Tulloch PA: Ordered assemblies

in fibrous proteins. In: Structural aspects of recognition and assembly in biological macromolecules. Balaban M, Sussman JL, Traub W, Yonath A (eds), Boston, International Science Services, 1981, 1, pp 327–340.
79. Fjolstad M, Helle O: A hereditary dysplasia of collagen tissues in sheep. J Pathol 112: 183–188, 1974.
80. Cahill JI, Jones BR, Barnes GRG, Craig AS: A collagen dysplasia in a greyhound bitch. NZ Vet J 28: 203–204; 213, 1980.
81. Belton JC, Crise N, McLaughlin RF, Tueller EE: Ultrastructural alterations in collagen associated with microscopic foci of human emphysema. Hum Pathol 8: 669–677, 1977.
82. Menton DN, Hess RA: The ultrastructure of collagen in the dermis of tight-skin (Tsk) mutant mice. J Invest Dermatol 74: 139–147, 1980.
83. Uitto J, Santa Cruz DJ, Starcher BC, Whyte MP, Murphy WA: Biochemical and ultrastructural demonstration of elastin accumulation in the skin lesions of the Buschke-Ollendorff Syndrome. J Invest Dermatol 76: 284–287, 1981.
84. Vogel A, Holbrook KA, Steinmann B, Gitzelmann R, Byers PH: Abnormal collagen fibril structure in the gravis form (type I) of Ehlers-Danlos Syndrome. Lab Invest 40: 201-206, 1979.
85. Dlugosz J, Gathercole LJ, Keller A: Cholesteric analogue packing of collagen fibrils in the cuvierian tubules of Holothuria forskali (Holothuroidea, Echinodermata). Micron 10: 81–87, 1979.
86. Bailey AJ, Gathercole LJ, Dlugosz J, Keller A, Voyle CA: Proposed resolution of the paradox of extensive cross-linking and low tensile strength of cuvierian tubule collagen from the sea cucumber Holothuria forskali. 4: 329–334, 1982.
87. Witschel H, Fine BS, Grutzner P, McTigue JW: Congenital hereditary stromal dystrophy of the cornea. Arch Opthalmol 96: 1043–1051, 1978.
88. Jakus MA: Further observations on the fine structure of the cornea. Invest Opthalmol 1: 202–225, 1962.
89. Black CM, Gathercole LJ, Bailey AJ, Beighton P: The Ehlers-Danlos Syndrome: an analysis of the structure of the collagen fibres of the skin. Br J Dermatol 102: 85–96, 1980.
90. Holbrook KA, Byers PH; Ultrastructural characteristics of the skin in a form of the Ehlers-Danlos Syndrome type IV. Lab Invest 44: 342–350, 1981.
91. Parry DAD, Flint MH, Gillard GC, Craig AS: A role for glycosaminoglycans in the development of collagen fibrils. FEBS Lett 149: 1–7, 1982.
92. Henkel W, Glanville RW: Covalent crosslinking between molecules of Type I and Type III collagen. The involvement of N-terminal, nonhelical regions of the α1(I) and α1(III) chains in the formation of intermolecular crosslinks. Eur J Biochem 122: 205–221, 1982.
93. Fraser RDB, MacRae TP, Miller A, Suzuki E: Molecular conformation and packing in collagen fibrils. J Mol Biol 167: 497–521, 1983.
94. Timpl R, Wiedemann H, van Delden V, Furthmayr H, Kuhn K: A network model for the organization of Type IV collagen molecules in basement membrane. Eur J Biochem 120: 203–211, 1981.
95. Eikenberry EF, Childs B, Sheren SB, Parry DAD, Craig AS, Brodsky B: Crystalline fibril structure of Type II collagen in lamprey notochord sheath. J Mol Biol (in press).
96. Zambrano NZ, Montes GS, Shigihara KM, Sanchez EM, Junqueira LCU: Collagen arrangement in cartilages. Acta Anat 113: 26–38, 1983.
97. Inouye H, Worthington CR: X-ray diffraction study of the cornea. Biophys J 41: 285a, 1983.
98. Sayers A: Koch MHJ, Whitburn SB, Meek KM, Elliott GF, Harmsen A: Synchrotron X-ray diffraction study of corneal stroma. J Mol Biol 160: 593–607, 1982.

Authors' address:
Department of Chemistry, Biochemistry and Biophysics
Massey University
Palmerston North, New Zealand

CHAPTER 3

Collagen distribution in tissues

GREGORIO S. MONTES, MARIA S.F. BEZERRA and LUIZ C.U. JUNQUEIRA

1. Introduction

Collagens are a closely related family of fibrous proteins with similar structure. This family of proteins provides an extracellular framework for all metazoan organisms. Besides, collagen also possesses non-structural functions.

All collagen molecules are long rods (about 300 nm long and only 1.4 nm in diameter) consisting of a triple helix of three α chains. To date, three different types of interstitial collagen have been characterized biochemically on the basis of the constitution of their polypeptide chains. The above-mentioned interstitial collagens are known as type I, type II, and type III collagens. Collagens have also been isolated from structures identified morphologically as basement membranes.

The morphological entity of interstitial collagens is the *fibril,* an ordered molecular polymer visible by electron microscopy (Figures 5, 6, 9, 10, 11, 12). Depending on the collagen type, and the tissue studied, fibrils can be seen as structures that are more or less clearly banded in longitudinal sections and with variable diameters in transverse sections, located in the extracellular matrix of connective tissues. The ordered fibrillar aggregates that can be seen under the light microscope are known as *fibers* (Figures 1, 5, 6, 15, 16).

The term collagen describes a family of specialized molecules, each a genetically distinct type that has evolved for a particular function. Each of the different collagen types is characterized by its chemical structure, morphology, distribution, functional properties, and pathology. The aim of this chapter is to review the ultrastructural aspects of collagen distribution in tissues and to correlate these observations with the available light microscopic studies, and also with the current biochemical data on the subject.

During the last decade collagen has been the subject of a large literature. As it is impossible to cite all the original contributions in this short review, readers are encouraged to refer to recent reviews on the different topics (1–10).

2. Methods

For the study of collagen type distribution in tissues three methods have been used, namely biochemical, immunohistologic, and histochemical methods. Although several reports have described the occurrence of collagen types in tissues and organs it has been difficult to obtain a distinct identification, specific localization, and precise quantitation of the different collagen types.

It is our intention to describe briefly the bases of these methods and of their limitations in order to provide a clue to understanding the results that will be discussed here.

2.1. Chemical methods

These include extraction of the collagens, with or without enzymatic digestions, followed by purification using differential salt precipitation and chromatography. Another chemical method used to assess collagen type occurrence consists of quantitating specific marker peptides generated by Cyanogen Bromide (CNBr) digestion of the whole tissue. Limitations of these techniques include preferential extraction of one collagen type over another (probably due to differences in cross-linking and in association with other macromolecules) and losses resulting from the purification procedures.

When using samples for biochemical assay, very few authors describe the structure used or perform histologic controls. Even if the material collected is carefully dissected to avoid inclusion of adjacent tissues in the samples, it is difficult to ascertain how much collagen of the entire organ is studied. For

Ruggeri, A and Motta, PM (eds): Ultrastructure of the connective tissue matrix. ISBN-13:978-1-4612-9789-5

instance, it is very difficult to seperate glandular epithelia and smooth muscle from the stroma of some organs, and, even when achieved, there is no method to ascertain the homogeneity of the connective tissue.

The above comments may explain why discrepancies are evident when comparing results reported by different authors. For example, collagen type II, which is known for its tissue selectivity for cartilages, has also been detected in some tendons and in the eye balls of birds and amphibians. Although these findings seem to be exceptions, they are not so if it is kept in mind that many tendons contain cartilaginous inclusions in their pressure-bearing areas and that histological studies have shown that amphibian and avian eye balls contain cartilaginous structures.

In spite of these limitations, better results have been obtained recently using more sophisticated and reliable biochemical methodologies (11, 12).

2.2. Immunohistology

The molecules of the various collagen types contain different antigenic determinants and can therefore be distinguished by specific antibodies.

Antibodies against one of the different types of collagen are raised in rabbits and purified by affinity methods. Monoclonal antibodies to different collagen types can also be produced, using hybridoma cells (5). Tissue sections are then incubated with the antibodies that can thus bind to tissue antigens.

To make sites of reaction evident, antibodies are linked to peroxidase and – after the incubation of tissue slices – bound antibodies are reacted with diaminobenzidine in the presence of hydrogen peroxide, prior to examination in the electron microscope. For light microscopy, cryostat sections are submitted to an indirect immunofluorescent staining and studied under a fluorescence microscope.

Although immunohistologic methods have been widely used to localize the different collagen types in tissue sections (2, 5), an unambiguous identification of each collagen type has been very difficult to achieve. The limitations of this method include insufficient purification of the antigens (it is difficult to remove all the contaminating non-collagenous proteins) and of the antibodies (all unspecific antibodies must be removed). In addition, localization of a given collagenous antigen requires more than just specific antibodies: fixation conditions suitable for preservation of antigenicity must be tested. Equally important, but often difficult to control or evaluate, is the fact that collagen fibers may be partially masked or may not react completely with the antibody reagents, since it has been shown that each collagen type interacts differently (considering both the quantitative and qualitative aspects) with the ground substance (13, 14, 15). It is interesting to remark that, despite the very well-known species specificity of immunological reactions of collagen (16), it is frequently observed in the reports of the literature that, for the study of human tissue sections, antibodies are obtained from rabbits that are immunized with bovine collagen. Among other artifacts, non-specific binding of, or failure to wash out unbound, antibodies should also be considered. Besides, immunohistologic methods are not readily adaptable to quantitation.

This technique has been particularly useful in assessing collagen type distribution in tissues displaying developmental or disease-associated changes, but its application is still largely empirical and subject to many artifacts; thus, non-immunological techniques must be used to corroborate the immunohistologic findings.

The above comments explain the discrepancies observed between results reported by different authors. For example, an immunohistologic study of the collagens of human liver (16) states that reticulin fibers within the liver lobule contain collagen type III and that they are not stained by collagen type I antibodies; these results are not consistent with a more recent report claiming that these same fibers contain both type I and type III collagens in human and bovine livers (17).

2.3. Histochemistry

As collagen molecules are orderly disposed in a parallel orientation, a normal birefringence is one of the classic characteristics of collagenous structures. Collagen molecules, being rich in basic aminoacids, strongly react with acidic dyes.

Sirius Red, an elongated strongly acidic dye, reacts with collagen and promotes an enhancement of its normal birefringence due to the fact that the dye molecules are attached to collagen molecules in such a way that their long axes are parallel (18). The enhancement of birefringence promoted by the Picrosirius-polarization method is specific for collagen. Thus, this method (being specific for collagen detection in tissue sections, 18) has proved to be more useful for studying collagen distribution in histological slides than the routinely used trichrome techniques (13).

Interstitial collagens display different interference

colors, and intensities of birefringence, in tissue sections studied by the Picrosirius-polarization method. The comparative study of vertebrate organs by this method disclosed a striking correlation between the localization of the different colors and intensities of birefringence, and the so far described biochemical distribution of collagen types I, II, and III, which led Junqueira et al. (19) to postulate that these different collagen types could be distinguished in tissue sections by the Picrosirius-polarization method. Their results demonstrated that collagen type I (which corresponds to what have been classically called *collagen fibers* by histologists) shows up as thick, strongly birefringent, yellow or red fibers; while collagen type III appears under the form of thin, weakly birefringent, greenish fibers that could be identified with *reticular fibers* (20). Collagen type II, present in hyaline and elastic cartilages does not form fibers and displays a weak birefringence of a variable color.

The observation that collagen types I, II, and III show different colors and intensities of birefringence in a same tissue section (19) can be explained by the widely known fact that these different interstitial collagens display distinct patterns of physical aggregation (1, 3, 4, 6, 20, 21).

Thus, collagen type I (*collagen fibers*) forms thick fibers, composed of closely packed thick fibrils (20, 22), and consequently displays an intense birefringence of yellow to red color. Collagen type III (*reticular fibers*) forms thin fibers composed of loosely packed thin fibrils (20, 22) and thus displays a weak birefringence of a greenish color. Collagen type II (present in hyaline and elastic cartilages) does not form fibers, and its very thin fibrils are disposed as a loose mesh that strongly interacts with the ground substance. This type of physical aggregation results in a weak birefringence, and a variable color, both of which depend on the interaction of collagen molecules with the ground substance (23).

It is thus evident that, in the 5–7-μm sections routinely used in histology, the different colors and intensities of birefringence displayed by collagen types I, II, or III are due to differences in their patterns of physical aggregation.

The Picrosirius-polarization method has proved to be useful for the study of collagen distribution in tissue sections (19–27). However, in the early stages of the development of this method, it became evident that tissue section thickness is of importance on the color and intensity of birefringence displayed by collagen, for folds in tissue sections are more strongly birefringent and display a different color when compared to the surrounding tissue. The fact

that color and intensity of birefringence depend on collagen fiber thickness has been shown using different methodologies (21, 28). These reports have demonstrated that the Picrosirius-polarization method is useful for studying the distribution of the different interstitial collagen types in tissue sections from normal adult organs. But care should be taken when interpreting pictures obtained studying embryonic or pathologic material, where processes involving intense collagen production or degradation are frequent.

However this method can also be of value for the study of collagen degradation (21, 26, 29) provided that, under these conditions, one should be cautious regarding the characterization of collagen types.

2.4. Electron microscopy

During the past decade the study of collagen by means of electron microscopic methods has proved useful for various purposes. Thus, electron microscopic observations on the collagen of dermatosparaxic cattle, published as early as 1970 (30), served as a basis for the discovery of procollagen (6). The electron microscope has also been used in the analysis of the primary structure of collagen, observing the highly reproducible banded pattern of segment-long-spacing crystallites (31) which has permitted the characterization of collagen types I, II, and III (32, 33). Scanning electron microscopy of the *in vitro* reconstitution of collagen polymers demonstrated the formation of thick bundles of pure collagen type I and thin isolated fibers of type III, showing that the chemical structure of collagen molecules determines their physical aggregation (34). In addition, freeze-fracture, thin-section, and negative-staining electron microscopic studies permitted the description of collagen fibrillogenesis and of the microfibrillar arrangement of collagen fibrils (35–37).

Ultrastructural observations on the connective tissue of several organs in representative species of the main vertebrate classes disclosed the presence of two distinct populations of collagen fibrils in all sites examined (20, 22, 24). These two distinct populations of collagen fibrils could be recognized on the basis of their diameter and of their either loose or compact arrangement.

One population is formed of thicker fibrils, with marked variation in diameter, arranged in densely packed blocks. The second population is composed of loosely packed thin fibrils of more uniform diameters (Figures 5, 6, 10, 12).

In certain anatomical sites (for example in nerves, arteries, smooth muscle, etc.) these two distinct fibril populations were found to be segregated into different compartments (20, 22, 24). The comparative study of these two collagen fibril populations in those anatomical sites by electron microscopy disclosed a striking correlation between the localization of the two different populations, and the so far described biochemical and immunohistologic distribution of collagen types I and III, which led Montes et al. to postulate that collagen type I forms thicker fibrils than collagen type III (20). Also a correspondence could be established between the localization of thicker fibrils and the presence of collagen fibers in histological slides; the same could be said for the distribution of thinner fibrils in the sites where reticular fibers could be localized by silver impregnation techniques (20).

Different collagen fibril diameters had been previously reported in nerves (38) and other organs (39), although the difference in fibril diameters was not related to the type of collagen. It is interesting to observe that, in the more recent papers where collagen fibril diameter was related to collagen type distribution in tissues (20, 22), the morphometric and statistical studies performed provided results which are similar to the approximate values available in the previous reports.

Ultrastructural research using freeze fracture techniques and scanning electron microscopy (34, 37) support the transmission electron microscopic findings, showing that collagen type III fibrils present a relatively small and uniform diameter when compared to collagen type I. No exception has so far been reported to this ultrastructural evidence that permits the characterization of interstitial collagen types by means of the routinely used transmission electron microscopy.

However, all papers published to date using electron microscopy to study collagen type distribution in tissues, have also used histochemical characterization of collagen types and have discussed the available chemical and immunohistologic data to corroborate the ultrastructural findings (20, 22–25, 27, 51).

3. Distribution of structurally distinct collagen types in tissues

Although all collagens have common structural features, each collagen type is characterized by its chemical structure, functional properties, interaction with the ground substance, pathology, morphology, and distribution (Table 1).

Collagen appears in some shape or form in virtually every tissue. Although each tissue may contain several types of collagen, the relative proportions of

Table 1. Main characteristics of the different collagen types*

Collagen type	Molecular formula	Tissue distribution	Optical microscopy	Ultrastructure	Site of synthesis	Interaction with glycosaminoglycans	Function
I	$[\alpha_1 \,(I)]_2\alpha_2$	Dermis, bone, tendon, dentin, fasciae, sclera, organ capsules, fibrous cartilage	Closely packed, thick non-argyrophilic, strongly birefr. yellow or red fibers. *Collagen fibers*	Densely packed, thick fibrils with marked variation in diameter	Fibroblast, osteoblast, odontoblast, chondroblast	Low level of interaction, mainly with *dermatan sulfate*	Resistance to tension
II	$[\alpha_1 \,(II)]_3$	Hyaline and elastic cartilages	Loose collagenous network only visible by aid of Picrosirius-polarization	No fibers, very thin fibrils embedded in abundant ground substance	Chondroblast	High level of interaction, mainly with *chondroitin sulfates*	Resistance to intermittent pressure
III	$[\alpha_1 \,(III)]_3$	Smooth muscle, endoneurium, arteries, uterus, liver, spleen, kidney, lung	Loose network of thin argyrophilic, weakly birefr., greenish fibers. *Reticular fibers*	Loosely packed, thin fibrils with more uniform diameters	Smooth muscle, fibroblast, reticular cells, Schwann cells, hepatocyte	Intermediate level of interaction, mainly with *heparan sulfate*	Structural maintenance in expansible organs
IV	$[\alpha_1 \,(IV)]_3$	Epithelial and endothelial basement membranes	Thin, amorphous, weakly birefr. membrane	Neither fibers nor fibrils are detected	Endothelial and epithelial cells	i.d.	Support and filtration
V	$[\alpha \,A]_3$ and $[\alpha \,B]_3$	Placental basement membranes	i.d.	i.d.	i.d.	i.d.	i.d.

* Reproduced with permission from (1).
birefr. = birefringent; i.d. = insufficient data.

each of the different collagen types are tissue-specific. These proportions may change with time in the same tissue, as will be seen below when comparing fetal and adult skin.

The different collagen types may be segregated into two major classes. Most familiar are the interstitial collagens that form the bulk of the major connective tissues. Less abundant, and as yet poorly defined, collagens can be assigned to a second group that includes basement membrane collagens and others that appear to be located in the immediate pericellular environment. Collagens of the latter group do not form fibrils and thus little information is obtained when they are observed under the electron microscope; besides, these collagen types are being actively studied at present and are still a matter of controversy. For these reasons, this review will deal mainly with the fibrillar, interstitial collagens.

3.1. Collagen type I

This type of collagen forms the bulk of the vertebrate body collagen, and it accounts for about 90% of the collagen in mammals, being the most abundant collagen in bones and tendons. Collagen type I is predominant in skin, organ capsules, fasciae, ligaments, dentin, perichondrium, and fibrous cartilages; fibroblasts, osteoblasts, and odontoblasts produce this type of collagen.

Collagen type I forms the thick (2–10 μm) *collagen fibers* observed by histologists under the light microscope. When studied by the Picrosirius-polarization method these fibers show up as thick, strongly birefringent, yellow or red fibers (Figure 1).

The electron microscopic picture of collagen type I fibers shows bundles of tightly packed, thick (average diameter: 78 nm) fibrils that leave little room for the ground substance (Figures 6, 10; Tables 1, 2, 3). When observed in longitudinal sections, these fibrils display a characteristic periodic transverse striation, resulting in a 64–67 nm banded staining pattern. This is typical of collagen and is particularly clear in this collagen type because of its scarce interaction with masking ground substances. These typical ultrastructural features differ from those of collagen types II and III, and permit their characterization by means of electron microscopy.

Collagen type I fibrils are present where tensile strength is necessary and thus possess a very important structural function.

3.2. Collagen type II

It is found in hyaline and elastic cartilages, and it is produced by chondroblasts. This type of collagen forms thin (average diameter: 25 nm) fibrils that form a meshwork in association with large amounts of proteoglycans (Figure 11; Table 1).

Because of these characteristics, this type of collagen is not visible by the usual optical microscopic methods. However, oriented populations of collagen type II fibrils become evident when studied by Picrosirius-polarization method (Figure 2) (23). With the exception of articular cartilage, in all the other hyaline cartilages as well as in all elastic cartilages, collagen type II is observed under the electron microscope as constituted of a loose meshwork of thin fibrils that show a very small and regular diameter (Figure 11). The characteristic banding pattern of collagen, clearly visible in fibrous cartilages (Figure 13), is less evident in hyaline and elastic cartilages, and this is probably due to a masking effect of the abundant interfibrillary material.

A comparative study of collagen type II fibrils in different cartilages showed that their diameters are extremely variable. In elastic and non-articular hyaline cartilages, collagen fibril diameters are small and more uniform; while in articular cartilages fibril diameters are larger and more variable (resembling the collagen type I fibril diameters of fibrous cartilages, which suggests that collagen fibrils in cartilages that support high pressures such as articular and fibrous cartilages differ from the collagen fibrils of the hyaline and elastic cartilages not subjected to high pressures). These features are illustrated in Figure 26. These observations suggest that cartilage collagen fibril diameter might be related to its functionally important resistance to the intermittent action of pressure.

The study of the collagen fibril diameter in the cartilages studied showed that in this tissue three gradients can occur. The first case observed in the articular cartilage whose fibril diameter increases steadily from the articular surface to the bone tissue. The second gradient occurs in all cartilages that possess a perichondrium, where a distinct gradient of fibril diameters can be observed decreasing from the perichondrium inward. The third example of gradient of fibril diameter can be observed increasing centrifugally from each cell.

3.3. Collagen type III

This type of collagen has been found only in organs

70

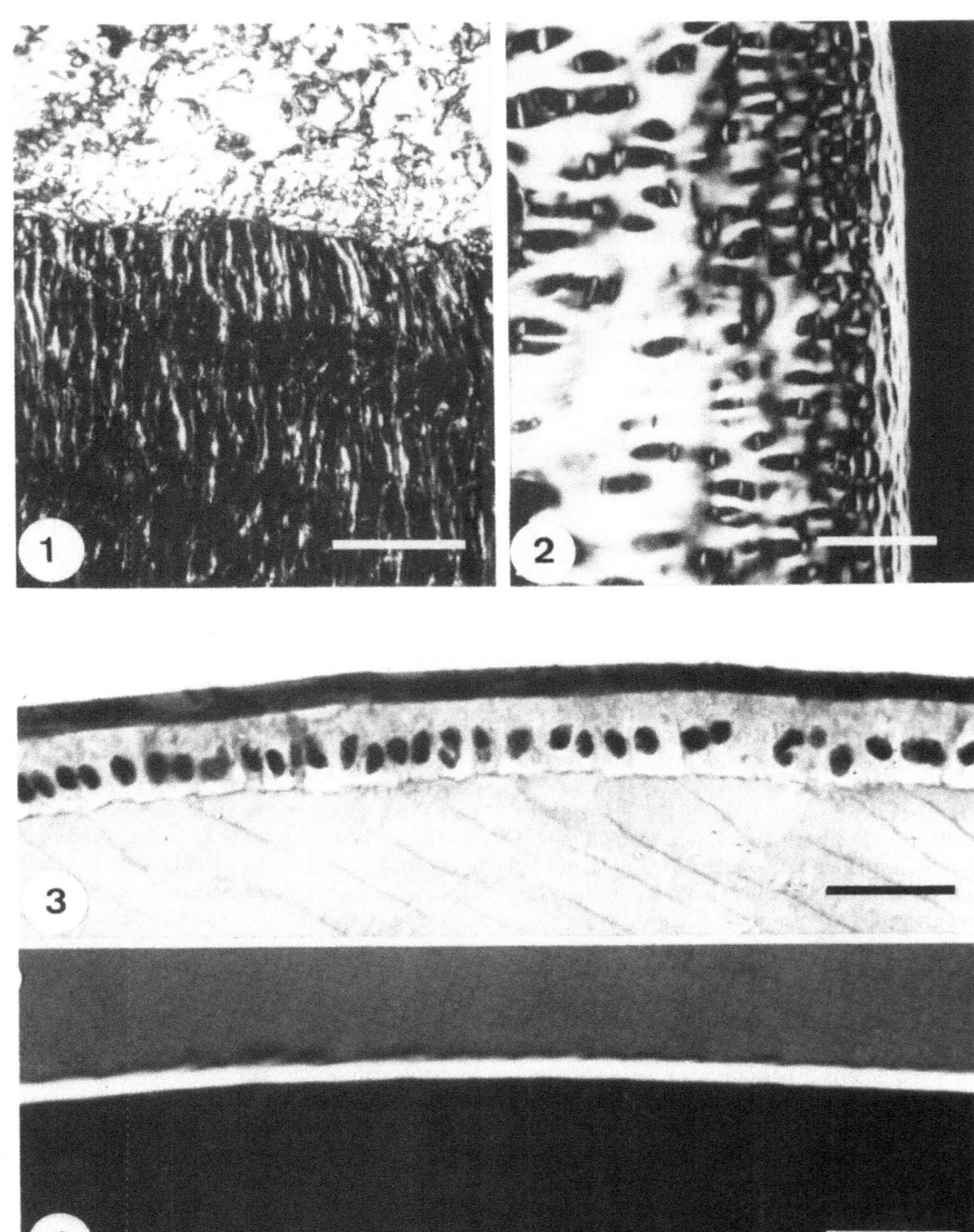

that also contain collagen type I (such as uterus, arteries, skin, intestines, lung, spleen, liver, kidney, etc.). It is present in many organs and is mainly, but not exclusively, related to smooth muscle cells. Thus, in arteries it is located in the tunica media of these structures (in the extracellular spaces of the concentric layers of muscle cells) (24); while the adventitial layer and veins are composed mainly of collagen type I. In intestines, it is localized to the muscular layers and lamina propria. In the uterus, an organ composed mainly of fibrous and muscular tissues, it is probably associated with the muscular component.

Preliminary immunohistologic results have been presented suggesting that thin reticular fibers react with fluorescent antibody against collagen type III (40, 41).

Histochemical evidence has been reported demonstrating that collagen type III can be observed by aid of the light microscope in the form of thin (0.5–2 μm) argyrophilic, weakly birefringent, greenish fibers (Figures 1, 7, 8) that correspond to the *reticular fibers* described by histologists. In all tissues studied the distribution of these argyrophilic fibers shows a striking correlation with the localization of collagen type III by means of biochemical and immunohistologic methods (19, 20).

When observed under the electron microscope, reticular fibers are composed of a loosely packed bunch of thin (average diameter: 45 nm) fibrils (Figure 5). The interfibrillary space contains abundant thin and short, irregular, electron-dispersing bridges that join these fibrils (Figure 12) (20). The differences between the arrangement and diameter of the fibrillar components in the various types of interstitial collagen are clearly evident in Tables 2 and 3 and Figures 10–12.

Recently, reticular fibers have been found to be the main extracellular fibrillary structure in four different locations: the connective tissue surrounding the axon of nerves (endoneurium); the inner sheaths of tendons (endotenonium and peritenonium); the tunica media of rat aorta, and the inner muscular layer of dog small intestine (20, 24, 42). Biochemical studies on this material should give convincing results regarding the collagen type present in *reticular fibers*. The argyrophilia of these fibers, by the way, has been shown to be due to its carbohydrate, and not its protein (collagen), moiety (20, 39).

Collagen type III is produced by smooth muscle cells, fibroblasts, reticular cells of hemopoietic organs, liver cells and Schwann cells of the peripheral nervous system.

The study of Schwann cell tumors, and of the infection of Schwann cells by *Mycobacterium leprae*, when associated with other observations in the literature (43, 44), strongly suggests that Schwann cells produce collagen, very probably of type III (22, 25, 27, 45). Biochemical studies on this material would be of interest to give stronger support to these results.

Collagen type III is always present in tissues and organs that require a motile structural scaffolding, such as arteries, muscular layers of the intestine, uterus, etc. Reticular fibers, whose fibrils are loosely bound by mobile bridges of protein-carbohydrate

Table 2. Area occupied by the fibrillary component in cross sections of reticular and collagen fibers*

Class	Species	Reticular fibers	Collagen fibers
Amphibia	Bufo ictericus (toad)	21.93%	43.45%
Reptilia	Xenodon merremii (snake)	27.53%	52.28%
Mammalia	Rattus norvegicus (rat)	28.17%	49.09%
	Oryctolagus cuniculus (rabbit)	29.08%	49.14%
	Human	28.81%	56.15%

* Reproduced with permission from (20).

←

Figure 1. Collagen types I and III. Picrosirius-polarization method. When studied by this method oriented collagen molecules can be seen as brightly birefringent structures against a dark background. In this photomicrograph of a chicken aorta the adventitia (upper region) is composed of closely packed, thick, strongly birefringent fibers of collagen type I (collagen fibers), whereas the tunica media (lower region) is composed of a loose network of thin, weakly birefringent fibers of collagen type III (reticular fibers). Bar = 50 μm; × 400.

Figure 2. Collagen type II. Photomicrograph of a guinea-pig articular cartilage. Observe (on the right-hand side) that collagen is disposed parallel to the articular surface, while the collagen in the deeper layers (left) shows a perpendicular orientation. No 'fibers' can be distinguished as such; however, the preferential orientation of collagen fibrils promotes a distinct birefringence that is characteristic of collagen type II. Picrosirius-polarization method. Bar = 50 μm; × 370. (Reproduced with permission from (23)).

Figure 3. Basement membrane collagen. Section of a guinea-pig lens surface showing that the lens capsule, overlying the epithelial monolayer, is darkly stained by Sirius Red. Picrosirius-haematoxylin. Bar = 25 μm; × 800 (reproduced with permission from (48)).

Figure 4. Basement membrane collagen. The same structure as in Figure 3 observed with polarization microscopy. The intense birefringence of the lens capsule (horizontal white band) denotes that collagen molecules are oriented in this basement membrane. Picrosirius-polarization method. Bar = 25 μm; × 800 (reproduced with permission from (48)).

72

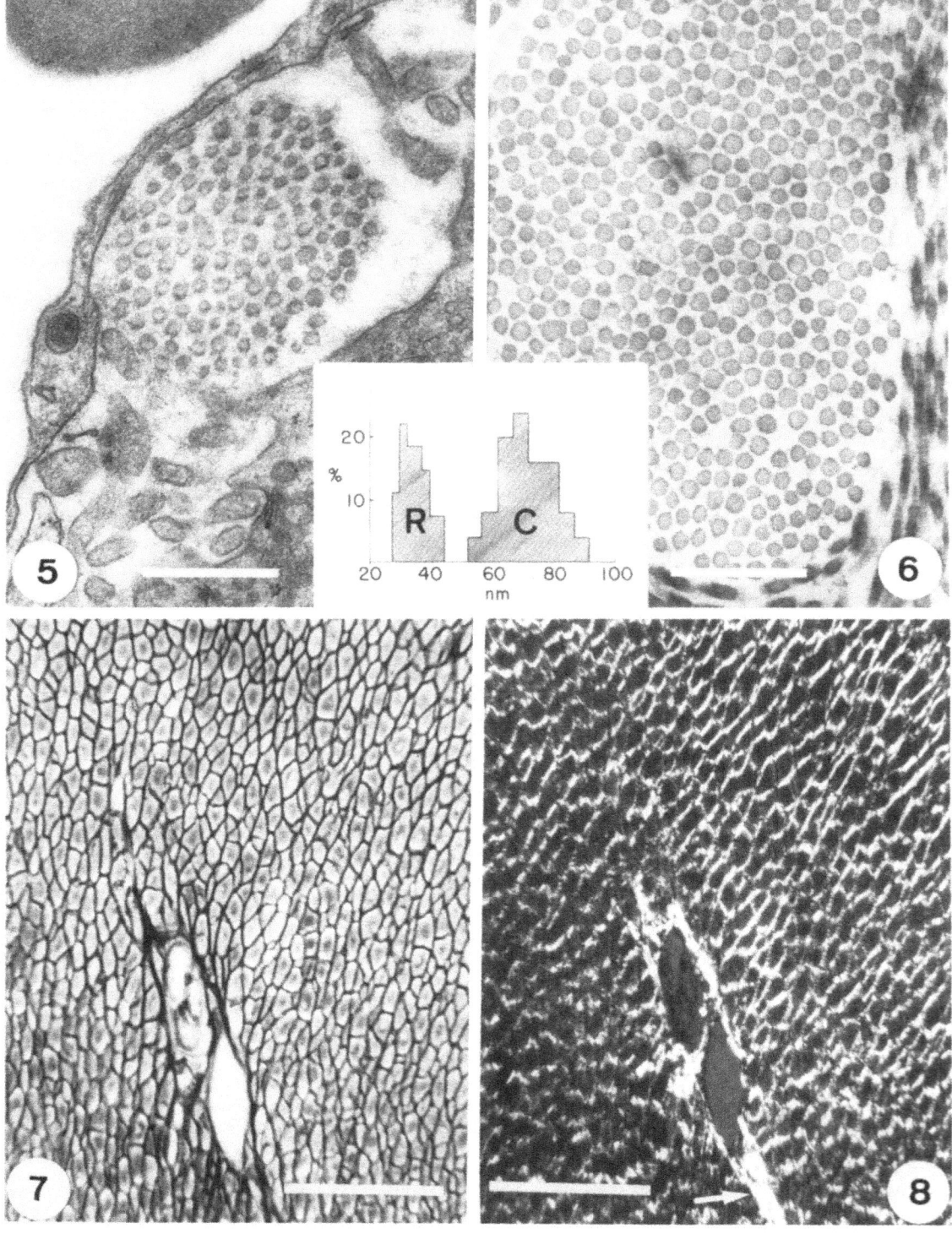

Table 3. Average fibril diameter from reticular and collagen fibers of different vertebrates*

Class	Species	Organ	Reticular fiber: fibril diameter (average in nm)	Collagen fiber: fibril diameter (average in nm)	Ratio between fibril diameter from collagen- and reticular fibers
Osteichthyes	*Tilapia melanopleura* (bony fish)	Endoneurium	26.63 ± 4.49	58.96 ± 16.23	1.94
		Aortic media	34.15 ± 5.21		
Amphibia	*Bufo ictericus* (toad)	Pancreas	43.94 ± 6.60	72.35 ± 15.95	1.72
		Smooth muscle	32.75 ± 5.00		
		Endoneurium	41.86 ± 4.58		
		Aortic media	49.20 ± 6.15		
Reptilia	*Xenodon merremii* (snake)	Liver	57.84 ± 9.90	82.03 ± 22.15	1.46
		Kidney	55.52 ± 8.80		
		Smooth muscle	53.56 ± 15.56		
		Endoneurium	58.12 ± 8.28		
		Aortic media	55.59 ± 15.75		
Aves	*Gallus gallus* (chicken)	Endoneurium	36.05 ± 4.65	52.45 ± 12.44	1.52
		Aortic media	32.83 ± 2.39		
Mammalia	*Rattus norvegicus* (rat)	Spleen	33.38 ± 6.03	71.03 ± 11.38	1.88
		Kidney	39.58 ± 6.94		
		Smooth muscle	40.20 ± 7.00		
		Endoneurium	37.49 ± 4.24		
		Aortic media	38.43 ± 6.90		
	Callithrix jacchus (marmoset)	Liver	33.01 ± 5.79	53.11 ± 9.04	1.66
		Kidney	34.69 ± 5.85		
		Smooth muscle	28.16 ± 4.70		
	Human	Smooth muscle	44.27 ± 6.58	76.61 ± 7.91	1.63
		Endoneurium	50.01 ± 4.24		

* Adapted with permission from (20).

complexes, very probably possess biomechanical properties compatible with these requirements.

3.4. Basement membrane collagens

At least two types of collagen (type IV and type V collagens) have been characterized in structures identified morphologically as basement membranes (Table 1). The chemical structure, and other characteristics, of the types of collagen associated with glycoproteins, and also with proteoglycans, in basement membranes is still a subject of controversy (8).

However, it is accepted that basement membrane

←

Figure 5. Electron micrograph of a reticular fiber in the space of Disse of a rat liver. Observe the loose disposition and thin diameter of the constituent collagen type III fibrils. Note the abundance of interfibrillary material. Bar = 0.5 μm; × 43,000. (Reproduced with permission from (13)).

Figure 6. Electron micrograph of part of a collagen fiber from rat connective tissue (same enlargement as in Figure 5). Observe the large diameter and close packing of the constituent collagen type I fibrils that assemble to form a large collagen fiber. Bar = 0.5 μm; × 43,000. The inset summarizes the results obtained from measurements performed when studying the diameters of the collagen fibrils in reticular (R) and in collagen (C) fibers from rat connective tissue of various organs. Observe that each type of fiber is the site of a different fibril population, for fibrils in collagen fibers are thicker and show a marked variation in diameter, when compared to the thin fibrils of more uniform diameters that are present in reticular fibers.

Figure 7. Photomicrograph of a section of the inner muscular layer of the dog small intestine. A network of reticular fibers surrounds the smooth muscle cells. Silver impregnation technique. Bar = 50 μm; × 500. (The corresponding electron microscopic picture can be seen in Figure 32.)

Figure 8. An adjacent section of the same material as in Figure 7, observed by the Picrosirius-polarization method. Note the presence of thin, weakly birefringent collagen type III fibers surrounding the smooth muscle cells. The adventitia of blood vessels is composed of thick, strongly birefringent, collagen type I fibers (arrow). Observe the remarkable correspondence in the distribution of the thin, weakly birefringent, collagen type III fibers with the reticular fibers in the serial section of Figure 7. Bar = 50 μm; × 500.

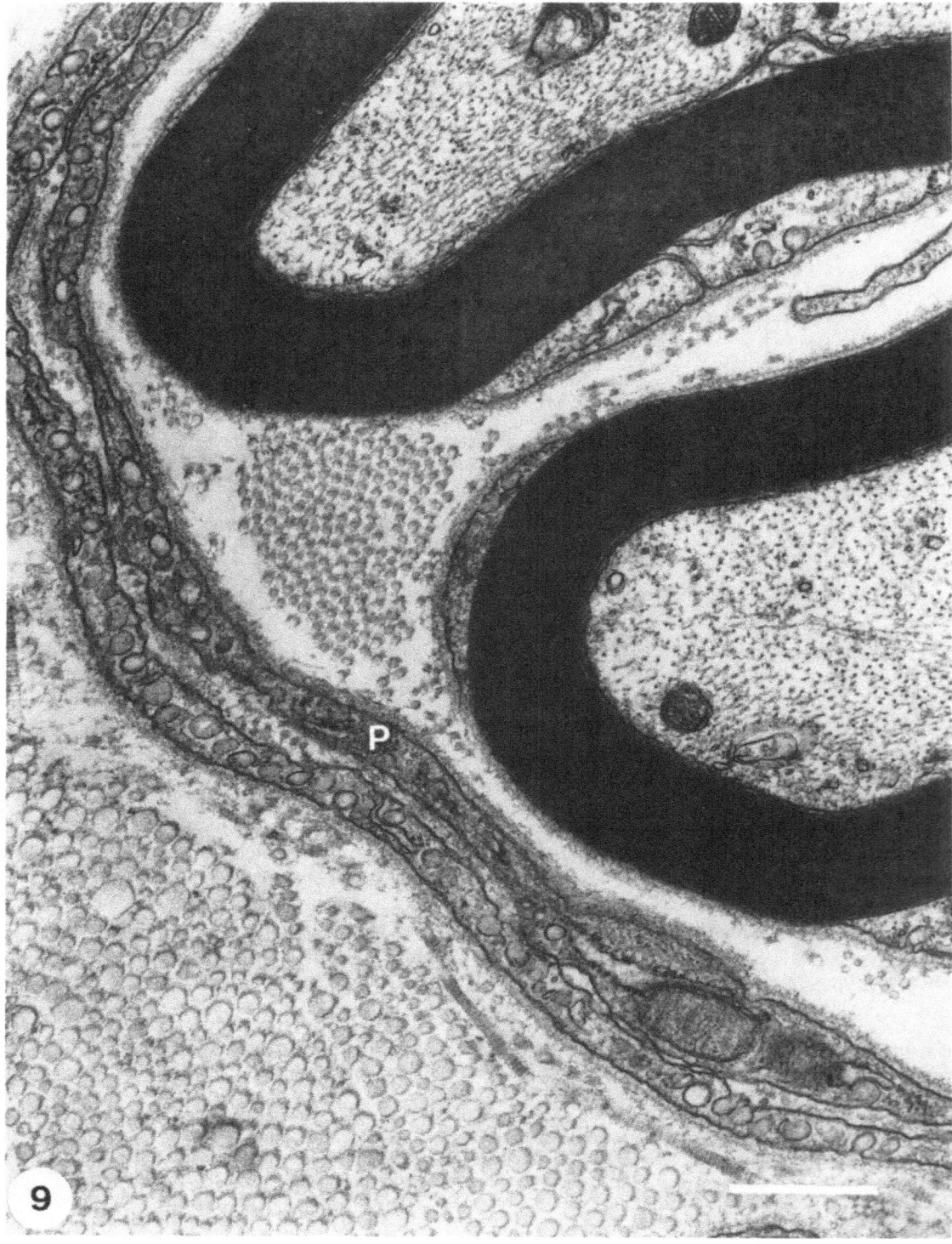

Figure 9. Electron micrograph from a thin nerve branch in the mouse skin. Observe the perineurial cells (P) separating the thin endoneurial collagen fibrils surrounding the two myelinated axons, which contrast with the thicker epineurial collagen fibrils in the lower left corner. Bar = 0.5 μm; × 46,500.

collagens do not form visible fibrils or fibers. Very thin microfibrils can be seen under the electron microscope (Figure 12). Several authors are of the opinion that these microfibrils are arranged randomly forming a tight feltwork-like structure (46, 47). In spite of this, a clear birefringence can be detected in many basement membranes studied by means of the Picrosirius-polarization method, suggesting that there are collagen molecules orderly disposed in basement membranes (Figures 3, 4, 17) (48).

Two main functions are widely accepted for basement membrane collagens. The first relates to a. structural support, and separation, for epithelial cells. The second refers to the function of basement membranes as filters; this function has been well studied in the renal glomerulus [see Chapter 11].

4. Collagen distribution in organs

Summing up, the results discussed so far strongly suggest that collagen type I forms the *collagen fibers* and that collagen type III forms the *reticular fibers* observed by histologists; while collagen type II is identified with hyaline and elastic cartilages and type IV is localized to basement membranes.

Therefore, in all sites where collagen fibers have been described (such as tendons, fasciae, organ capsules, adventitial layer of blood vessels, dermis, and submucous layer of the digestive tube) closely packed, thick collagen fibrils of variable diameters with a clearly evident banded staining pattern have been observed under the electron microscope (Figure 10). On the other hand, in all structures to which *reticular fibers* have been localized (such as endoneurium, endotenonium, arterial tunica media, mucosal lamina propria, and smooth muscle layers of the gastrointestinal tract, and the reticular network of parenchymatous organs such as the liver, kidney, and lung) loosely disposed, thin collagen fibrils of more uniform diameters have been detected by means of electron microscopy (Figure 12).

These features provide a general background for understanding the distribution of the different collagen types in the various organs. In addition, the ultrastructural features combined with the available histochemical, biochemical, and immunohistologic data of the literature, regarding some of the systems, will be discussed below in order to corroborate the electron microscopic findings.

4.1. Digestive system

The gastrointestinal tract possesses collagen types I and III that form the collagen and reticular fibers respectively, whose most obvious role is to provide the tensile strength that is necessary to hold the different tissues together as a functional unit, yet permitting high levels of compliance. Thus collagen type I provides its tensile strength to the dense connective tissues of the serous covering and submucous layer, while collagen type III is present in the mucosal lamina propria and in the smooth muscle layers that require higher levels of compliance (Figure 15).

With exception of a few collagen fibers present in the adventitia of blood vessels, the inner muscular layer of the dog small intestine is devoid of collagen-fiber-containing trabeculae, its stroma being rich in orderly disposed reticular fibers (20). Histochemical, immunohistologic, and ultrastructural evidence has been reported demonstrating that the distribution of these reticular fibers shows a striking correlation with the localization of collagen type III (Figures 7, 8) (20).

In the parenchymatous glands associated with the digestive tube, collagen type distribution is also characteristic. Collagen type I is present in the organ capsule and its septa, and in the adventitia of blood vessels; while collagen type III is localized to the network of reticular fibers that surround the epithelial cells. Thus, in the liver for example, collagen type I is present in the Glisson's capsule and in the portal spaces; while collagen type III is localized to the network of reticular fibers that lie in the space of Disse, between the sinusoid lining and the hepatic cell plates (Figures 5, 16). Immunohistologic data corroborate these histochemical and ultrastructural findings (16).

4.2. Skin

Biochemical studies showed that collagen type III constitutes more than 60% of the collagen in fetal skin, but makes up less than 20% of the collagen in adult skin (49). Although immunohistologic observations suggested that in the adult skin type III collagen is present primarily in the papillary dermis just beneath the epidermis, while the deeper dermal regions contain mainly collagen type I (40); a biochemical study reported that the ratio of type I to type III collagens was the same at all levels of the dermis (50).

Ultrastructural and histochemical results from this laboratory show that the fine-woven meshwork of

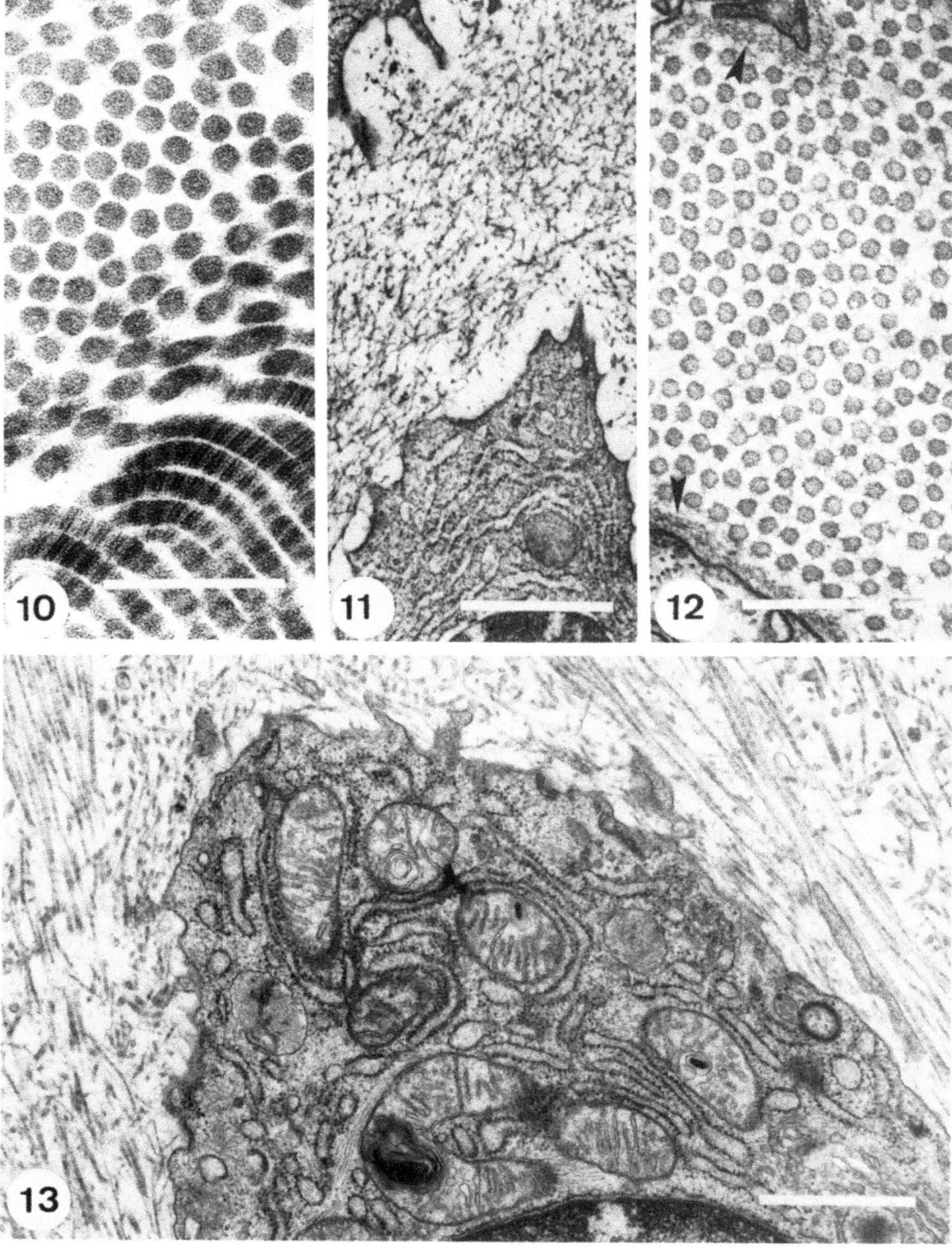

reticular fibers in the papillary layer can be distinctly differentiated from the coarse collagen fibers present in the deeper layers (51). Thus, in the papillary layer, reticular fibers display a distinct argyrophilia when studied by means of silver impregnation techniques and show up as thin, weakly birefringent greenish fibers when observed by the Picrosirius-polarization method. In addition, electron microscopic studies disclosed the presence of thin fibrils in the papillary dermis just beneath the epidermis, while thicker fibrils could be localized ultrastructurally to the deeper layers where coarse collagen type I fibers had been characterized by the Picrosirius-polarization method (Figure 14). Our morphometric data express quantitatively the differences between the two distinct collagen fibril populations (Figure 27).

Thus, our results give strong support to the immunohistologic localization of collagens in dermis (40), suggesting that the collagen in the papillary dermis is mainly of the type III, while collagen type I is predominant in the deeper layers. The foregoing findings do not support the previous biochemical report (50), however they are consistent with a more recent biochemical study (52) that has found a higher relative content of type III collagen in the upper dermis.

4.3. Nervous system

Collagen types I and III had been shown by chemical methods to coexist in human femoral nerves (53). However, the localization of these collagen types in the peripheral nervous system was only described recently (22). In this report, nerves and ganglia from a variety of fish, amphibian, reptilian, and mammalian species were studied by optical and electron microscopy. Observations using the Picrosirius-polarization method strongly suggested that two different types of collagen are present in the connective tissues of nerves and ganglia. Electron microscopy of these organs showed two different collagen fibril populations, distinguishable on the basis of diameter, located in different compartments of these structures (Figure 9). Thicker fibrils are present in nerve and ganglionic epineurium. Thinner fibrils are present in the endoneurium, surrounding nerve fibers and ganglionic cells, and between the concentric layers of perineurial cells (Figure 28). These results, consistently observed in all species studied and very probably represent a general phenomenon in vertebrates, strongly suggested that collagen type I is present primarily in the epineurium, and collagen type III is localized to the peri- and endoneurium. An immunohistologic report strongly supported the histochemical and ultrastructural findings discussed above (54).

4.4. Arteries

Biochemical observations have shown that arteries contain collagen types I and III (55, 56). The study of the distribution of these interstitial collagens by immunohistologic and histochemical methods have shown that collagen types I and III are located mainly in the adventitia and tunica media of these structures respectively (19, 57). Recently, the fact that collagen types I and III are segregated into different compartments of the arterial wall has been demonstrated by means of light- and electron microscopy (24). In this report, the collagen of the adventitia has been shown to present features which are characteristic of collagen type I; while the collagen in the tunica media revealed histochemical and ultrastructural aspects which are typical of collagen type III (Figures 1, 29). As these results were consistently observed in all arteries of representative species from the main vertebrate classes, they suggest the presence of a general structural pattern of collagen distribution in vertebrate arteries.

←

Figure 10. Collagen type I. Electron micrograph of a collagen fiber in the epineurium of a human sural nerve. Observe the close packing and large diameter of collagen type I fibrils. In the lower region, longitudinal sections of these fibrils show the characteristic banded staining pattern that is particularly evident in this type of collagen. Bar = 0.5 μm; × 56,000.

Figure 11. Collagen type II. Electron micrograph of a hyaline cartilage. Observe that the collagen type II fibrils are unbanded and present a very small diameter. Bar = 1 μm; × 24,000.

Figure 12. Collagen type III and basement membrane collagen. Electron micrograph of a reticular fiber in the endoneurium of a human sural nerve. Same magnification as in Figure 10. Observe the loose disposition, small fibril diameter, and the abundance of interfibrillary material (forming bridges that connect these collagen type III fibrils). Surrounding the cellular portions present in this picture, typical basement membranes, with no visible collagen fibrils, can be seen (arrowheads). Bar = 0.5 μm; × 56,000.

Figure 13. Electron micrograph of a rat knee joint meniscus. The magnification of this picture of a fibrocartilage is the same as that of the hyaline cartilage in Figure 11. Observe that the collagen type I fibrils in the upper region are thicker, present a more visible cross striation, and are assembled to form fibers, these characteristics differing from those of the fibrils in hyaline cartilages. Bar = 1 μm; × 24,000.

78

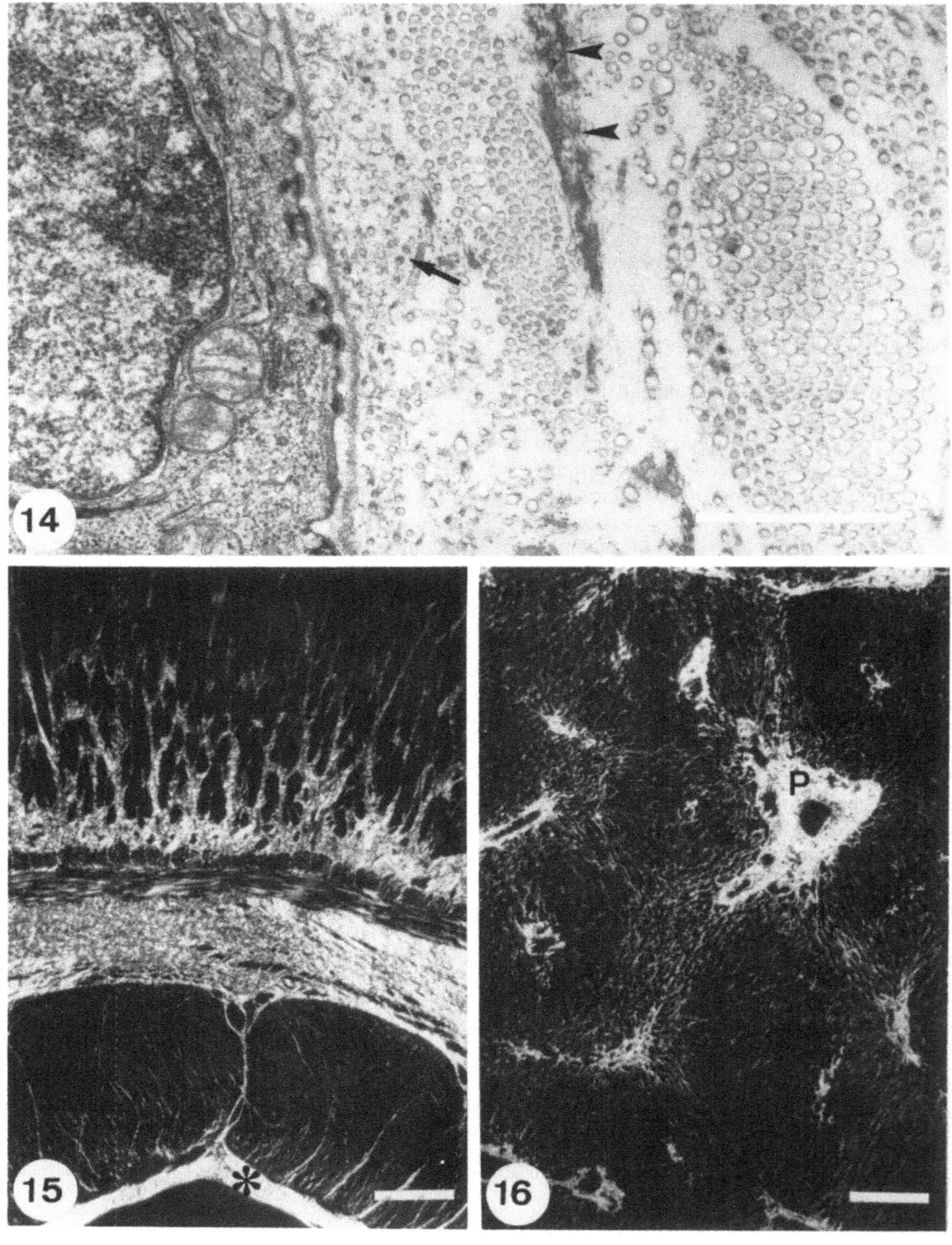

5. Tissue distribution of specific collagen–proteoglycan interactions

It is known that collagen interacts with proteoglycans and that this interaction has a biological importance that explains, to a certain extent, some of the functional characteristics of the different types of collagen (13) [see also Chapter 6].

Collagen–proteoglycan interaction has been the subject of several studies (for reviews, see 13, 58, 59). Methods were developed to permit the observation of proteoglycans by aid of electron microscopy. Not only did these methods permit the observation of isolated proteoglycan molecules (60), but is was also possible to localize structures that could be identified as proteoglycans in connective tissues (61–65).

Recently a method has been developed that permits the quantitative estimation of this interaction in tissue sections (14). This method is based on the quantitation of the collagen amino groups that are blocked by the interaction with the proteoglycan sulfate radicals. These blocked groups cannot, therefore, bind to the anionic dye Sirius Red, thus reducing the stainability of tissue sections. When adjacent sections are treated with papain, the proteoglycans are hydrolyzed, and the previously blocked amino groups are liberated, thus increasing the stainability of the sections. In a next step, the Sirius Red bound to both control- and papain-treated sections is eluted in dilute sodium hydroxyde and measured in a spectrophotometer. Thus, the difference in the amount of dye present in control- and papain-digested sections measures the amino groups blocked due to collagen–proteoglycan interaction.

When this method was applied to different tissues, containing different collagen types, from sections of representative species of vertebrates, it showed in all cases studied a close relation between the type of collagen and the intensity of its interaction with proteoglycans (14). Thus collagen type II shows high levels of interaction with proteoglycans (an average of 56% of its amino groups are exposed by papain treatment) while collagen type I displays very low levels of interaction (its average exposure being of 16%). Collagen type III shows intermediate levels of interaction with proteoglycans (30 % of its amino groups being exposed by papain digestion). Further, the type and amount of proteoglycans present in tissue also varies according to the type of collagen contained in the sample (15). Thus, tissues composed of collagen type I possess small amounts of proteoglycans (about 2.8 mg/gm of dry weigth) that contain almost exclusively dermatan sulfate, while tissues containing only collagen type II have high amounts (an average of 15.0 mg/gm of dry tissue) of chondroitin sulfates AC. Tissues containing collagen type III possess intermediate levels of proteoglycans (average: 11.5 mg/gm of dry weight) that are composed mainly of heparan sulfate.

These studies show that not only does the amount of proteoglycans vary according to the type of collagen present in the tissues, but also the type of the constituent glycosaminoglycan is specifically related to the type of interstitial collagen present in the tissue studied.

These conclusions are important for interpreting the morphological aspects observed by electron microscopy of the different types of collagen and can contribute to distinguish collagen type distribution (13).

Thus, one can easily understand why collagen fibers (collagen type I), which contain a small amount of proteoglycans and display low levels of interaction with these compounds, show their fibrils compactly disposed and possess very few branched structures connecting these fibrils (Figures 6, 10). It also explains why hyaline and elastic cartilages that contain collagen type II and are rich in proteoglycans (which display high levels of interaction with the collagen present) show their fibrils widely separated from one another and the space between them is filled with coarse granules and multiform branched structures that have been identified as proteoglycans (61–65) (Figure 11).

←

Figure 14. Electron micrograph of mouse skin. At left, an epithelial cell showing hemidesmosomes and its basement membrane. Observe that the fibrils in the papillary dermis, just beneath the epithelial cell (arrow), are thinner than the collagen fibrils in the deeper layer (on the right-hand side). Arrowheads point an elaunin fiber of the elastic system. Bar = 1 μm; × 33,000 (reproduced with permission from (51)).

Figure 15. Section of dog stomach. Thin, weakly birefringent collagen type III fibers (reticular fibers) can be seen in the lamina propria (upper region) and also in the smooth muscle layers. Observe that the thick, strongly birefringent collagen type I fibers (collagen fibers) present in the serous covering (asterisk) penetrate the outer muscular layer forming collagenous septa. The submucous layer is also composed of collagen type I. Picrosirius-polarization method. Bar = 200 μm; × 60.

Figure 16. Section of dog liver. Thick, strongly birefringent collagen type I fibers (collagen fibers) can be seen in the portal spaces (P), while thin reticular fibers (composed of collagen type III) form a network surrounding the hepatic cell plates within the lobules. Picrosirius-polarization method. Bar = 200 μm; × 60.

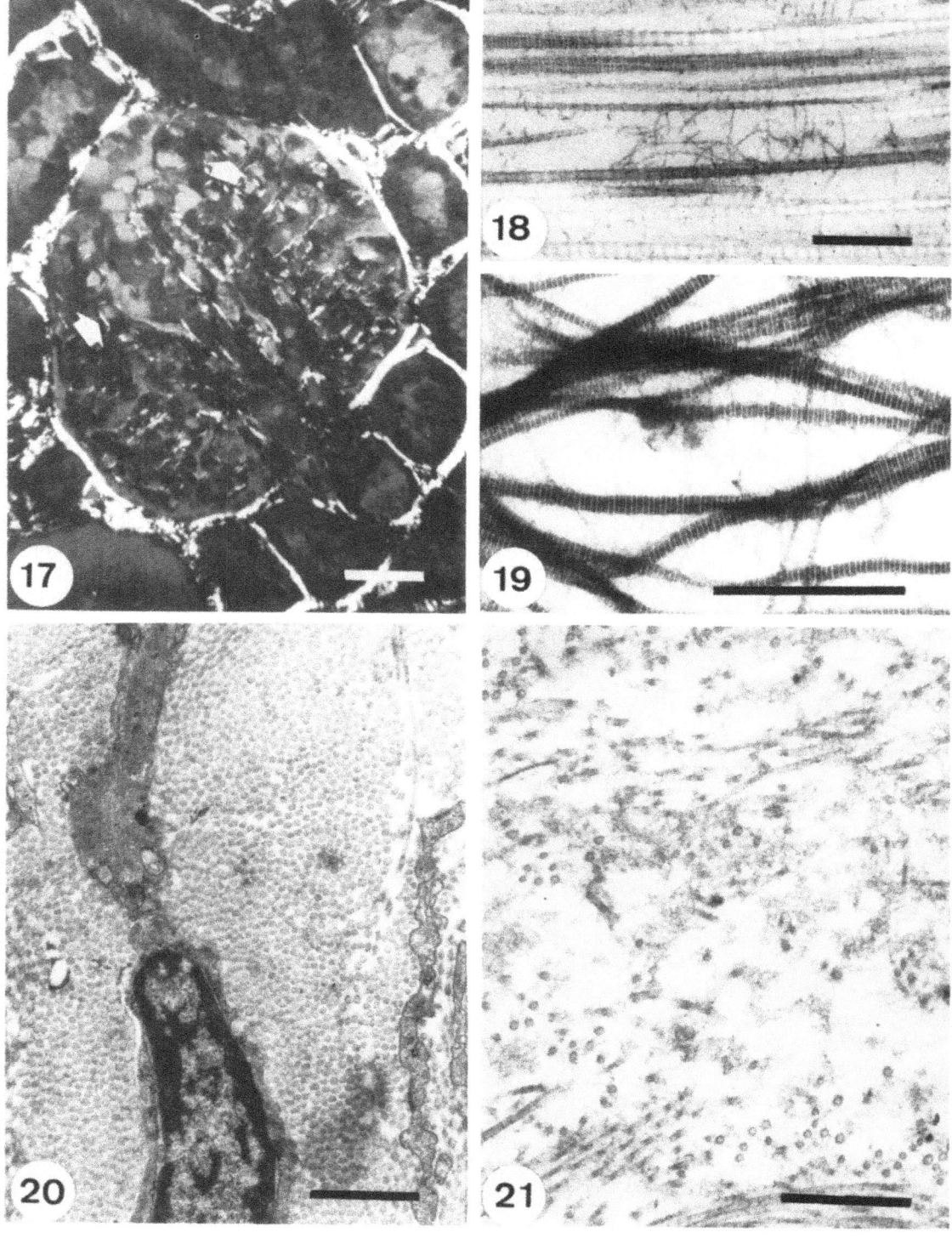

It is interesting to observe that the cross striation of collagen fibrils is hardly recognizable in normal cartilage and other proteoglycan-rich tissues and that this striation becomes clearly evident after digestion with papain (Figures 18, 19). This fact strongly suggests that, in cartilage, the 'stainability' of collagen cross striation is inhibited by the interaction with proteoglycans that not only block reactive (amino) groups but also form a highly hydrophilic viscous enveloping layer that may make difficult the access of 'staining' material to the collagen fibrils.

This possibility should be kept in mind for it can be an important factor in hindering efforts to obtain results with immunohistologic methods in the study of the distribution of the different collagen types.

Collagen type III, present in the reticular fibers, which shows intermediate levels of interaction with proteoglycans, displays a less compact disposition of its fibrils when compared to that of collagen fibers (20) and shows a net of multiform branched structures binding its fibrils (Figure 12).

The above-mentioned aspects strongly suggest that collagen–proteoglycan interaction has an obvious structural role in tissues, lending to them a characteristic elasticity by imparting resistance to pressure, that acts like a biological spring. The importance of collagen–proteoglycan interactions has been observed in several biological models of which three examples will be cited here. In all of them the hydrolysis of proteoglycans, or of collagen, lead to clear-cut changes in the characteristics of the tissue studied. In the first two examples the hydrolysis of proteoglycans occur while, in the third, collagen is hydrolyzed:

a) *The effect of intravenous papain injection on the rabbit ear cartilage*. In this model (66) the injection of papain promotes the hydrolysis of the ear cartilage proteoglycans, resulting in a loss of turgidity of the ear which then collapses. The resynthesis of the proteoglycans occurs in 72 hr with concomitant recuperation of the ear turgidity.

b)*The phenomenon of autotomy in holothurians*. It was recently suggested (67) that the softening of these animals body wall, which occurs during autotomy, is due to the activity of a protease that hydrolyzes the proteoglycans bound to the abundant collagen fibers present in this structure. The electron microscopy of this material presented morphological evidence in favor of this hypothesis (Figures 18, 19).

c) *Cervical dilation*. Cervical dilation during partum occurs due to the conspicuous softening of the uterine cervix, a tissue that is normally very hard. Evidence has been presented that this phenomenon is due to an intense process of local collagenolysis, without parallel changes in the proteoglycans present in this tissue (26). In this paper, clear-cut quantitative morphological evidence for collagenolysis could be obtained from the electron microscopic observations performed (Figures 20, 21).

These examples illustrate the value of electron microscopy on the study of collagen–proteoglycan interaction in normal and pathological tissues, a field that undoubtedly should be explored further.

6. Collagen distribution in pathological models

As consequence of the recent findings related to the biology of collagen, a few well-known syndromes presenting with connective tissue disorders could be characterized as primary diseases of collagen. Defects in collagen biosynthesis and processing can be due to genetic defects or environmental causes. The latter (mainly the action of drugs, or dietary deficiencies) are responsible for the acquired diseases of collagen. The molecular bases of some inherited collagen disorders have been reviewed recently (68).

Figure 17. Section of a human kidney. The central region is occupied by a glomerulus. The distinct birefringence of the basement membranes (arrows) shows that they contain orderly disposed collagen molecules. Picrosirius-polarization method. Bar = 20 μm; × 620 (reproduced with permission from (48)).

Figure 18. Electron micrograph of a sea cucumber (*Stichopus badionotus*) body wall. Collagen fibrils are irregularly impregnated by uranyl acetate and lead citrate. Observe that the collagen fibrils are bound by thin irregular microfibrils of the ground substance. Bar = 0.5 μm; × 31,000 (reproduced with permission from (67)).

Figure 19. Electron micrograph of papain-digested body wall of the sea cucumber *Stichopus badionotus*. Observe the gradual separation of collagen fibrils into thinner units, and the absence of microfibrillar bridges in the ground substance. The removal of interfibrillar material by enzymatic digestion promotes a better visualization of the banded staining pattern of collagen fibrils. Positive staining by uranyl acetate. Bar = 0.5 μm; × 60,000 (reproduced with permission from (67)).

Figure 20. Electron micrograph of a human nonpregnant cervix. Observe the close packing of the collagen fibrils forming collagen fibers. Bar = 1 μm; × 17,000.

Figure 21. Electron micrograph of a human intrapartum cervix. Great reduction in the amount of collagen fibrils; a diffuse granular deposit occupies most of the area normally filled by collagen fibrils. Observe the variable diameters of the transversely sectioned fibrils that are being degraded. The electron microscopic picture of collagen degradation shows a corroded aspect, particularly obvious in transverse sections where fibrils show a ragged and irregular outline. Bar = 1 μm; × 20,000.

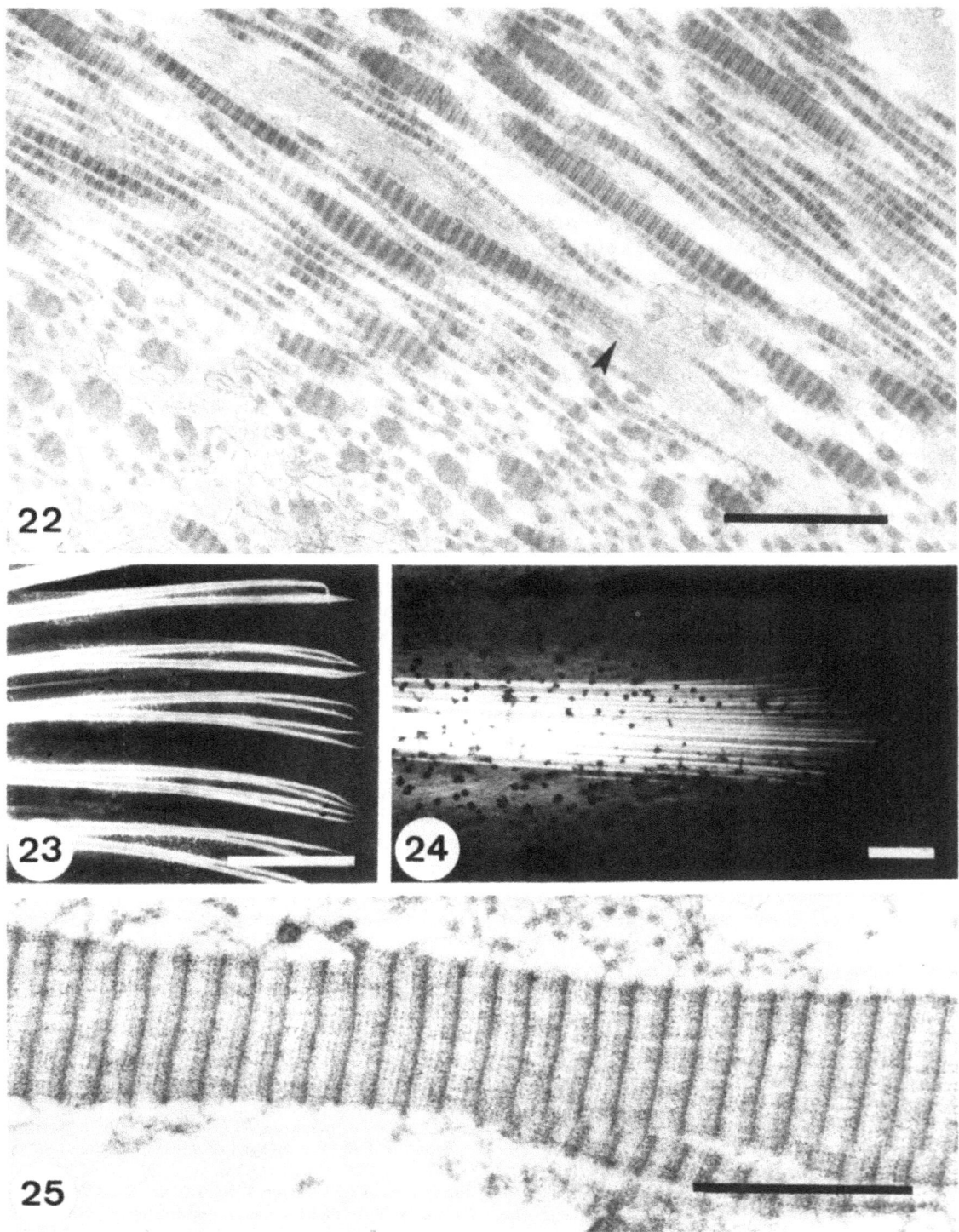

22

23

24

25

In addition to the above-mentioned diseases of collagen, there is a full range of diseases affecting collagen. These diseases affect collagen locally, and are usually the consequence of tissue injury promoted by various causes such as trauma, inflammation, etc.

Consequently the number of diseases of collagen, or affecting collagen, is very large. This section, however, will only deal briefly with some typical illustrative examples concentrating mainly on those that have been studied with the aid of the electron microscope.

It must be kept in mind, however, that collagen pathology is not limited to heritable and acquired disorders in the synthesis and extracellular processing of this protein; changes in amounts of collagen production and degradation are also responsible for many diseases involving connective tissue disorders. Thus, an excess of collagen synthesis is certainly a major problem in parenchymatous organs such as the liver. A well-known example is the cirrhotic response to prolonged injury to hepatocytes. Ultrastructural aspects of the collagen in hepatic fibrosis have been reported (69, 70).

The invasion of Schwann cells by *Mycobacterium leprae* promotes leprous neuritis that is also characterized by the proliferation of collagen; in this disease it was recently demonstrated that, despite the remarkable overall increase in the amount of nerve collagen, its other characteristics such as fibril diameters and collagen type distribution show no difference when compared with normal nerves (25).

Collagenases are important for understanding chronic inflammatory diseases such as rheumatoid arthritis and other pathologic conditions in which collagenous tissues are specifically degraded. Collagen degradation has been comprehensively reviewed elsewhere (10); nevertheless the ultrastructural aspects of collagen degradation are also illustrated in this chapter (Figures 20, 21) and have been in part discussed in Section 5c.

Among the non-structural functions of collagen,

its influence on cell differentiation and proliferation has important implications in developmental biology and in many diseases, including cancer. Other non-structural functions of collagen, which have special pathologic significance, include the role of collagen in platelet aggregation and thrombus formation (57), its involvement in cell attachment to a substratum (9), and the progressive changes in the physical and chemical properties of collagen that are related to the aging process.

Malignant cells are capable of synthesizing collagen, and this fact has permitted a differential diagnosis of some tumors based on the study of their collagen populations (27).

The ultrastructural characteristics of abnormal collagen fibrils have been reviewed recently (71). Among collagen fibril abnormalities, variation in diameters is the most common; this may be due mainly to either collagen degradation (Figures 20, 21) or hyperpolymerization (Figure 22), a feature that is very common in several pathologic processes such as hepatic fibrosis (70) and tumors (27).

However, hyperpolymerization of collagen is not limited to pathologic conditions. As a matter of fact, histochemical, biochemical, and ultrastructural results from this laboratory have shown that actinotrichia (the thin spicules that stretch out beyond the distal end of teleost fin rays, Figures 23, 24) are composed of hyperpolymerized collagen molecules (Figure 25) (72). Indeed, each actinotrichium is a single giant collagen fibril that measures an average 3 μm in diameter and 650 μm in length.

7. Conclusions

The ultrastructural aspects of collagen distribution described here, and the correlation of these observations with the available light microscopic studies, and also with the current biochemical and immunohistologic data on the subject, enable the following conclusions to be drawn.

←

Figure 22. Electron micrograph of a human neurofibroma showing an unusual aspect of collagen polymerization. Agglomerates of microfibrils hyperpolymerize to form very thick collagen fibrils (arrowhead). Bar = 0.5 μm; × 50,000 (reproduced with permission from (27)).
Figure 23. Negative image of a whole mount of a part of a teleost tail fin. Observe that the rays originating from the base of the fin (left), extend distally, dichotomously branching towards the margin of the fin (right). Bar = 0.5 cm; × 4.
Figure 24. A higher magnification of the margin of a teleost fin shows a cluster of strongly birefringent thin spicules (actinotrichia) that stretch out beyond the distal end of each ray. Polarization microscopy. Bar = 100 μm; × 100.
Figure 25. Electron micrograph of an actinotrichium from the false mouthbreeder teleost *Tilapia melanopleura.* The banded staining pattern of actinotrichia in this species measures an average 64.2 nm, ranging from 49.0 to 75.7 nm in different specimens. Observe that each actinotrichium consists of hyperpolymerized collagen molecules which constitute a single ultrastructural entity, which can measure up to 5 μm in diameter and up to 850 μm in length. These aspects suggest that we may be dealing with a new type of collagen. Bar = 0.5 μm; × 69,000 (reproduced with permission from (72)).

84

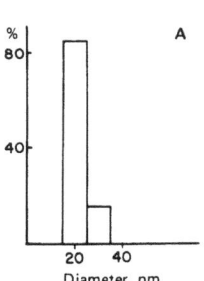 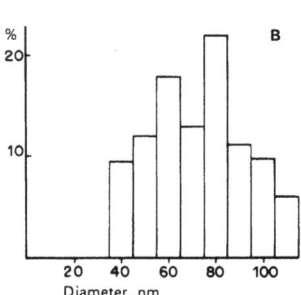 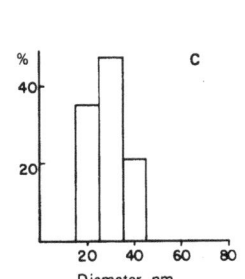 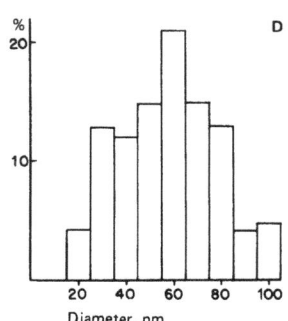

Figure 26. Histograms illustrating the distribution of the collagen fibril diameters measured in different cartilages of the rat: (A) ear elastic cartilage, (B) femoral articular cartilage, (C) tracheal hyaline cartilage and (D) knee joint meniscus fibrocartilage. Observe that the cartilages normally submitted to strong pressures (articular and fibrous cartilages) present larger and more variable diameters as compared to the elastic and hyaline cartilages, which are not subjected to high pressures (reproduced with permission from (23)).

The electron microscopic picture of the fibrillar, interstitial collagens shows that each collagen type displays typical ultrastructural features that permit their characterization by means of the routinely used transmission electron microscopy. On the contrary, basement membrane collagens, which are being actively studied at present and are still subject to controversies, do not form fibrils and little information can be obtained when observing them under the electron microscope.

All tissues in which *collagen fibers* have been described by histologists, contain collagen type I. This collagen type forms closely packed, thick collagen fibrils of variable diameters with a clearly evident banded staining pattern that can be readily recognized under the electron microscope.

Collagen type III could be detected in all struc-

tures to which *reticular fibers* have been localized. The electron microscope discloses the presence of loosely disposed, thin collagen fibrils of more uniform diameters in such locations.

Collagen type II, which is known for its tissue selectivity for hyaline and elastic cartilages, forms a meshwork of very thin collagen fibrils in association with large amounts of proteoglycans; instead of assembling to constitute ordered fibrillar aggregates (fibers), as type I and type III collagens do.

The foregoing conclusions were drawn from exhaustive morphometric and statistical ultrastructural studies performed on several representative species of the main classes of vertebrates (1, 20, 22–27, 51), and no exception has so far been found to the above-mentioned electron microscopic evidences that permit the characterization of interstitial collagen types.

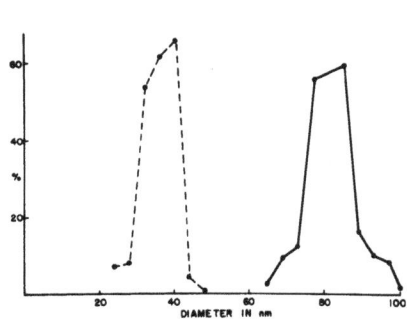 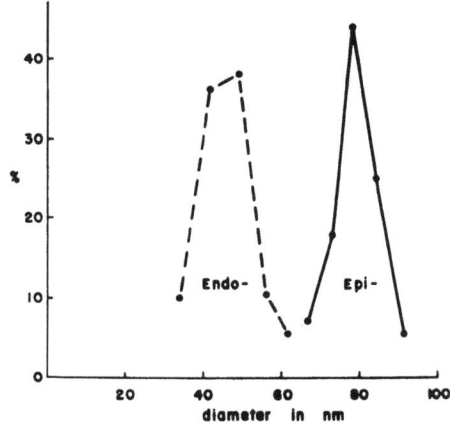

Figure 27. Graph showing the frequence of collagen fibril diameters in the papillary dermis just beneath the epidermis (-----), and in the deeper layers of the dermis (———), of human skin. Observe the two clearly distinct populations of fibrils that are segregated into different compartments of the dermis (reproduced with permission from (51)).

Figure 28. Graph illustrating the distribution of collagen fibril diameters in the epi- and endoneurium of a human sural nerve, showing that each compartment is the site of a different population. The electron microscopic picture of the epi- and endoneurium of this same nerve can be seen in Figures 10 and 12.

However, it has been claimed that it is not possible to identify collagen types on the basis of their electron microscopic appearance (5). This statement was based on erroneous premises such as the fact that type II collagen formed fibrils which had the characteristics of type I collagen (the error is that calcified cartilage of growing bone was considered, and it should have been kept in mind that chondroblasts undergo a shift in collagen synthesis from type II to type I preceding calcification of the cartilaginous matrix, 73). The fact that purified types I and III collagens form fibrils of similar size when reconstituted *in vitro* was acknowledged as a second evidence (however a more recent report on the scanning electron microscopic images of the *in vitro* reconstitution of collagen polymers demonstrated the formation of thick bundles of pure collagen type I and thin isolated fibers of type III, showing that the chemical structure of collagen molecules determines their psysical aggregation, 34). Indeed, there is ample evidence that collagen type I forms thicker fibrils than collagen type III, and this fact has been repeatedly stated in the abundant literature on the topic (1, 3, 4, 6, 20). However, it should be kept in mind that the characterization of collagen types, based on fibril diameters, is not reliable when studying embryionic material or cell cultures where intense collagen production and deposition are responsible for the gradual increase in width of each fibril.

It is noteworthy that, later on, the same article quoted in the previous paragraph (5) states that more recent ultrastuctural studies have shown that type III collagen assembles to thin fibrils while the thicker fibrils seemed to be type I collagen as revealed by the use of ferritin-labeled antibodies (74).

The reconstitution of collagen polymers *in vitro* may not necessarily reflect the pattern found in the intact tissue. As a matter of fact, despite the detailed knowledge of the structure of the distinct collagen molecules, it is still unclear how the assembly of the different types of collagen molecules to higher structures is controlled (5). The fact that tissues containing only one collagen type have been used to show that each collagen type interacts differently (considering both the qualitative and quantitative aspects) with the ground substance (14, 15), suggests that (besides the role played by the chemical structure of collagen molecules in determining there physical aggregation) the assembly of collagen molecules *in vivo* may also be regulated by other tissue-specific components such as proteoglycans and glycoproteins (75, 76). As results of this discussion it is felt that further ultrastructural studies on the *in vitro* and *in vivo* assembly of collagen molecules, and

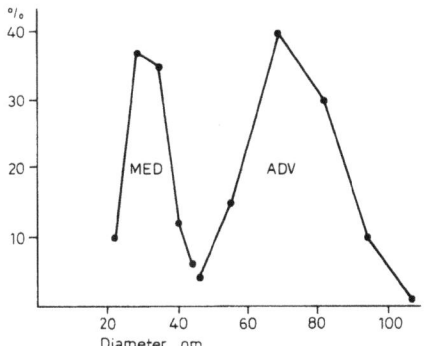

Figure 29. Distribution of collagen fibril diameters measured in the adventitia (ADV) and tunica media (MED) of rat thoracic aorta, showing that each compartment is the site of a different population (reproduced with permission from (24)).

on the influence of non-collagenous macromolecules on such assembly would be of interest to expand the knowlegde regarding this topic.

Electron microscopic characterization of the different collagen types has proved useful for localizing them in places where it has not been possible to detect them by biochemical techniques. In fact, collagen type III has been identified ultrastructurally (42) and by immunohistology (2, 40) in the endotenonium (Figure 33), while no biochemical evidence for type III collagen in the tendon is available. Another question that remains unsettled is whether several collagen types are intermingled in one fibril or assemble to separate structural entities. Electron microscopy may prove useful to solve this problem too. Indeed, unpublished results from this laboratory show that in certain anatomical sites (such as liver and skin) occasionally reticular fibers may be composed of a mixture of thick and thin fibrils assembled to constitute a single ordered fibrillar aggregate, whereas in other locations (namely in the endoneurium, around smooth muscle cells, and around adipose cells) each reticular fiber is formed of a homogeneous bundle of thin fibrils with uniform diameters (Figures 30–33). These findings suggest that some reticular fibers in skin and liver may be composed of both type I and type III fibrils, whereas endoneurial reticular fibers contain exclusively collagen type III fibrils (Figure 9). It is noteworthy that an increase in the collagen type I fibril population of hepatic reticular fibers seems to correlate directly with the aging process. These results speak in favor of the hypothesis that each of the different interstitial collagen types assemble to seperate structural entities which are homogeneously composed of only one type of collagen molecules. The foregoing find-

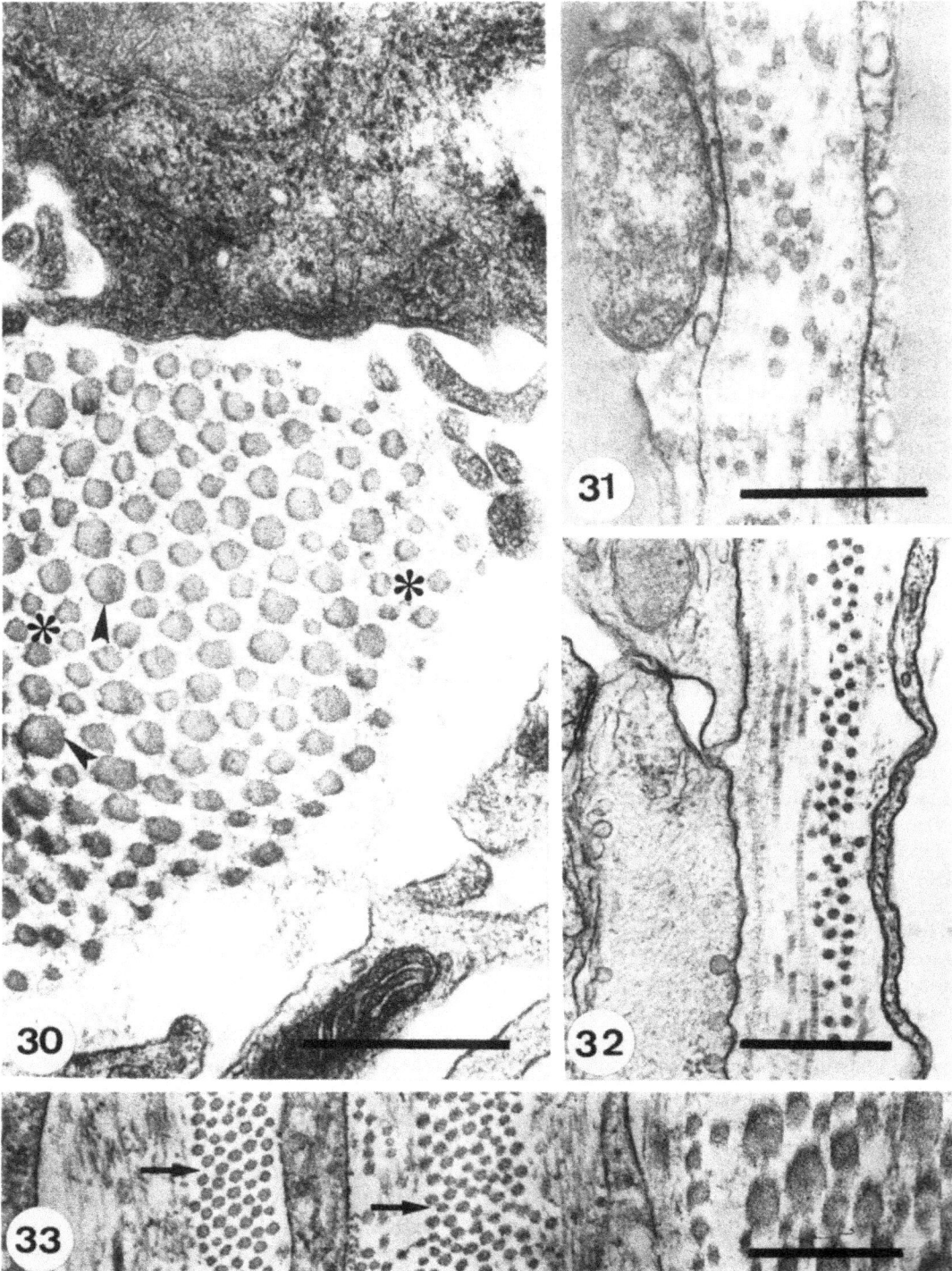

Figure 30. Electron micrograph of a mouse liver showing part of a reticular fiber in the space of Disse. Observe that the thin collagen type III fibrils (surrounding asterisks) are interspersed with thicker fibrils (arrowheads). Bar = 0.5 μm; × 63,000.

Figure 31. Electron micrograph of mouse subcutaneous adipose tissue showing that the reticular fibers in the intercellular space are homogenously composed of loosely packed, thin collagen type III fibrils. Bar = 0.5 μm; × 56,000.

Figure 32. Electron micrograph of reticular fibers surrounding smooth muscle cells. Observe the uniform diameters of the transversely sectioned collagen type III fibrils. Bar = 0.5 μm; × 45,000.

Figure 33. Electron micrograph of the sheath that surrounds each tendon (peritenonium) showing that the thin collagen fibrils (arrows) of uniform diameters are arranged in layers alternating to rows of flattened cells. At right, thicker collagen fibrils of variable diameters. Bar = 0.5 μm; × 46,000.

ings may also explain why discrepancies between results reported by different authors using immunohistologic methods are so evident. Thus, both studies of the collagens of human liver, one stating that reticulin fibers contain collagen type III and that they are not stained by collagen type I antibodies (16) and the other claiming that the same fibers contain both type I and type III collagens (17), may be correct in view of the electron microscopic observations discussed here.

On the other hand, the fact that the electron microscope discloses the existence of a gradient of fibril diameters increasing steadily from type II collagen in the cartilaginous matrix to type I collagen in the perichondrium (23); and from the endoneurium (type III collagen), through the perineurium, to the epineurium (type I collagen) may indicate that, contrasting with what was suggested in the previous paragraph, two types of genetically distinct collagen molecules are intermingled in a single fibril. It is hoped that this short discussion will have served to show how much more remains to be studied about collagen and that at the same time it will have illustrated the importance of electron microscopy for the study of collagen biology.

To give stronger support to these observations, it would be interesting to use immunoelectron microscopic techniques. Ferritin-labeled antigens, however, may attach non-specifically to collagen and other tissue components. Fewer non-specific reactions have been observed with peroxidase-labeled antibodies, although its electron-dense oxidation product does not allow the resolution of ferritin-labeled antibodies. However, once such problems have been overcome, immunoelectron microscopic techniques may provide a means of corroborating the ultrastructural findings summarized in this chapter that seem to be so accurate, in view of the sensitivity of the electron microscope for localizing and measuring small structures in tissues.

Acknowledgements

Studies cited from the authors' laboratory have been aided by grants from Fundação de Amparo à Pesquisa do Estado de São Paulo (FAPESP) and Financiadora de Estudos e Projetos (FINEP-FNDCT). G.S. Montes is supported by the Department of Histology and Embryology, Institute of Biomedical Sciences, University of São Paulo, and by Conselho Nacional de Desenvolvimento Científico e Tecnológico (CNPq). L.C.U. Junqueira is supported by CNPq.

References

1. Montes GS, Junqueira LCU: Biology of collagen. Rev Can Biol Exp 41: 143–156, 1982.
2. Gay S, Miller EJ: Collagen in the physiology and pathology of connective tissues. Stuttgart, Gustav Fischer, 1978.
3. Bornstein P, Sage H: Structurally distinct collagen types. Ann Rev Biochem 49: 957–1003, 1980.
4. Eyre DR: Collagen: molecular diversity in the body's protein scaffold. Science 207: 1315–1322, 1980.
5. von der Mark K: Localization of collagen types in tissues. Int Rev Connect Tissue Res 9: 265–324, 1981.
6. Minor RR: Collagen metabolism. Am J Pathol 98: 225–280, 1980.
7. Adams E: Invertebrate collagens. Science 202: 591–598, 1978.
8. Kefalides NA, Alper R, Clark CC: Biochemistry and metabolism of basement membranes. Int Rev Cytol 61: 167–228, 1979.
9. Kleinman HK, Klebe RJ, Martin GR: Role of collagenous matrices in the adhesion and growth of cells. J Cell Biol 88: 473–485, 1981.
10. Pérez-Tamayo R: Pathology of collagen degradation. Am J Pathol 92: 509–566, 1978.
11. Tanaka S, Hata R-I, Nagai Y: Two-dimensional electrophoresis patterns of the CNBr peptides of collagen types I, II, III, and V. Coll Res 5: 445–452, 1981.
12. Reiser KM, Last JA: Quantitation of specific collagen types from lungs of small mammals. Anal Biochem 104: 87–98, 1980.
13. Junqueira LCU, Montes GS: Biology of collagen-proteoglycan interaction. Arch Histol Jpn 46: 589–629, 1983.
14. Junqueira LCU, Bignolas G. Mourão PAS, Bonetti SS: Quantitation of collagen-proteoglycan interaction in tissue sections. Connect Tissue Res 7: 91–96, 1980.
15. Junqueira LCU, Toledo OMS, Montes GS: Correlation of specific sulfated glycosaminoglycans with collagen types I, II, and III. Cell Tissue Res 217: 171–175, 1981.
16. Biempica L, Morecki R, Wu CH, Giambrone M-A, Rojkind M: Immunocytochemical localization of type B collagen. Am J Pathol 98:591–602, 1980.
17. Konomi H, Sano J, Nagai Y: Immunohistochemical localization of type I, III and IV (basement membrane) collagens in the liver. Acta Pathol Jpn 31: 973–978, 1981.
18. Junqueira LCU, Bignolas G, Brentani R: Picrosirius staining plus polarization microscopy, a specific method for collagen detection in tissue sections. Histochem J 11: 447–455, 1979.
19. Junqueira LCU, Cossermelli W, Brentani R: Differential staining of collagens type I, II, and III by Sirius Red and polarization microscopy. Arch Histol Jpn 41: 267–274, 1978.
20. Montes GS, Krisztán RM, Shigihara KM, Tokoro R, Mourão PAS, Junqueira LCU: Histochemical and morphological characterization of reticular fibers. Histochemistry 65: 131–141, 1980.
21. Pérez-Tamayo R, Montfort I: The susceptibility of hepatic collagen to homologous collagenase in human and experimental cirrhosis of the liver. Am J Pathol 100: 427–442, 1980.
22. Junqueira LCU, Montes GS, Krisztán RM: The collagen of the vertebrate peripheral nervous system. Cell Tissue Res 202: 453–460, 1979.
23. Zambrano NZ, Montes GS, Shigihara KM, Sanchez EM, Junqueira LCU: Collagen arrangement in cartilages. Acta Anat 113: 26–38, 1982.
24. Carrasco HF, Montes GS, Krisztán RM, Shigihara KM, Carneiro J, Junqueira LCU: Comparative morphologic and histochemical studies on the collagen of vertebrate arteries. Blood Vessels 18: 296–302, 1981.
25. Junqueira LCU, Montes GS, Almeida Neto A, Barros C, Tedesco-Marchese AJ: The collagen of permanently damaged nerves in human leprosy. Int J Leprosy 48: 291–297, 1980.
26. Junqueira LCU, Zugaib M. Montes GS, Toledo OMS, Krisztán RM, Shigihara KM: Morphologic and histochemical evidence for the occurrence of collagenolysis and for the role of neutrophilic polymorphonuclear leukocytes during cervical dilation. Am J Obstet Gynecol 138: 273–281, 1980.
27. Junqueira LCU, Montes GS, Kaupert D, Shigihara KM, Bolonhani

TM, Krisztán RM: Morphological and histochemical studies on the collagen in neurinomas, neurofibromas, and fibromas. J Neuropathol Exp Neurol 40: 123–133, 1981.

28. Junqueira LCU, Montes GS, Sanchez EM: The influence of tissue section thickness on the study of collagen by the picrosirius-polarization method. Histochemistry 74: 153–156, 1982.

29. Kuttan R, Di Ferrante N: Sirius Red–collagen interaction: a method for the measurement of collagen and bacterial collagenase activity. Biochem Int 1: 455–462, 1980.

30. O'Hara PJ, Read WK, Romane WM, Bridges CH: A collagenous tissue dysplasia of calves. Lab Invest 23: 307–314, 1970.

31. Bruns RR, Gross J: Band pattern of the segment-long-spacing form of collagen. Its use in the analysis of primary structure. Biochemistry 12: 808–815, 1973.

32. Stark M, Miller EJ, Kühn K: Comparative electron-microscope studies on the collagens extracted from cartilage, bone, and skin. Eur J Biochem 27: 192–196, 1972.

33. Wiedemann H, Chung E, Fujii T, Miller EJ, Kühn K: Comparative electron-microscope studies on type-III and type-I collagens. Eur J Biochem 51: 363–368, 1975.

34. Lapière ChM, Nusgens B, Piérard GE: Interaction between collagen type I and type III in conditioning bundles organization. Connect Tissue Res 5: 21–29, 1977.

35. Trelstad RL, Hayashi K, Gross J: Collagen fibrillogenesis: intermediate aggregates and suprafibrillar order. Proc Natl Acad Sci USA 73: 4027–4031, 1976.

36. Ruggeri A, Benazzo F, Reale E: Collagen fibrils with straight and helicoidal microfibrils: a freeze-fracture and thin-section study. J Ultrastruct Res 68: 101–108, 1979.

37. Reale E, Benazzo F, Ruggeri A: Differences in the microfibrillar arrangement of collagen fibrils. Distribution and possible significance. J Submicrosc Cytol 13: 135–143, 1981.

38. Thomas PK: The connective tissue of peripheral nerve: an electron microscope study. J Anat 97: 35–44, 1963.

39. Snodgrass MJ: Ultrastructural distinction between reticular and collagenous fibers with an ammoniacal silver stain. Anat Rec 187: 191–206, 1977.

40. Becker U, Nowack H, Gay S, Timpl R: Production and specificity of antibodies against the aminoterminal region in type III collagen. Immunology 31: 57–65, 1976.

41. Takyo C, Peyrol S, Cardier JF, Grimaud JA: Connective matrix organization in human pulmonary fibrosis. VIII FECTS Meeting. Copenhagen, 1982, p 61.

42. Strocchi R, Marchini, Leonardi L, Guízzardi S, Ruggeri A: Ultrastructural aspects of rat tail tendon sheats. J Anat. (in press) 1984.

43. Nathaniel EJH, Pease DC: Collagen and basement membrane formation by Schwann cells during nerve regeneration. J Ultrastruct Res 9: 550–560, 1963.

44. Bunge MB, Williams AK, Wood PM, Uitto J, Jeffrey JJ: Comparison of nerve cell and nerve cell plus Schwann cell cultures, with particular emphasis on basal lamina and collagen formation. J Cell Biol 84: 184–202, 1980.

45. Junqueira LCU, Montes GS, Bezerra MSF: Do Schwann cells produce collagen type III ? Experientia 35: 114, 1979.

46. Farquhar MG: Structure and function in glomerular capillaries. In: Biology and Chemistry of Basement Membranes. Kefalides NA (ed), New York, Academic Press, 1978, pp 43–80.

47. Schwartz D, Veis A: Characterization of bovine anterior-lens-capsule basement-membrane collagen. Eur J Biochem 103: 29–37, 1980.

48. Junqueira LCU, Montes GS, Toledo OMS, Bexiga SRR, Gordilho MA, Brentani RR: Evidence for collagen molecular orientation in basement membranes. Histochem J 15: 785–794, 1983.

49. Epstein EH: [α1 (III)]₃ human skin collagen. J Biol Chem 249: 3225–3231, 1974.

50. Epstein EH, Munderloh NH: Human skin collagen. Presence of type I and type III at all levels of the dermis. J Biol Chem 253: 1336–1337, 1978.

51. Junqueira LCU, Montes GS, Martins JEC, Joazeiro PP: Dermal collagen distribution. An ultrastructural and histochemical study. Histochemistry (in press).

52. Tajima S, Nagai Y: Distribution of macromolecular components in calf dermal connective tissue. Connect Tissue Res 7: 65–71, 1980.

53. Seyer JM, Kang AH, Whitaker JN: The characterization of type I and type III collagens from human peripheral nerve. Biochim Biophys

Acta 492: 414–425, 1977.

54. Shellswell GB, Restall DJ, Duance VC, Bailey AJ: Identification and differential distribution of collagen types in the central and peripheral nervous systems. FEBS Lett 106: 305–308, 1979.

55. Chung E, Miller EJ: Collagen polymorphism: characterization of molecules with the chain composition [α1 (III)]₃ in human tissues. Science 181: 1200–1201, 1974.

56. Trelstad RL: Human aorta collagens: evidence for three distinct species. Biochem Biophys Res Commun 57: 717–725, 1974.

57. Gay S, Balleisen L, Remberger K, Fietzek PP, Adelmann BC, Kühn K: Immunohistochemical evidence for the presence of collagen type III in human arterial walls, arterial thrombi, and in leukocytes incubated with collagen in vitro. Klin Wochenschr 53: 899–902, 1975.

58. Lindahl U, Höök M: Glycosaminoglycans and their binding to biological macromolecules. Ann Rev Biochem 47: 385–417, 1978.

59. Podrazký V: Interactions between connective tissue components. In : Connective Tissue Research: Chemistry, Biology, and Physiology. Deyl Z, Adam M (eds), New York, Alan R Liss, 1981, pp 151–162.

60. Rosenberg L, Hellmann W, Kleinschmidt AK: Electron microscopic studies of proteoglycan aggregates from bovine articular cartilage. J Biol Chem 250: 1877–1883, 1975.

61. Thyberg J, Lohmander S, Friberg U: Electron microscopic demonstration of proteoglycans in guinea pig epiphyseal cartilage. J Ultrastruct Res 45: 407–427, 1973.

62. Yamada K, Hoshino M: Digestion with chondroitinases of acid mucosaccharides in rabbit cartilages as revealed by electron microscopy. Histochem J 5: 195–197, 1973.

63. Ruggeri A, Dell'Orbo C, Quacci D: Electron microscopic visualization of proteoglycans with Alcian Blue. Histochem J 7: 187–197, 1975.

64. Shepard N, Mitchell N: The localization of proteoglycan by light and electron microscopy using Safranin O. J Ultrastruct Res 54: 451–460, 1976.

65. Shepard N, Mitchell N: The localization of articular cartilage proteoglycan by electron microscopy. Anat Rec 187: 463–476, 1977.

66. Thomas L: Reversible collapse of rabbit ears after intravenous papain, and prevention of recovery by cortisone. J Exp Med 104: 245–252, 1956.

67. Junqueira LCU, Montes GS, Mourão PAS, Carneiro J, Salles LMM, Bonetti SS: Collagen–proteoglycans interaction during autotomy in the sea cucumber, Stichopus badionotus. Rev Can Biol 39: 157–164, 1980.

68. Byers PH, Barsh GS, Holbrook KA: Molecular mechanism of connective tissue abnormalities in the Ehlers-Danlos syndrome. Coll Res 1: 475–489, 1981.

69. Peyrol S, Grimaud JA, Borojevic R: Les septa fibreux actif et passif dans les fibroses hépatiques sévères chez l'homme: mise en évidence d'hyperfibres de collagène en microscopie électronique C R Acad Sci Paris 282: 333–336, 1976.

70. Mancini AM, Ruggeri A, Riva R, Ferri M, Marchini M: Pseudoelastica in alcoholic hepatic fibrosis: histochemical and ultrastructural aspects. Bas Appl Histochem 26: 241–248, 1982.

71. Staubesand J, Fischer N: The ultrastructural characteristics of abnormal collagen fibrils in various organs. Connect Tissue Res 7: 213–217, 1980.

72. Montes GS, Becerra J, Toledo OMS, Gordilho MA, Junqueira LCU: Fine structure and histochemistry of the tail fin ray in teleosts. Histochemistry 75: 363–376, 1982.

73. Reddi AH: Cell biology and biochemistry of endochondral bone development. Coll Res 1: 209–226, 1981.

74. Fleischmajer R, Gay S, Perlish JS, Cesarini J-P: Immunoelectron microscopy of type III collagen in normal and scleroderma skin. J Invest Dermatol 75: 189–191, 1980.

75. Scott JE, Orford CR: Dermatan sulfate-rich proteoglycan associates with rat tail-tendon collagen at the d band in the gap region. Biochem J 197: 213–216, 1981.

76. Scott JE, Orford CR, Hughes EW: Proteoglycan-collagen arrangements in developing rat tail tendon: an electron-microscopical and biochemical investigation. Biochem J 195: 573–581, 1981.

Authors' address:
Laboratório de Biologia Celular
Faculdade de Medicina de USP
Av. Dr. Arnaldo 455
01246 São Paulo, Brazil

CHAPTER 4

Ultrastructural aspects of freeze-etched collagen fibrils

MAURIZIO MARCHINI and ALESSANDRO RUGGERI

Among the various morphological techniques used in the ultrastructural study of collagen, freeze-etching has recently been used succesfully thus supplying new and interesting data.

This technique is particularly advantageous because it does not necessitate the use of the chemical agents for fixing, dehydrating, embedding and staining material. On the other hand, rapid freezing guarantees the reproduction of the real form.

According to the procedures used so far, fresh specimens are frozen at $-150\,C$ in monochlorodifluoromethane (freon) and stored in liquid nitrogen at $-110\,C$. Freezing may be preceeded by immersion in a cryoprotectant (30% glycerol solution) in order to avoid the formation of large ice crystals during freezing. Thereafter, the frozen specimens are fractured (freeze-fracturing); the fracture surfaces may be submitted to varying periods of ice sublimation (freeze-etching) and then, whether etched or not, are replicated with carbon and platinum.

It is well known that, after having stained collagen fibrils with heavy metal, an intrafibrillar filamentous aspect or a fibrillar banding pattern may appear indicating the characteristic microfibrillar arrangement and molecular packing of the collagen fibril. These two aspects, filamentous aspect and banding, visualized on replicas of freeze-fractured or freeze-etched collagen fibrils, may be of particular importance if well interpreted.

Filamentous aspect

When a cryoprotectant is used before freezing, most of the freeze-fractured and replicated collagen fibrils appear composed of filamentous subunits which are comparable to those observed in the 'filament zones' of isolated and negatively stained collagen fibrils or to those visualized in thin sections on samples either inertly dehydrated or treated with extracting chemicals.

These filamentous subunits (Figure 1) were initially interpreted by some authors as being extrafibrillar or artefact material (1–4). Other authors , on the other hand, considered them to be microfibrillar subunits of collagen fibrils (5–9). These, in fact, have a 4–5 nm diameter which is close to that deduced for the five-stranded (10) or four-stranded (11) substructures called microfibrils.

It was shown by submitting samples of rat tail tendon to the effect of a 30% glycerol solution for different periods that swelling is induced according to the time of immersion (12). By combining data obtained by means of light microscopy and, subsequently, of electron microscopy on thin sections and replicas, it has been shown that this swelling is due to an increase in interfibrillar spacing and to microfibrillar dissociation of the collagen fibrils themselves into subunits (Figures 2, 3). Furthermore, X-ray diffraction analysis, carried out on the same samples, has shown that the treated fibrils maintain both the period of 67 nm and the molecular spacing of 1.5 nm, thus demonstrating that the molecular packing in the microfibrils remains unchanged. These observations lead to the suggestion that the intermicrofibrillar and the intramicrofibrillar charge distributions are different and give further evidence of the existence of the microfibril as a supramolecular substructure of the collagen fibril.

Freeze-etching has also shown that the microfibrils are arranged in the collagen fibrils in different ways (7, 8, 13):

(a) A helicoidal arrangement of right-turning microfibrils with an 18° spiral angle (Figure 4) was observed in papillary dermis, tunica intima and media of the aorta wall, peri- and endotendineum, perimysium, perichondrium, interstitial collagen of lung, liver and lymph node, lamina propria of oesophagus and trachea, ligamentum nucae and corneal stroma. Collagen fibrils with this type of microfibrillar arrangement are characterized by an uniform and relatively small diameter.

Ruggeri, A and Motta, PM (eds): Ultrastructure of the connective tissue matrix. ISBN-13:978-1-4612-9789-5

90

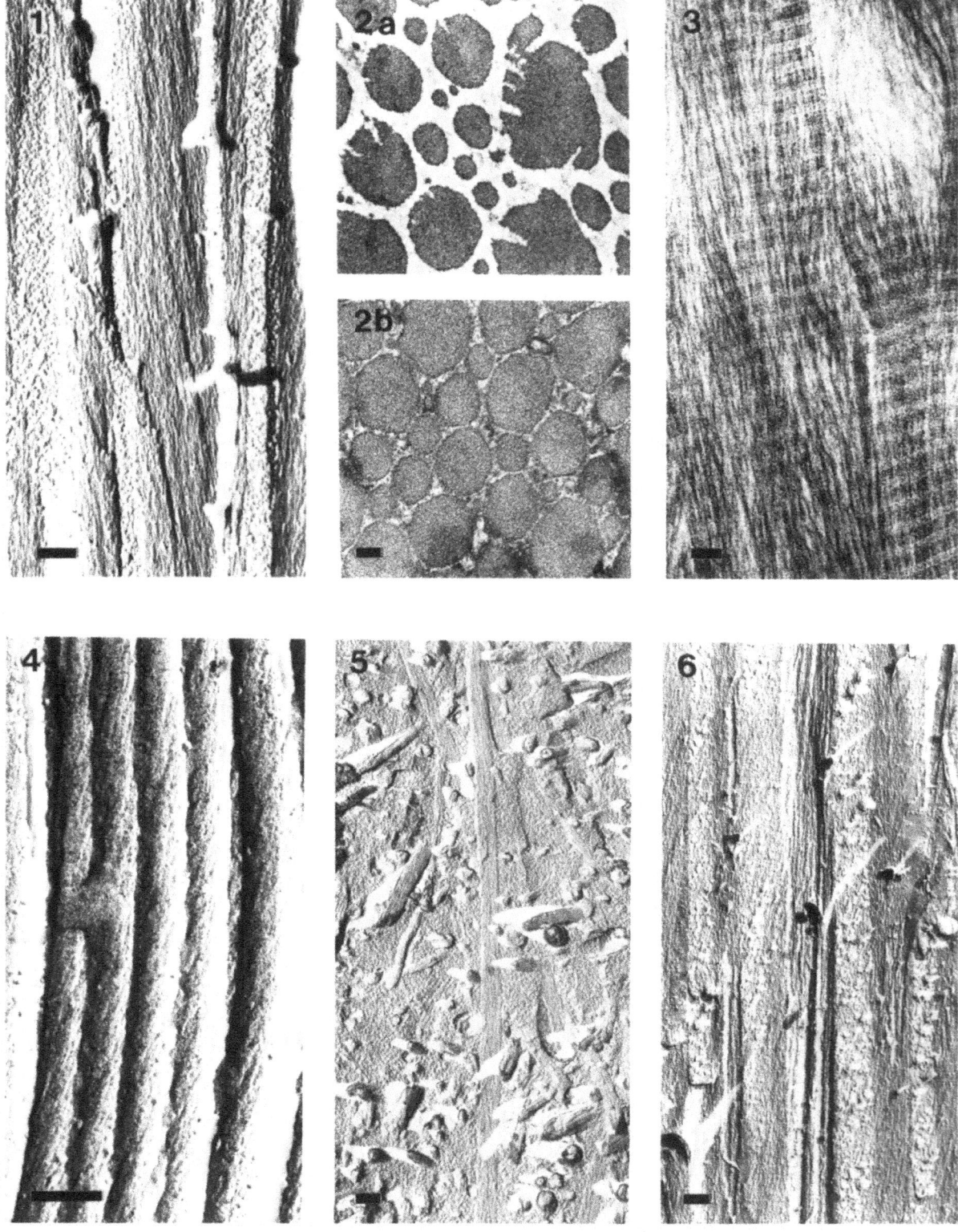

(b) A straight arrangement with microfibrils strictly parallel to the axis of the collagen fibril (Figure 5) was observed in the articular, hyaline and elastic cartilage and in the inner zone of the annulus fibrosus of the intervertebral disc. Collagen fibrils of this type usually have a small diameter varying within restricted range.

(c) A straight arrangement with slightly wavy microfibrils, the inclination angles of which never exceed 10°, (Figure 6) was observed in tendons, sclera, capsule of liver and limph node, chorda tendinea of heart and outer zone of the annulus fibrosus of the intervertebral disc. The fibrils characterized by this type of microfibrillar arrangement have diameters varying from 20 to 400 nm.

An analysis of the tissue distribution of these different kinds of collagen fibrils suggests that collagen fibrils with a microfibrillar helicoidal arrangement correspond to type III collagen (except in the case of type I corneal collagen), whereas collagen fibrils with a straight microfibrillar arrangement correspond to type I and type II collagen.

As far as collagen fibrils with helicoidal microfibrillar arrangement are concerned, it has been noticed that the 18° spiral angle is constant. If all these microfibrils present the same degree of coiling (which should correspond to the fourth supercoiling level (14)), one may deduce that in these fibrils the D-period is shorter than that observed in tendons where microfibrils are straightly arranged. Such an assumption is consistent with recent X-ray diffraction studies (15) showing the existence, in skin, spleen reticulum and aorta collagen fibrils (helicoidal arrangement), of a repeating period of 65.2 nm instead of 66.9 nm measured in tendon collagen fibrils (straight arrangement).

Banding

The collagen fibrils of freeze-etched and replicated samples appear to be cross-striated when a cryoprotectant is not used. The banding is determined by an alternate succession of elevated and depressed segments along the collagen fibril, with a period close to 64–70 nm reported for the heavy-metal-stained collagen fibrils or for those examined by X-ray diffraction analysis (Figure 7).

The periodical variability in the diameter of the fibril has been interpreted as the effect of various degrees of shrinkage caused by ice sublimation during etching, before replication of the sample (2, 4). This interpretation was confirmed by the observation that the frequency and the characteristics of the banding pattern are modified according to the length of etching time (16). It is reasonable to believe that this phenomenon is more accentuated in the more hydrated regions with a lower molecular density, thus putting the regions with a high molecular density in relief. On the basis of this consideration, it was suggested that the depressed segments correspond to the gap zones and the elevated segments to the overlap zones (2). This interpretation was also given by collimating pictures of freeze-etched collagen fibrils and of isolated and negatively stained collagen fibrils (17). The above-mentioned data lead to the conclusion that, following adequate standardization of the techniques of preparation, two important banding parameters may be assessed on replicas i.e. the intraperiod distance (D) and the period percentages occupied respectively by gap and overlap zones.

It has been demonstrated that samples submitted to ten or more minutes etching show collagen fibrils with two ridges located at the margins of the elevated

←

Figure 1. Replica of a cryoprotected and freeze-fractured collagen fibril of rat tail tendon. A filamentous aspects is observable. Bar = 0.1 μm.

Figure 2. Thin cross-section of rat tail tendon collagen fibril (a) treated for 1 h with a 30% glycerol solution and then fixed with glutaraldehyde or (b) fixed with glutaraldehyde without treatment. In (a) both an increase of fibrillar spacing and fibrillar swelling are evident. Bars = 0.1 μm.

Figure 3. Thin longitudinal section of rat tail tendon collagen fibrils treated as in Figure 2a. The collagen fibrils are swollen and both microfibrillar and cross banding are evident. Bar = 0.1 μm.

Figure 4. Replica of freeze-fractured collagen fibrils of rat thoracic aorta. The fibrils are composed of right-turning microfibrils. Bar = 0.1 μm. (reprinted from Ruggeri et al. (7) with permission of J Ultrastruct Res).

Figure 5. Replica of freeze-fractured collagen fibrils of bovine nasal septum cartilage. The microfibrils show a straight arrangement and run parallel to the main axes of the collagen fibrils. Bar = 0.1 μm (reprinted from Ruggeri et al. (7) with permission of J Ultrastruct Res).

Figure 6. Replica of freeze-fractured collagen fibrils of rat tail tendon. The microfibrils show a straight arrangement and are slightly wavy. Bar = 0.1 μm (reprinted from Ruggeri et al. (7) with permission of J Ultrastruct Res).

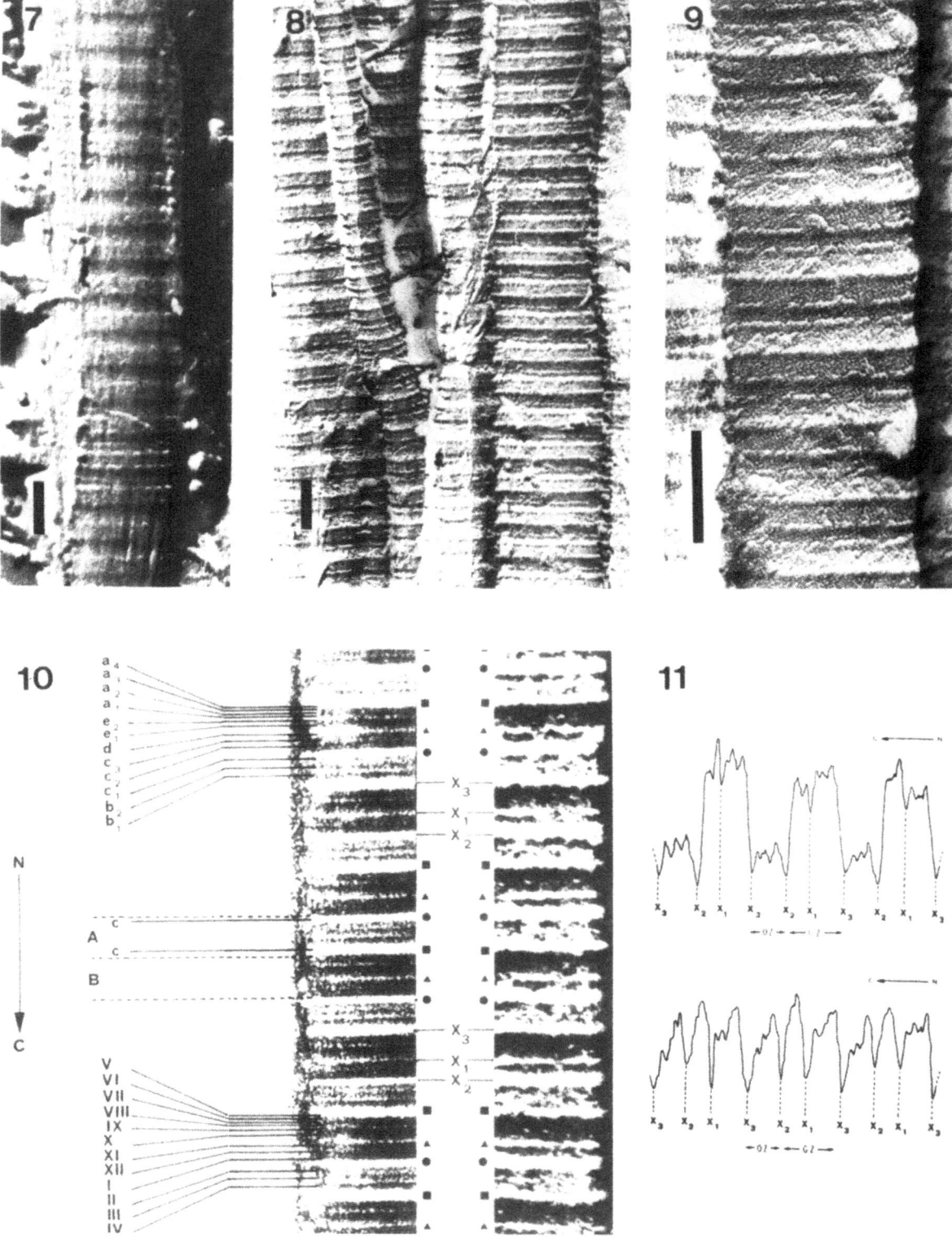

segments and a third ridge situated at an intermediate point of the depressed segments (Figures 8, 9). Therefore, longer sublimation times entail the visualization of differences in molecular density even within the gap and overlap zones. These three ridges correspond to the three widest light bands observable in negatively stained collagen fibrils, which are considered as bands of higher molecular density (for terminology see Figure 10).

The identification of the three ridges in the three main dense bands is confirmed by the similarity between the densitometric diagrams of deep-etched collagen fibril and negatively stained collagen fibril pictures (Figure 11). Furthermore, the first type of diagram, due to the relative regularity of the period, has given reliable preliminary data for the study of banding of fresh and unstained collagen fibrils (16).

Concluding remarks

As may be deduced from this brief chapter, ultrastructural research on freeze-fractured and freeze-etched collagen fibrils has contributed to the study of the various types of collagen packing. This method is particularly effective when complemented with other techniques. Thus, ultrastructural data obtained from replicas and supported by conventional embedding, sectioning and staining and by X-ray diffraction have given further confirmation of the existence of the microfibril as a supramolecular substructure of the collagen fibril. These techniques have also supplied information on differences in microfibrillar arrangement, helicoidal or straight, which may be correlated to the genetical type of collagen and may also imply a variability in the D-period distance, as assumed already on the basis of traditional electron microscopy observations. Data obtained from the study of freeze-fractured and deep-etched collagen fibrils have given interesting results allowing the detection, on replicas, of the gap and overlap zones and of the three bands corresponding to the intraperiod segments with higher molecular density. It has been shown that pictures of banding patterns of collagen fibrils, obtained from replicas, were suitable for accurate densitometric analysis which, especially if supported by X-ray diffraction analysis, may give a new perspective to the study of collagen based on banding. However, it is important to specify that even though freeze-etching offers the advantage of being applicable to fresh material, like all morphological techniques, it is not completely free from artefact. This mainly depends on the use of a cryoprotectant, on the rate of cooling and on etching. At present frequent technical innovation and research on the chemical–physical processes which accompany the artefact events, guarantee a short-term improvement in the reliability of the pictures of replicas and in their interpretation.

Acknowledgements

The personal investigations mentioned in this paper have been supported by grants of the Italian National Research Council.

References

1. Berger H: Elektronenmikroscopische Befunde mit der Gefrierätztechnik am kollagen Bindegwebe menschlicher. Hautartzt 20: 225–228, 1969.
2. Reed R: Freeze-etched connective tissue. Int Rev Connect Res 6: 257–305, 1973.
3. Breathnach AS: Application of freeze-fracture replication technique to investigative dermatology. Br J Dermatol 88: 563–574, 1973.
4. Stolinski G, Breathnach AS: Freeze-fracture replication and surface sublimation of frozen collagen fibrils. J Cell Sci 23: 325–334, 1977.
5. Szirmai JA, van Raamdonk W, Galavazi G: Freeze-etch studies of the extracellular matrix of connective tissues. J Cell Biol 47: 209a, 1970.
6. Rayns DG: Collagen from frozen fractured glycerinated beef heart. J Ultrastruct Res 48: 59–66, 1974.

Figure 7. Replica of a freeze-etched collagen fibril, after 1-min etching, showing a sequence of elevated and depressed segments. Bar = 0.1 μm (reprinted from Marchini et al. (16) with permission of Connect Tissue Res).

Figure 8. Replica of freeze-etched collagen fibrils after 20-min etching. All fibrils show three intraperiod ridges. Bar = 0.1 μm (reprinted from Marchini et al. (16) with permission of Connect Tissue Res).

Figure 9. Replica of a freeze-etched collagen fibril, after 20-min etching. Bar = 0.1 μm.

Figure 10. Collimation between an isolated and negatively stained collagen fibril picture (magnified 230,000 and a deep-etched collagen fibril picture (magnified 212,000: a correspondence between the widest light bands and the three ridges is evident. They are indicated as x_3, x_2 and x_1 adopting the terminology introduced by Bairati et al. (18). On the left the terminologies introduced by Schmitt and Gross (19), Tromans et al. (20) and Bruns and Gross (21) are indicated (modified from Marchini et al. (16) with permission of Connect Tissue Res).

Figure 11. (a) Densitometric diagram obtained from an isolated and negatively stained collagen fibril micrograph. (b) Densitometric diagram obtained from a deep-etched collagen fibril micrograph. In these diagrams there is a correspondence between the main negative peaks (x_3, x_2 and x_1). The arrows indicate the molecular N terminal end – C terminal end direction. GZ: gap zone; OZ overlap zone, modified from Marchini et al. (16) with permission of Connect Tissue Res).

94

7. Ruggeri A, Benazzo F, Reale E: Collagen fibrils with straight and helicoidal microfibrils: A freeze-fracture and thin section study. J Ultrastruct Res 68: 101–108, 1979.

8. Marchini M, Strocchi R, Castellani PP, Riva R: Ultrastructural observations on collagen and proteoglycans in the annulus fibrosus of the intervertebral disc. Basic Appl Histochem 23: 137–148, 1979.

9. Itoh T, Klein L, Geil PH: Age Dependence of collagen fibril and subfibril diameters revealed by transverse freeze-fracture and etching technique. J Microsc 125: 343–357, 1982.

10. Smith JW: Molecular pattern in native collagen. Nature 219: 157–158, 1968.

11. Veis A, Yuan L: Structure of the collagen microfibril. A four-strand overlap model. Biolpolymers 14: 895–900, 1975.

12. Leonardi L, Ruggeri A, Roveri N, Bigi A, Reale E: Light microscopy, electron microscopy and X-ray diffraction analysis of glycerinated collagen fibers. J Ultrastruct Res (in press).

13. Reale E, Benazzo F, Ruggeri A: Differences in the microfibrillar arrangement of collagen fibrils. Distribution and possible significance. J Submicrosc Cytol 13: 135–143, 1981.

14. Piez KA: Structure and assembly of the native collagen fibril. Connect Tissue Res 10: 25–36, 1982.

15. Brodsky B, Eikenberry EF: Characterization of fibrous form of collagen. Methods Enzimol 82: 127–174, 1982.

16. Marchini M, Morocutti M, Castellani PP, Leonardi L, Ruggeri A: The banding pattern of rat tail tendon freeze-etched collagen fibrils. Connect Tissue Res 11: 175–184, 1983.

17. Hashimoto K. Ultrastructure of freeze-cleaved dermal collagen. Acta Derm Venerol 54: 241–248, 1974.

18. Bairati A, Petruccioli MG, Torri Tarelli L: Studies on the ultrastructure of collagen fibrils. 1. Morphological evaluation of the periodic structure. J Submicrosc Cytol 1: 113–141, 1969.

19. Schmitt FO, Gross J: Further progress in the electron microscopy of collagen. J Am Leather Chem Assoc 43: 658–675, 1948.

20. Tromans WJ, Horne RW, Gresham GA, Bailey AJ: Electron microscope studies on the structure of collagen fibrils by negative staining. Z Zellforsch 59: 798–802, 1963.

21. Bruns RR, Gross J: High-resolution analysis of the modified quarter-stagger model of the collagen fibril. Biopolymers 13: 931–941, 1974.

Authors' address:
Università di Bologna
Istituto di Anatomia Umana Normale
Via Irnerio, 48
40126 Bologna, Italy

CHAPTER 5

Electron microscopy of proteoglycans

C. JOHAN O. THYBERG

1. Introduction

In addition to the fibrillar proteins collagen and elastin, proteoglycans constitute a major macromolecular component in the extracellular matrix of connective tissues. They have also been identified as components of basement membranes and cell surface coats. Chemically, the proteoglycans consist of a protein core to which glycosaminoglycan chains are covalently attached. The latter are composed of alternating uronic acid and hexosamine residues and are all polyanions with acidic sulfate and/or carboxyl groups (1). A particularly high content of proteoglycans is found in cartilage and much of the information that is available today derives from studies on this tissue. An average cartilage proteoglycan has a molecular weight of about 2.5 af0910^6 and is composed of a protein core to which about 100 chondroitin sulfate and 50–60 keratan sulfate chains are bound. Within the cartilaginous matrix, most of these monomers occur in large aggregates, formed by noncovalent interaction with hyaluronic acid and link proteins (2–4). The structure of proteoglycans in other connective tissues, in basement membranes, and in cell surface coats is less well known. The type, number, and size of glycosaminoglycan chains per molecule have been found to vary considerably, but the supramolecular organization (e.g. aggregate formation) is still poorly defined.

In recent years, considerable interest has been paid not only to the structure and metabolism of proteoglycans but also to their influence on the functional properties of cells and tissues and their role in cell differentiation and cell–cell and cell–matrix interaction (5–7). In this context, electron microscopy has provided an important experimental tool by making it possible to identify the molecules in thin tissue sections and to study their distribution and spatial relation to cells and other matrix components. Moreover, application of protein monolayer technique of the type originally devised for DNA has

enabled visualization and structural characterization of isolated molecules. The purpose of this chapter is to shortly review the contribution of electron microscopy to our present knowledge of proteoglycans.

2. Electron microscopy of tissue proteoglycans

2.1. Methodology

In contrast to collagen and elastin, proteoglycans lack distinct morphological characteristics in tissue sections prepared for electron microscopy by conventional methods. This has necessitated the development of cytochemical procedures for their fine structural indentification. Electron microscopic autoradiography with radioactive inorganic sulfate, a precursor of sulfated glycosaminoglycans, was one of the first techniques to be used for this purpose (8, 9). It made it possible to determine the site of sulfation and to follow intracellular transport and extracellular discharge and spreading of the molecules. However, this technique lacks the sensitivity and resolution required to resolve structural detail.

At about the same time, the principle of ionic interaction between negatively charged glycosaminoglycans and positively charged reagents was introduced. Among the various compounds that have been utilized in this context, colloidal iron (10), colloidal thorium dioxide (11), ruthenium red (12, 13), alcian blue (14), lanthanum (15), lysozyme (16), safranin O (17), toluidine blue O (18), cuprolinic blue (19) and high iron diamine (20) can be mentioned. Lysozyme is used as a vital stain, colloidal thorium dioxide and high iron diamine to stain thin sections, and the others to stain the tissue en bloc before embedding and sectioning. The most detailed information with regard to chemical structure and reactivity with different tissue components is available for ruthenium red (12, 13) and alcian blue (21, 22). Commercial availability, ease of use, fine grain

Ruggeri, A and Motta, PM (eds): Ultrastructure of the connective tissue matrix. ISBN-13:978-1-4612-9789-5

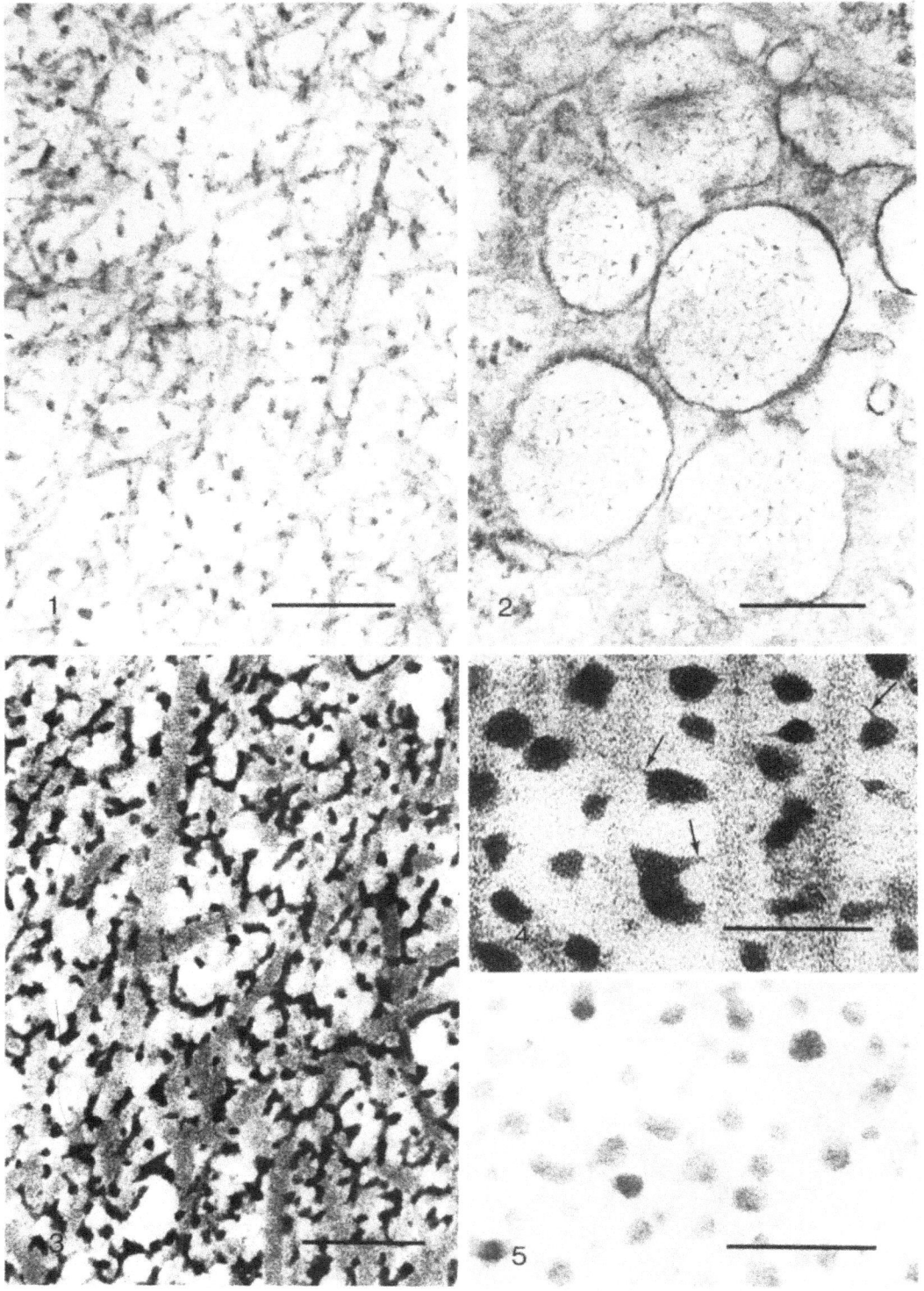

and high electron density are other factors which have made these reagents particularly popular.

Polyvalent cationic probes of the type mentioned above will react with polyanions in general. It is therefore necessary to combine the staining procedure with one or more additional tests in order to ascertain the proteoglycan nature of the stained material. Such tests usually aim at removing the whole molecules or parts of them and to examine the effects thereof on the stainability of the tissue. Three different approaches will be mentioned here. First, there exists enzymes of varying specificity that degrade glycosaminoglycans. Some of these enzymes or groups of enzymes, like hyaluronidases and chondroitinases, are commercially available. Others, like heparinases and heparitinases, can be prepared relatively easily. Their combined action make them powerful tools for identification of glycosaminoglycans (23). A second possibility is to use proteolytic enzymes like trypsin or papain, which attack the protein core of the proteoglycans and release diffusible peptide-glycosaminoglycan fragments (2, 4). A third alternative is to employ the dissociative extraction method developed for isolation of cartilage proteoglycans. It is based on the principle that salt solutions of high concentration, like 4 M guanidinium chloride, 2 M calcium chloride or 3 M magnesium chloride, dissociate the proteoglycan aggregates into monomers, which diffuse out of the tissue (2, 4).

In practice, unfixed or fixed tissue is digested with enzymes or extracted with salt solutions of high concentration, stained and examined by electron microscopy. Parallel control samples are exposed to inactivated enzymes, kept in buffer without enzymes or immersed in salt solutions of low concentration. A simple chemical control of the effectiveness of the various treatments is obtained by determining the uronic acid and/or hexosamine content of the samples. A more detailed analysis of the glycosaminoglycan content of the tissue before and after the treatments provides additional information.

The critical electrolyte concentration method represents another means of analyzing staining specificity. It is based on the principle that a cationic stain bound to an anionic substrate can be displaced by other cations at concentrations that are directly related to the strength of binding between the stain and the substrate. Above a critical concentration of the competing cation, the stain is displaced from its binding sites and no staining will be obtained. This method was originally worked out for light microscopic histochemistry (22) but has later been utilized also in electron microscopy (24). Chemical blocking of free anionic groups by methylation is another procedure that has been employed in polysaccharide histochemistry. This treatment causes loss of sulfate groups and esterification of carboxyl groups but has an unfavorable effect on tissue fine structure which limits its usefulness.

Demonstration of intracellular proteoglycans constitutes an inherent problem in the methodology outlined above. Most of the cationic probes are relatively large molecules that do not easily penetrate into the cells. This is also true for the enzymes that are used to test staining specificity. Moreover, extraction with salts of high concentration destroys cell fine structure. One way to overcome the limited penetration is to do staining and enzyme digestions after embedding and sectioning of the tissue (11, 20, 25). However, this is technically difficult and connected with risks for displacement and/or unspecific adsorption of stain to the sections as well as an uncertainty concerning the susceptibility of plastic-embedded tissue components to enzymatic digestion. It is also difficult to check the effectiveness of enzymes on thin sections chemically.

←

Figure 1. Extracellular matrix of mouse hyaline cartilage after routine preparation including fixation with glutaraldehyde and osmium tetroxide and section staining with uranyl acetate and lead citrate. The matrix consists of a network of thin collagen fibrils without distinct cross striations and small polygonal granules which, however, are difficult to distinguish from cross-sectioned fibrils. Bar = 0.25 μm; × 74,000.

Figure 2. Golgi vacuoles in a chondrocyte from mouse hyaline cartilage after routine preparation (see legend to Figure 1). The vacuoles contain a flocculent material partly consisting of granules that are similar in appearance to those in the extracellular matrix (cf. Figure 1). Bar = 0.25 μm; × 76,000.

Figures 3, 4. Extracellular matrix of guinea pig hyaline cartilage after fixation with glutaraldehyde and osmium tetroxide in the presence of ruthenium red and section staining with uranyl acetate and lead citrate. The matrix granules are positively stained (cf. Figure 1) and stand out in clear contrast to the collagen fibrils, with which they are closely associated, sometimes by filamentous projections (Figure 4, arrows). Adjacent granules are also connected to each other, thus forming ribbon-like structures of varying length (Figure 3). As discussed in the text, such structures may be a representation of the aggregates in which the proteoglycans normally occur in the matrix. Bars = 0.25 and 0.1 μm, respectively. Figure 3, × 74,000; Figure 4, × 224,000.

Figure 5. Extracellular matrix in an unstained section of guinea pig hyaline cartilage fixed in the presence of ruthenium red. Granules stained by the latter reagent are the only clearly visible structures in the matrix. Bar = 0.2 μm; × 143,000.

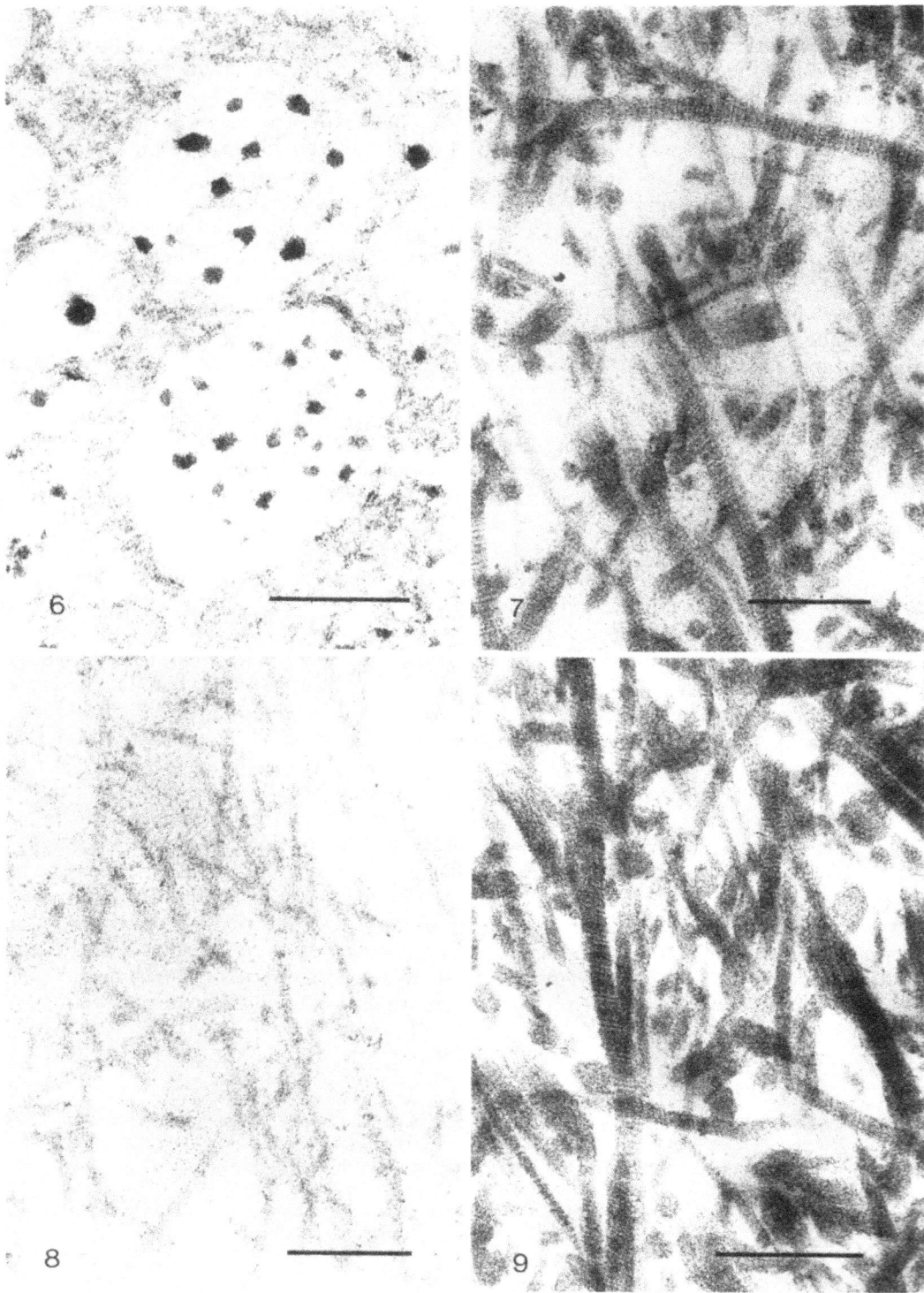

Recently, immunological methods have been applied in proteoglycan histochemistry as well. The potency of this approach lies in the possibility to visualize the protein part of the molecules, which cannot be demonstrated by other means. Antibodies directed against the protein core and the link proteins of cartilage proteoglycans (26–29) and the protein core of aortic proteoglycans (28) have been utilized for this purpose. The use of antibodies further enables investigations on the coordination of production and localization of different matrix components. This possibility was recently exploited in a study on the synthesis and extracellular deposition of collagen, fibronectin and proteoglycans by cultured chondrocytes (30). So far, the immunohistochemical studies have largely been restricted to the light microscopic level, but some preliminary electron microscopic observations have been presented (26, 29). In the future, immunoelectron microscopy can be expected to widen our knowledge on the intracellular routes of secretion of different parts of the proteoglycans, their assembly into supramolecular aggregates, and the interaction of the molecules with the cell surface and other matrix components.

Finally, the application of lectins in carbohydrate histochemistry should be mentioned. Lectins are proteins of widespread occurrence, first discovered in plants, that bind specifically to mono- or oligosaccharides. They can be linked covalently to electron microscopic tracers like horseradish peroxidase and ferritin. Such conjugates have been utilized for cytochemical detection of cell surface carbohydrates (31). They can be expected to find an increased use also in studies on extracellular matrix components.

2.2. Cartilage proteoglycans

In 1960 Godman and Porter (32) reported that the matrix of hyaline cartilage consists of fine fibrils and small floccules or granules. They tentatively suggested that proteinpolysaccharide complexes reside within the latter structures. One of the first attempts to demonstrate such complexes electron microscopically by histochemical means was made by Revel (11), who used colloidal thorium dioxide at low pH to stain thin sections of hyaline and elastic cartilage embedded in methacrylate. This resulted in precipitation of thorium throughout the extracellular matrix. The heavy reaction and the coarse granularity of the colloidal metal particles thus prevented resolution of finer detail. Nevertheless, digestion of the tissue with hyaluronidase completely removed the stainable material. Methylation of the sections similarly led to loss of stainability. In the cells, a positive reaction was restricted to Golgi vacuoles. This finding agreed with the simultaneous identification of the Golgi complex as the site of sulfation in the chondrocytes (8, 9).

It was later established that polygonal granules of varying size (about 10–100 nm in diameter) represent an integral structural component of the cartilaginous matrix, visible by conventional electron microscopy. Granules of similar appearance are also present in the Golgi vacuoles of the chondrocytes (33–35; Figures 1, 2). It was further demonstrated that the matrix granules were removed by digestion of the tissue with hyaluronidase or trypsin and by extraction with 4 M guanidinium chloride or 1.9 M calcium chloride, indicating that they contain proteoglycans (33, 35). At about the same time, the matrix granules were shown to bind cationic reagents of the type referred to above (13–20; Figures 3–5). A positive staining of the granules in the Golgi vacuoles was also evident (Fig. 6)). Moreover, stainable granules were found to be removed by digestion of the tissue with proteolytic or glycosaminoglycan-degrading enzymes (15, 17, 36; Figures 7, 8) or by extraction with 4 M guanidinium chloride (37; Fig. 9). Finally, if chondrocytes are enzymatically freed from surrounding matrix and cultivated *in vitro*, a new extracellular matrix including granules stained by ruthe-

←

Figure 6. Golgi vacuoles in a chondrocyte from guinea pig hyaline cartilage stained en bloc with ruthenium red. Positively stained granules (cf. Figure 2), similar in appearance to those in the extracellular matrix (cf. Figures 3, 4) are demonstrated. Section stained with uranyl acetate and lead citrate. Bar = 0.2 μm; × 106,000.
Figures 7, 8. Guinea pig hyaline cartilage digested with the proteolytic enzyme papain (Figure 7) or the glycosaminoglycan-degrading enzyme chondroitinase ABC (Figure 8), fixed, and stained en bloc with ruthenium red. Collagen fibrils are evident but the matrix granules have been removed (cf. Figures 3, 4). The hexosamine content of the tissue was reduced by 90% and 50–60%, respectively, by the enzyme treatments, indicating degradation of proteoglycans. Sections stained with uranyl acetate and lead citrate. Bars = 0.2 μm; × 94,000.
Figure 9. Guinea pig hyaline cartilage extracted with 4 M guanidinium chloride, fixed and stained en bloc with ruthenium red. Cross-striated collagen fibrils but no matrix granules are evident (cf. Figures 3, 4). The hexosamine content of the tissue was reduced by 70–75% by the treatment, indicating efficient extraction of proteoglycans. Section stained with uranyl acetate and lead citrate. Bar = 0.25 μm; × 89,000.

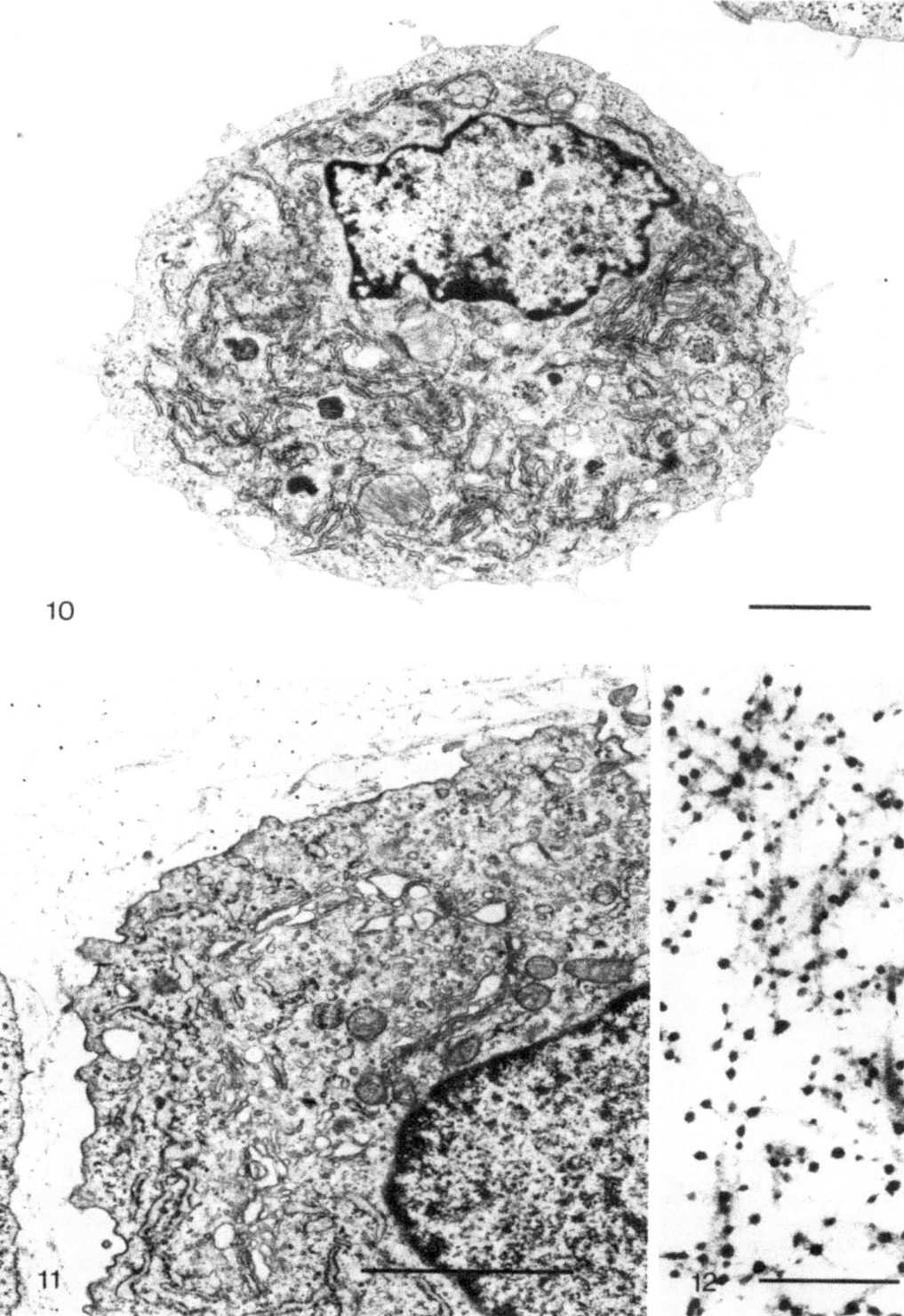

Figures 10–12. Regeneration of extracellular matrix after enzymatic isolation of chondrocytes from guinea pig hyaline cartilage. Figure 10 shows a freshly isolated chondrocyte. No remnants of the matrix are seen. Routine preparation. Bar = 2 μm; × 9,000. Figure 11 is from an 1.5-day-old culture of isolated chondrocytes. The cells are surrounded by small amounts of matrix. Routine preparation. Bar = 2 μm; × 16,000. Figure 12 demonstrates regenerated matrix in a 4-day-old culture of isolated chondrocytes after block staining with ruthenium red. The matrix has an appearance similar to that of intact cartilage (cf. Figures 3, 4) and consists of a network of thin collagen fibrils with attached ruthenium red-positive granules. Bar = 0.3 μm; × 70,000.

nium red is reformed (Figures 10–12).

Studies of this type verify that the matrix granules represent cartilage proteoglycans and that ruthenium red, alcian blue and other cationic probes can be used to visualize these macromolecules electron microscopically. However, the matrix granules do not give a true picture of how the proteoglycans are organized *in situ* but demonstrates a fixation-precipitated and dehydrated form of the molecules. This is illustrated by the finding that the granules change into filamentous structures if magnesium chloride is added in connection with fixation and staining (24, 38). Moreover, in cartilage fixed with glutaraldehyde and osmium tetroxide in the presence of ruthenium red and embedded in water-soluble embedding medium without prior dehydration, no matrix granules were observed. Instead, there was a diffuse staining between the collagen fibrils, which appeared in negative contrast (Figure 13). Similarly, unfixed or shortly glutaraldehyde-fixed cartilage, rapidly frozen and sectioned by ultracryotomy, did not show any evident structures in addition to the fibrils (Figure 14).

The removal of matrix granules by dissociative extraction suggest that they do not delineate proteoglycan monomers but aggregates. However, the granules are markedly smaller than isolated aggregates spread into a monomolecular film (see below). On the basis of such size comparisons, it was suggested that they represent one or a few monomers rather than entire aggregates (39). Moreover, as shown in Figure 3, the granules are often connected into long ribbons which could set forth the latter.

2.3. Proteoglycans in other connective tissues

In interstitial and synovial connective tissue, material stained by ruthenium red and alcian blue, has been found to be closely associated with the collagen fibrils (13, 14, 40, 41). This material occurred as an amorphous coat on the fibrils, transverse regularly spaced bands, and fine filaments connecting adjacent fibrils. Sensitivity to enzymes suggested that these structures consisted of proteoglycans. Similar observations have been made in dense connective tissues like tendons (19) and cornea (42, 43).

In arteries and other blood vessels, matrix granules similar to those of cartilage, stained with ruthenium red, and removed after hyaluronidase and chondroitinase digestion or guanidinium chloride extraction have been demonstrated (44–46; Figure 15). Histochemically demonstrable proteoglycans of this type are present in all layers of the arterial wall, that is the intima, media and adventitia. Recent chemical studies have shown that arteries contain chondroitin sulfate and dermatan sulfate proteoglycans (47–49). At present, considerable interest is paid to the role of these molecules in the pathogenesis of atherosclerosis.

In bone tissue, newly formed osteoid matrix contains granules that stain with ruthenium red but are smaller and more widely dispersed than in cartilage (Figure 16). Calcified bone matrix demonstrates small clusters of ruthenium red-positive material in the spaces between the collagen fibrils (Figure 17). It seems likely that the ruthenium red-stainable material of bone represents proteoglycans, but more strict histochemical evidence remains to be presented. Combined electron microscopic and histochemical studies have similarly indicated presence of proteoglycans in predentin and dentin (50, 51).

2.4. Basement membrane proteoglycans

The basement membrane is another extracellular structure that has been shown to contain proteoglycans. Using ruthenium red and alcian blue in combination with enzyme digestions, Trelstad et al. (43) demonstrated the presence of sulfated glycosaminoglycans in the basement membrane of embryonic chick cornea. Structurally, the stained material appeared as granules arranged in a bilaminar configuration, i.e. on both sides of the lamina densa. Granules of similar type were found in the basement membranes of the embryonic lens capsule and neural tube (52). A somewhat more diffuse distribution of ruthenium red-positive material was observed in the basement membrane of embryonic mouse submandibular glands (53).

Using cationized ferritin and ruthenium red, Kanwar and Farquhar (54) demonstrated a partly regular network of anionic sites in the glomerular basement membrane of the kidney. Morphologically, these sites had a structure similar to the matrix granules in cartilage and other connective tissues. Subsequent studies with enzymes of varying specificity made evident that the anionic sites of the glomerular basement membrane are mainly made up of heparan sulfate linked to protein (55; Figures 18–20). Accordingly, extraction of isolated glomerular membranes with 4 M guanidinium chloride yielded a major fraction of heparan sulfate proteoglycans and a minor fraction of chondroitin sulfate proteoglycans (56). Moreover, it was shown that enzymatic removal of heparan sulfate from the glomerular membrane leads to a dramatic increase in its permeability to

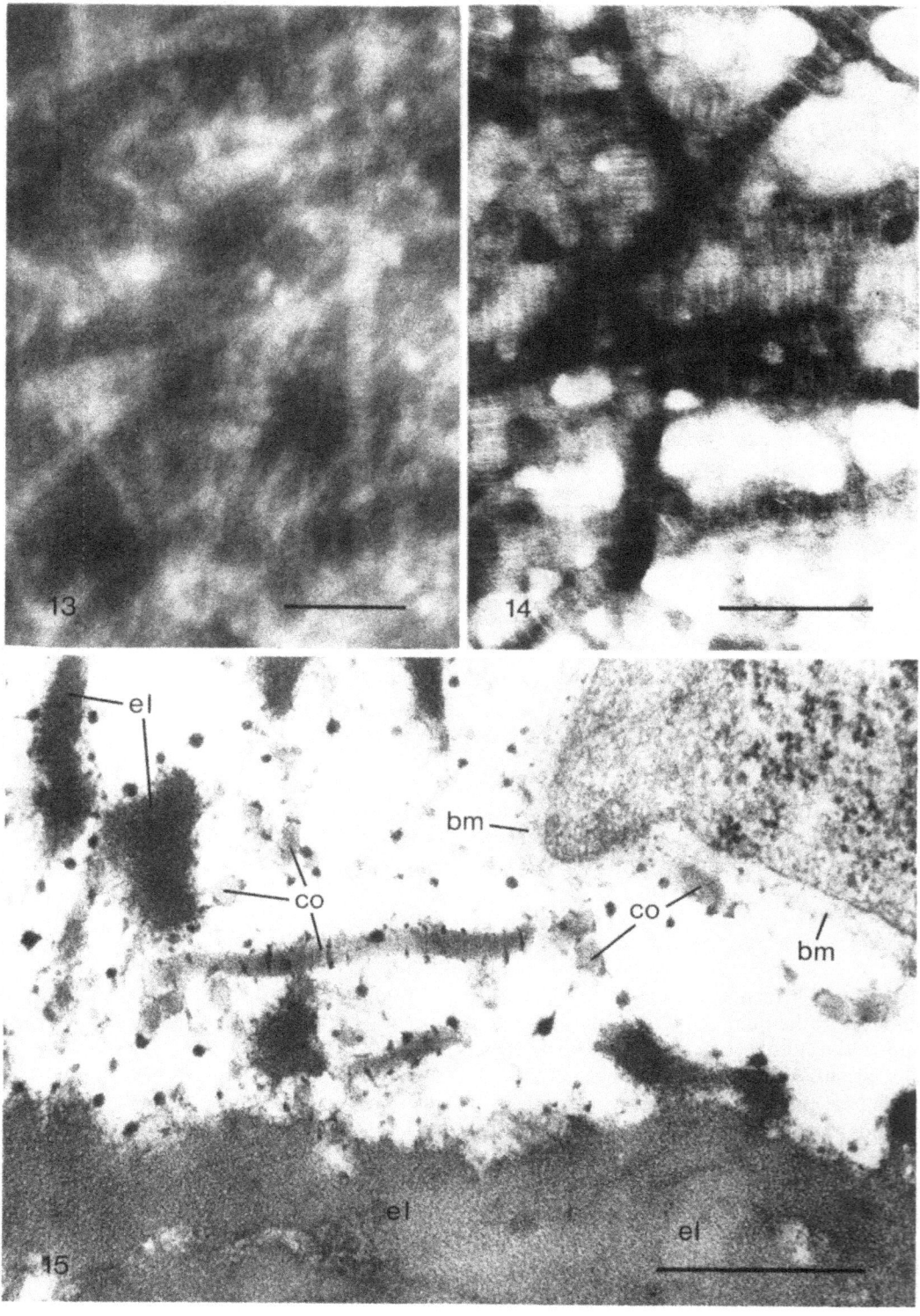

native, anionic ferritin. On the basis of these findings, it was suggested that the proteoglycans play an important role in establishing the permeability properties of the glomeruli to plasma proteins (57).

2.5. Cell surface proteoglycans

Most, if not all, cells are covered by a layer consisting of glycolipids, glycoproteins and proteoglycans. The lipid or protein part of the molecules are anchored in the membrane and the carbohydrate part extends from the outer aspect of the membrane. It has become increasingly clear that the cell surface coat is involved in such major functions as cell growth and differentiation, malignant transformation, expression of antigenicity, interaction with hormones and other specific ligands, and cell–cell and cell–matrix interaction (6, 7). Nevertheless, the unraveling of the structure and function of the cell surface coat is still only in its beginning.

Chemically, glycoproteins and glycolipids, both of which frequently contain sialic acid in their carbohydrate part, seem quantitatively to be the major components of the cell surface coat. However, proteoglycans with side chains of heparan sulfate, chondroitin sulfate and dermatan sulfate have also been identified as cell surface components (58–60). With the electron microscope, a few cell types, like intestinal epithelial cells, demonstrate a prominent fuzzy coat. In other cells, no such coat is evident in routine preparations. Improved preservation and visualization of the surface coat can be obtained by treatment of specimens fixed by conventional methods (glutaraldehyde and osmium tetroxide) with tannic acid (61).

More direct methods for demonstration of the cell surface coat are also available (62, 63). Visualization of proteoglycans with their acidic glycosaminoglycans usually includes the application of cationic probes like ruthenium red (13), alcian blue (14) or cationized ferritin (64). Figure 21 demonstrates the surface coat of cultured macrophages as revealed by ruthenium red staining and Figure 22 shows the binding of cationized ferritin to the surface of glutaraldehyde-fixed macrophages at pH 2. Under these conditions, only strongly acidic groups like carboxyl groups of sialic acid and sulfate groups of glycosaminoglycans are dissociated and able to bind the cationic probe. Prior digestion of the cells with neuraminidase partly removed the stainability with cationized ferritin at pH 2 (Figure 23). These observations suggest that sialic acid-containing glycoproteins as well as proteoglycans are present on the macrophage surface (cf. 58). Recently, Simionescu et al. (65) used cationized ferritin in combinations with enzymes of varying specificity to characterize proteoglycans on the luminal surface of the capillary endothelium. Their findings indicate that the anionic sites of the fenestral diaphragms are contributed primarily by heparan sulfate proteoglycans, whereas the anionic sites of the other parts of the plasma membrane are of a mixed chemical nature.

3. Electron microscopy of isolated proteoglycans

Electron microscopy has been used as a complement to biochemical and physicochemical methods to study the structure of isolated proteoglycans. The advantage of this technique is that it enables observation and measurement of indivudual molecules in polydisperse systems. If combined with specific fragmentation and purification methods, electron microscopy thus provides an important tool for testing models of proteoglycan structure set up from chemical data.

3.1. Methodology

Electron microscopy of isolated proteoglycans has

←

Figure 13. Extracellular matrix of guinea pig hyaline cartilage stained *en bloc* with ruthenium red and embedded in water-soluble resin without prior dehydration. Section stained with uranyl acetate. The collagen fibrils, some of which show evident cross-striations, stand out in negative contrast. A diffuse positive staining is seen in the interstices of the fibrillar network (cf. Figures 3, 4). Bar = 0.25 μm; × 74,000.

Figure 14. Extracellular matrix of guinea pig hyaline cartilage fixed for 10 min in buffered glutaraldehyde, frozen in the presence of glycerol and sectioned by ultracryotomy. Section stained with uranyl acetate. Cross-striated collagen fibrils are the only clearly recognizable components of the matrix. Bar = 0.25 μm; × 94,000.

Figure 15. Part of the media in a 5-day-old rat aorta stained *en bloc* with ruthenium red. Section contrasted with uranyl acetate and lead citrate. Ruthenium red-positive granules similar to those of the cartilaginous matrix (cf. Figures 3, 4) are demonstrated. Most of the granules are associated with collagen fibrils seen in longitudinal or cross-section (co), elastic fibers (el), or the basement membrane surrounding the smooth muscle cells (bm). Bar = 0.5 μm; × 63,000.

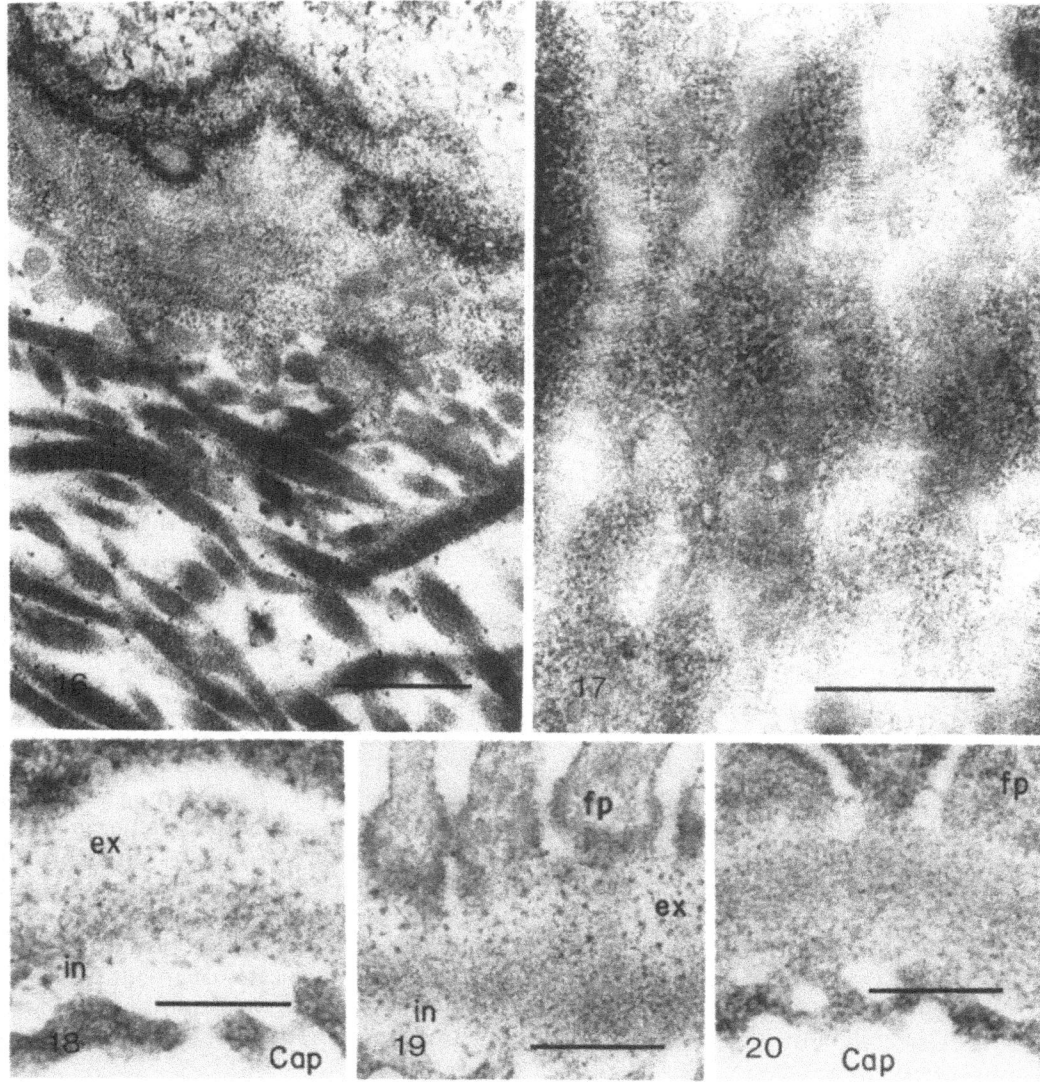

Figure 16. Metaphyseal spongiosa in a decalcified guinea pig bone specimen stained *en bloc* with ruthenium red. Section stained with uranyl acetate and lead citrate. From the top, this picture shows part of a cartilaginous septum delimited by a strongly stained lamina onto which a layer of diffuse, flocculent material and an osteoid matrix has been deposited. The latter is made up of collagen fibrils, wider and more clearly cross-striated than those of the cartilage matrix, and small scattered ruthenium red-positive granules. Bar = 0.25 μm; × 73,000.

Figure 17. Bone matrix in a decalcified guinea pig specimen stained *en bloc* with ruthenium red. Section stained with uranyl acetate. Cross-striated collagen fibrils coalescing into a dense network is the major component of the organic part of the bone tissue shown here. The meshes of this network is filled up with clusters of small ruthenium red-positive granules. Bar = 0.25 μm; × 96,000.

Figures 18–20. Details of glomerular basement membranes in rat kidneys perfused with buffer, chondroitinase ABC and heparitinase respectively, followed by perfusion fixation and block staining with ruthenium red. In the control (Figure 18) and in the specimen perfused with chondroitinase ABC (Figure 19), ruthenium red-positive granules, similar in appearance to those in the matrix of cartilage (cf. Figures 3, 4) and other connective tissues (cf. Figures 15–17), are found in the internal (in) and external (ex) light layers (laminae rarae) of the basement membranes. In the specimen perfused with heparitinase (Figure 20), no staining of this type is observed. The findings suggest that the above mentioned granules are mainly made up of heparan sulfate proteoglycans. Cap, capillary lumen; fp, foot processes of the glomerular epithelium. Sections stained with uranyl acetate and lead citrate. Courtesy of M.G. Farquhar (from 55). Bars = 0.3 μm; × 60,000.

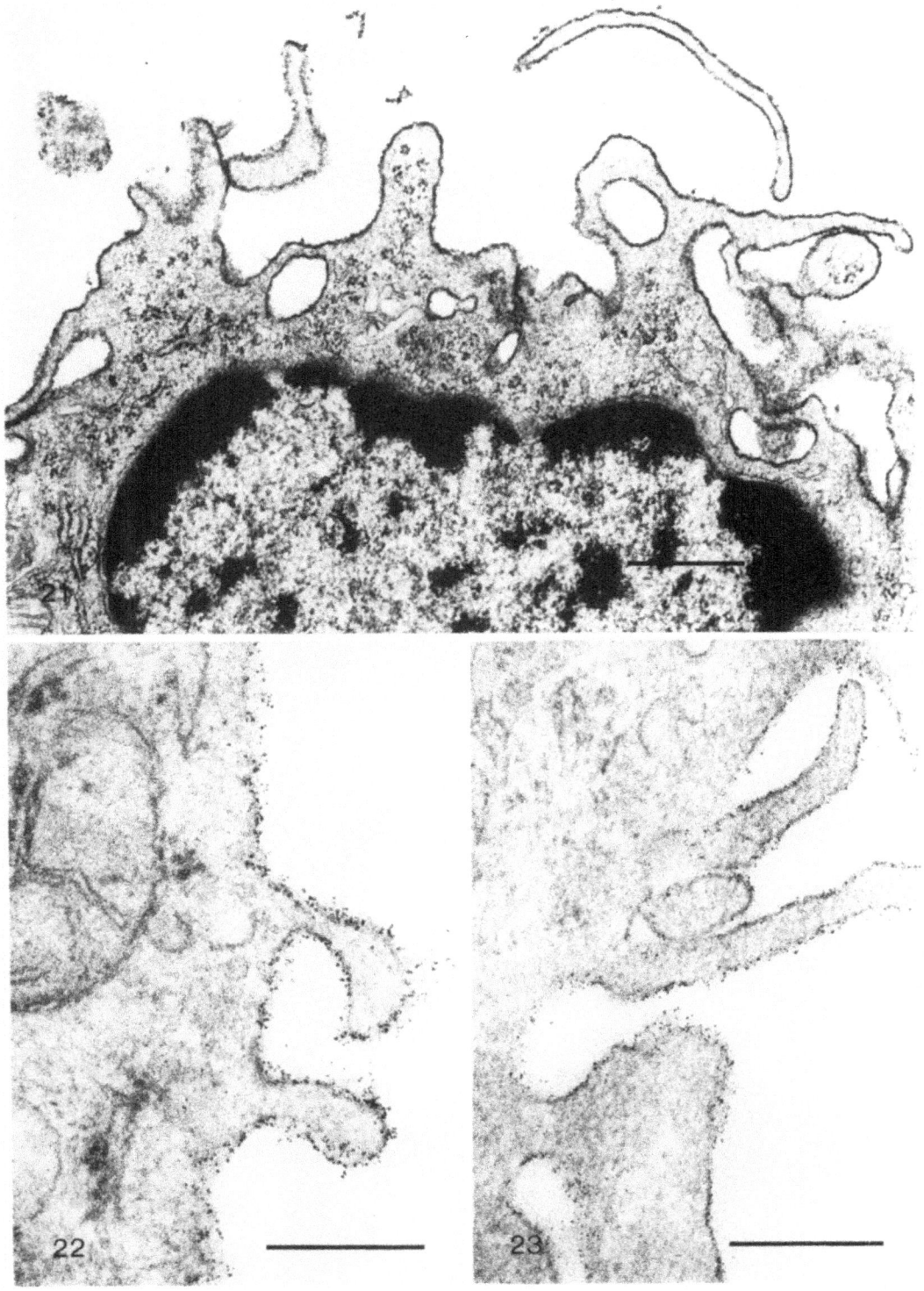

Figure 21. Cultured mouse peritoneal macrophage stained with ruthenium red. A thin layer of positively stained material covers the cell surface. Section stained with lead citrate. Bar = 0.5 μm; × 36,000.

Figure 22. Mouse peritoneal macrophage from a culture fixed in glutaraldehyde and stained with cationized ferritin in glycine-HCl buffer at pH 2.0. Ferritin particles are diffusely scattered over the cell surface. No section-staining. Bar = 0.2 μm; × 120,000.

Figure 23. Mouse peritoneal macrophage from a culture fixed in glutaraldehyde, digested with neuraminidase isolated from Vibrio cholerae and stained with cationized ferritin in glycine-HCl buffer at pH 2.0. Only few ferritin particles have bound to the cell surface as compared to Figure 22. Section briefly stained with lead citrate. Bar = 0.25 μm; × 94,000.

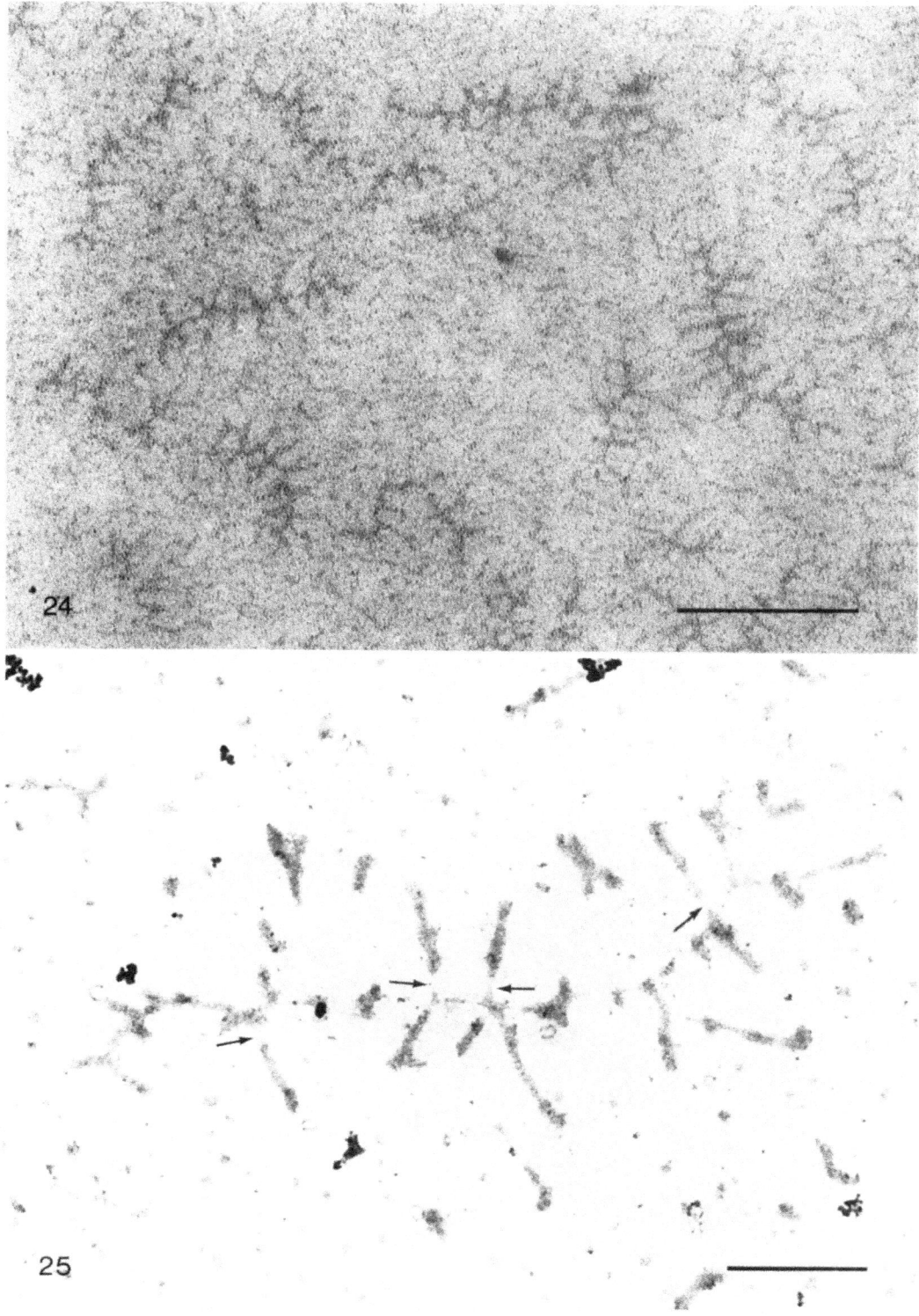

largely utilized the basic protein film technique originally developed for visualization of DNA (66). This technique is based on the following principle. Proteoglycans are mixed with a basic protein, usually cytochrome c, and the solution is allowed to flow down a glass ramp onto a suitable hypophase. By denaturation of cytochrome c at the air-liquid interface, a positively charged, monomolecular film forms on the hypophase. The proteoglycans with their negatively charged glycosaminoglycan chains interact with the basic protein and become embedded and spread out in the abovementioned film (67). The spontaneous adsorption method (68) represents a simplified variety of the above principle, which also minimizes the amount of material needed. Cytochrome c and proteoglycans are mixed in an ammonium acetate solution, and small droplets of this solution are placed on a hydrophobic surface. Formation of a cytochrome c film and adsorption and spreading out of the proteoglycans then occur spontaneously (69). The cytochrome c-proteoglycan film is picked up on a grid covered by a support film, stained with uranyl acetate, possibly in combination with phosphotungstic acid, dehydrated and dried in air. Additional contrast can be obtained by rotary shadowing the grids with heavy metals. Dark-field electron microscopy is another alternative which makes it possible to observe the molecules even without staining (70).

Recently, methods for electron microscopic examination of hydrated proteoglycans have also been described (71). For this purpose, a wet-cell was used together with a pulsed plasma soft X-ray source to provide contact replicas for subsequent examination by conventional scanning electron microscopy. This allowed observation of fully hydrated molecules with good contrast and a resolution of around 30 nm.

3.2. Cartilage proteoglycans

Electron microscopic studies of isolated proteoglycans has so far largely been restricted to material from cartilage. However, the first application of the basic protein film technique to components of the extracellular matrix concerned hyaluronic acid from normal human synovial fluid (72). The findings indicated that hyaluronic acid is almost exclusively a straight-chain polymer with a considerable range of chain lengths.

Subsequently, the structure of proteoglycan monomers isolated from hyaline cartilage was investigated (67, 69, 73–75). These studies showed that the molecules were made up of a central filament, about 300 nm long, and 25–30 side chain filaments with a mean length of about 40–50 nm (Figure 24). The interpretation of this picture was that the central filament represents the chondroitin sulfate-binding region of the protein core and that the hyaluronic acid-binding region, responsible for aggregate formation, is not clearly visible with this technique. The former part has a molecular weight of about 110,000 and should have a length of about 400 nm if fully extended. This value agrees well with the measured length of the spread monomers. Furthermore, the side-chain filaments are believed to represent chondroitin sulfate, the predominant glycosaminoglycan in cartilage proteoglycans. However, they do not correspond to single but clusters of chondroitin sulfate molecules (69). This conclusion was reached by correlating electron microscopic observations to chemical data indicating that each monomer contains about 100 chondroitin sulfate chains spaced along the protein core in clusters with an average of four chains per cluster (2–4). Moreover, isolated clusters appeared as single chains independent of whether they came from fractions containing an average of 1, 4 or 6 chondroitin sulfate molecules (69). The average molecular weight of chondroitin sulfate from hyaline cartilage is about 20,000, which

←

Figure 24. Proteoglycan monomers isolated from bovine nasal cartilage and spread into a monomolecular film using the spontaneous adsorption method. A solution containing 0.3 M ammonium acetate, pH 5.0, 2 µg/ml of cytochrome c, and 0.5 µg/ml of isolated proteoglycans was prepared. Small droplets of this solution was put on a plate of Teflon and a surface film of cytochrome c-proteoglycans was allowed to form. After 30 min this film was picked up on a grid, rinsed in distilled water, air-dried and stained with uranyl acetate in acetone (see ref. 69). The monomers have a bottlebrush-like structure with numerous side chain-filaments extending from a central filament. Bar = 0.2 µm; × 138,000.

Figure 25. Proteoglycan aggregates isolated from bovine nasal cartilage and spread into a monomolecular film using the spontaneous adsorption method (see legend to Figure 24). The cytochrome c-proteoglycan film was rinsed in 90% ethanol after being picked up on the grid and after being stained with uranyl acetate dissolved in either acetone or ethanol. This modification improved the visualization of the aggregates but lead to collapse of the side chain-filaments of the monomers (see 78). In the proteoglycan aggregate shown here, the monomers therefore appear as rod-like structures attached to a central filament. A thin, poorly contrasted hinge region is evident in several places where the monomers attach to the central filament (arrows). Bar = 0.2 µm; × 106,000.

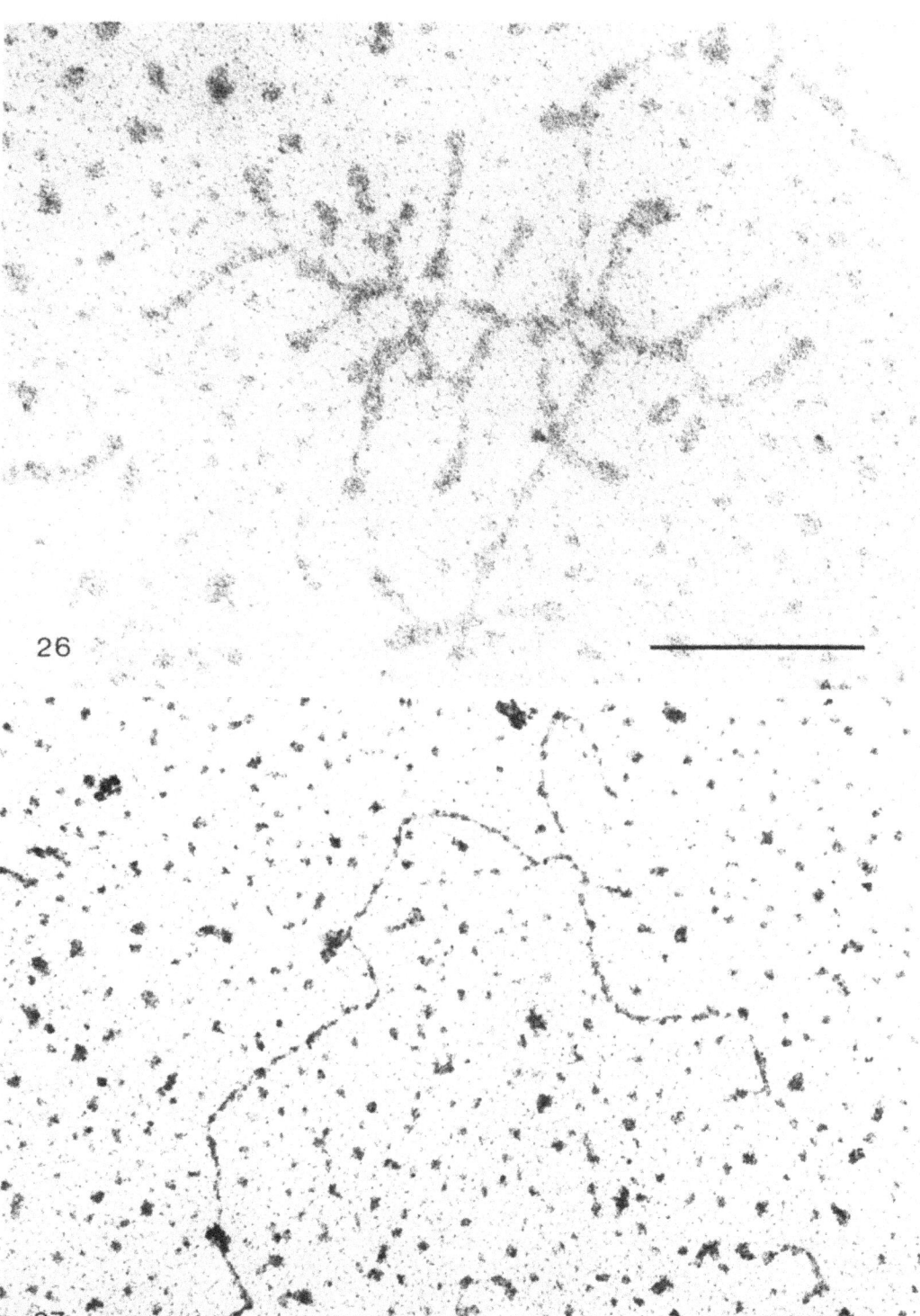

26

27

corresponds to about 40 disaccharide units. Since each unit is about 1 nm long, the length of a fully extended chondroitin sulfate chain should be about 40 nm. This value is in good agreement with the measured length of the side-chain filaments of the spread molecules. The keratan sulfate chains of the proteoglycans are of low molecular weight as compared to chondroitin sulfate. This is probably the reason why the keratan sulfate part of the monomers has not been possible to identify in the electron microscope.

The studies were then extended to aggregate fractions. In this work material isolated from normal hyaline cartilage (75–78), chondrocyte cultures (77) and chondrosarcoma tissue (39, 79) has been examined. The proteoglycan aggregates appeared electron microscopically as a central filament with a mean length of about 1,000–2,000 nm from which a variable number of monomers extended laterally on both sides (Figure 25). The side-chain filaments of the monomers were usually collapsed in these spreadings and therefore difficult to distinguish as separate entities. In the region of attachment of the monomers to the central filament, the former were poorly stained and not clearly delineated. This hinge region could correspond to the hyaluronic acid-binding region of the monomers, known to have a low content of glycosaminoglycans, mainly short chains of keratan sulfate.(2–4, 78).

A further confirmation of the proteoglycan aggregate structure by means of enzymatic modification and chemical purification in combination with electron microscopy was provided by Heinegård et al. (78). After removal of chondroitin sulfate with chondroitinase under conditions that left hyaluronic acid only moderately affected, the spread molecules retained a structure comparable to that described above (Figure 26). This was expected since the chondroitin sulfate clusters, as mentioned above, were collapsed in the spreadings of the aggregates. When the molecules were treated with trypsin, the largest fragments containing hyaluronic acid, the hyaluronic acid-binding part of the monomers and link proteins,

appeared electron microscopically as straight filaments with a length comparable to the central filament in the intact aggregates (Figure 27). Hyaluronic acid isolated from cartilage by papain digestion, precipitation with cetylpyridinium chloride and chromatography on a cetylpyridinium chloride-cellulose column gave a similar picture. There was a good correlation between electron microscopic and biochemical data with respect to the molecular weight of the hyaluronic acid chain and the number and spacing of monomers along this chain (79).

Taken together, biochemical and electron microscopic studies on isolated cartilage proteoglycans suggest that the molecules are built up as schematically shown in Figure 28.

4. Concluding remarks

Proteoglycans are complex molecules that occur in connective tissue matrices, basement membranes and cell surface coats. They consist of a protein core to which glycosaminoglycan chains are covalenty attached. Within the extracellular matrix most of these monomers exist in large aggregates, formed by noncovalent interaction with hyaluronic acid and link proteins. Electron microscopic and histochemical techniques have made it possible to identify proteoglycans in thin tissue sections and to study their distribution and spatial relation to cells and other matrix components. Because of the anionic character of the glycosaminoglycans, proteglycans can be stained with cationic dyes like ruthenium red and alcian blue. Such reagents demonstrate the presence of granular structures in the matrix of cartilage and other connective tissues and in basement membranes. These granules are removed by digestion of the tissue with proteolytic or glycosaminoglycan-degrading enzymes. Extraction of the tissue with salts of high concentration like 4 M guanidinium chloride, which dissociates the proteoglycans into diffusible monomers, similarly removes the granules. Such ob-

Figure 26. Proteoglycan aggregate isolated from bovine nasal cartilage and digested with chondroitinase ABC under conditions that allow degradation of most of the chondroitin sulfate chains, while leaving the hyaluronic acid as intact as possible (see 78). Spreading for electron microscopy by the spontaneous adsorption method (see legend to Figure 25). The modified aggregate has an appearance similar to that of the molecule shown in Figure 25. This could be expected since the side chain-filaments of the monomers are not clearly marked with the preparation method used for aggregates. Bar = 0.2 μm; × 159,000.

Figure 27. Proteoglycan aggregate isolated from bovine nasal cartilage, digested with trypsin and separated from low molecular weight fragments by ultracentrifugation in an associative cesium chloride gradient and gel chromatography (see 78). Spreading for electron microscopy by the spontaneous adsorption method (see legend to Figure 25). The molecules, which now consist of hyaluronic acid, the hyaluronic acid-binding region of the monomers and link proteins, appear as straight filaments. The remnants of the trypsin-digested monomers cannot be identified with certainty. Bar = 0.2 μm; × 100,000.

110

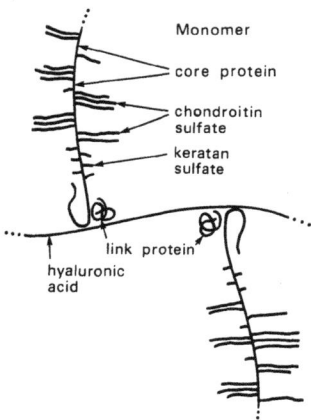

Figure 28. Schematic model of proteoglycan aggregate structure.

servations indicate that the matrix granules represent proteoglycans. However, they do not reveal the native but a fixation-precipitated and dehydrated form of the molecules. Moreover, size considerations suggest that each matrix granule represents one or several monomer portions of a proteoglycan aggregate rather than a whole aggregate.

Electron microscopy has also made it possible to observe and measure individual proteoglycans in isolated fractions. This has become a valuable complement to biochemical and physicochemical methods in the study of proteoglycan structure. The isolated molecules are mixed with a basic protein, usually cytochrome c, and the solution is let to flow down a glass ramp onto a hypophase. By denaturation of the basic protein at the air–liquid interface, a positively charged, monomolecular film is formed in which the proteoglycans with their negatively charged glycosaminoglycans become embedded and spread out. Formation of a cytochrome c-proteoglycan film will also occur spontaneously in a droplet of the above-mentioned solution. The film is picked up on a grid

covered by a support film, stained and studied in the electron microscope.

Examination of proteoglycan monomers isolated from hyaline cartilage with this technique has shown that the molecules consist of a central filament, about 300 nm long, and 25–30 side-chain filaments with a mean length of about 40–50 nm. The central filament is believed to represent the chondroitin sulfate-binding part of the protein core and the side-chain filaments are believed to represent clusters of chondroitin sulfate chains with an average of four molecules per cluster. Proteoglycan aggregates isolated from hyaline cartilage appear as a central filament with a mean length of about 1,000–2,000 nm from which a variable number of monomers extend laterally on both sides. In the region of attachment of the monomers to the central filament, the former were poorly contrasted. This hinge region is thought to correspond to the hyaluronic acid-binding part of the monomers, known to have a low content of glycosaminoglycans. After treatment with trypsin, the aggregates appear as a straight filament with a length comparable to the central filament of the intact molecules and to hyaluronic acid isolated from the cartilage.

In the future, new and improved methods will have to be developed for identification, isolation and characterization of proteoglycans from tissues other than cartilage. A more detailed knowledge concerning the secretory pathways of the different parts of the proteoglycans and the mechanism and site of assembly of the supramolecular aggregates in which they occur extracellularly is also needed. Information of this type will form the basis for further investigations on the function of the proteoglycans in the extracellular matrix and their role in cell–matrix and cell–cell interaction. The necessity to work at a subcellular and molecular level makes it important to attack the problems with a combination of chemical, electron microscopic and immunological methods.

References

1. Lindahl U, Höök M: Glycosaminoglycans and their binding to biological macromolecules. Ann Rev Biochem 47: 385–417, 1978.
2. Hascall VC, Heinegård DK: Structure of cartilage proteoglycans. In: Glycoconjugate Research, vol 1. Gregory JD, Jeanloz RW (eds), New York, Academic Press, 1979, pp 341–374.
3. Hascall VC, Hascall GK: Proteoglycans. In: Cell Biology of Extracellular Matrix. Hay ED (ed), New York, Plenum, 1981, pp 39–63.
4. Hascall VC, Kimura JH: Proteoglycans: isolation and characterization. Methods Enzymol 82: 769–800, 1982.
5. Slavkin HC, Greulich RC (eds): Extracellular Matrix Influences on Gene Expression. New York, Academic Press, 1975.
6. Lash JW, Burger MM (eds): Cell and Tissue Interactions. Society of general physiologists series 32, New York, Raven Press, 1977.
7. Lennarz WJ (ed): The Biochemistry of Glycoproteins and Proteo-
glycans. New York, Plenum Press, 1980.
8. Godman GC, Lane N: On the site of sulfation in the chondrocyte. J Cell Biol 21: 353–366, 1964.
9. Fewer D, Threadgold J, Sheldon H: Studies on cartilage. V. Electron microscopic observations on the autoradiographic localization of S^{35} in cells and matrix. J Ultrastruct Res 11: 166–172, 1964.
10. Curran RC, Clark AE, Lovell D: Acid mucopolysaccharides in electron microscopy. The use of the colloidal iron method. J Anat 99: 427–434, 1965.
11. Revel J-P: A stain for the ultrastructural localization of acid mucopolysaccharides. J Microsc 3: 535–544, 1964.
12. Luft JH: Ruthenium red and violet. I. Chemistry, purification, methods of use for electron microscopy and mechanism of action. Anat Rec 171: 347–368, 1971.
13. Luft JH: Ruthenium red and violet. II.Fine structural localization in animal tissues. Anat Rec 171: 369–416, 1971.

14. Behnke O, Zelander T: Preservation of intercellular substances by the cationic dye alcian blue in preparative procedures for electron microscopy. J Ultrastruct Res 31: 424–438, 1970.
15. Khan TA, Overton J: Lanthanum staining of developing chick cartilage and reaggregating cartilage cells. J Cell Biol 44: 433–438, 1970.
16. Eisenstein R, Arsenis C, Kuettner KE: Electron microscopic studies of cartilage matrix using lysozyme as a vital stain. J Cell Biol 46: 626–631, 1970.
17. Shepard N, Mitchell N: The localization of proteoglycan by light and electron microscopy using safranin O. A study of epiphyseal cartilage. J Ultrastruct Res 54: 451–460,1976.
18. Shepard N, Mitchell N: Simultaneous localization of proteoglycan by light and electron microscopy using toluidine blue O. A study of epiphyseal cartilage. J Histochem Cytochem 24: 621–629, 1976.
19. Scott JE, Orford CR, Hughes EW: Proteoglycan-collagen arrangements in developing rat tail tendon. An electron-microscopical and biochemical investigation. Biochem J 195: 573–581, 1981.
20. Takagi M, Parmley RT, Denys FR: Ultrastructural cytochemistry and radioautography of complex carbohydrates in secretory granules of epiphyseal chondrocytes. Lab Invest 44: 116–126, 1981.
21. Scott JE: Histochemistry of alcian blue. II. The structure of alcian blue 8GX. Histochemie 30: 215–234, 1972.
22. Scott JE: Histochemistry of alcian blue. III. The molecular basis of staining by alcian blue 8GX and analogous phthalocyanins. Histochemie 32: 191–212, 1972.
23. Linker A, Hovingh P: The uses of degradative enzymes as tools for identification and structural analysis of glycosaminoglycans. Fed Proc 36: 43–46, 1977.
24. Schofield BH, Williams BR, Doty SB: Alcian blue staining of cartilage for electron microscopy. Application of the critical electrolyte concentration principle. Histochem J 7: 139–149, 1975.
25. Hoshino M, Yamada K: Effects of digestion with chondroitinases upon mucosaccharide stainings of rabbit cartilage as revealed by electron microscopy. Histochemie 32: 221–229, 1972.
26. Dorfman A, Vertel BM, Schwartz NB: Immunological methods in the study of chondroitin sulfate proteoglycans. Curr Top Dev Biol 14: 169–198, 1980.
27. Poole AR, Pidoux I, Reiner A, Tang L-H, Choi H, Rosenberg L: Localization of proteoglycan monomer and link protein in the matrix of bovine articular cartilage. J Histochem Cytochem 28: 621–635, 1980.
28. Mangkornkanok-Mark M, Eisenstein R, Bahu RM: Immunologic studies of bovine aortic and cartilage proteoglycans. J Histochem Cytochem 29: 547–552, 1981.
29. Poole AR, Pidoux I, Reiner A, Rosenberg L: An immunoelectron microscope study of the organization of proteoglycan monomer, link protein, and collagen in the matrix of articular cartilage. J Cell Biol 93: 921–937, 1982.
30. Dessau W, Vertel BM, von der Mark H, von der Mark K: Extracellular matrix formation by chondrocytes in monolayer culture. J Cell Biol 90: 78–83, 1981.
31. Gonatas NK, Avrameas S: Detection of carbohydrates with lectin-peroxidase conjugates. In: Methods in Cell Biology vol XV. Prescott DM (ed), New York, Academic Press, 1977, pp 387–406.
32. Godman GC, Porter KR: Chondrogenesis studied with the electron microscope. J Biophys Biochem Cytol 8: 719–760, 1960.
33. Matukas VJ, Panner BJ, Orbison JL: Studies on ultrastructural identification and distribution of protein-polysaccharide in cartilage matrix. J Cell Biol 32: 365–377, 1967.
34. Smith JW: The disposition of proteinpolysaccharide in the epiphyseal plate cartilage of the young rabbit. J Cell Sci 6: 843–864, 1970.
35. Anderson HC, Sajdera SW: The fine structure of bovine nasal cartilage. Extraction as a technique to study proteoglycans and collagen in cartilage matrix. J Cell Biol 49: 650–663, 1971.
36. Thyberg J, Lohmander S, Friberg U: Electron microscopic demonstration of proteoglycans in guinea pig epiphyseal cartilage. J Ultrastruct Res 45: 407–427, 1973.
37. Thyberg J, Nilsson S, Friberg U: Electron microscopic studies on guinea pig rib cartilage. Structural heterogeneity and effects of extraction with quanidine-HCl. Z Zellforsch Mikrosk Anat 146: 83–102, 1973.
38. Ruggeri A, Dell'Orbo C, Quacci D: Electron microscopic visualization of proteoglycans with ruthenium red. Histochem J 9: 249–252, 1977.
39. Hascall GK: Cartilage proteoglycans: comparison of sectioned and spread whole molecules. J Ultrastruct Res 70: 369–375, 1980.
40. Myers DB, Highton TC, Rayns DG: Acid mucopolysaccharides closely associated with collagen fibrils in normal human synovium. J Ultrastruct Res 28: 203–213, 1969.
41. Takusagawa K, Ariji F, Shida K, Sato T, Asoo N, Konno K: Electron microscopic observations on pulmonary connective tissue stained by ruthenium red. Histochem J 14: 257–271, 1982.
42. Smith JW, Frame J: Observations on the collagen and proteinpolysaccharide complex of rabbit corneal stroma. J Cell Sci 4: 421–436, 1969.
43. Trelstad RL, Hayashi K, Toole BP: Epithelial collagens and glycosaminoglycans in the embryonic cornea. Macromolecular order and morphogenesis in the basement membrane. J Cell Biol 62: 815–830, 1974.
44. Wight TN, Ross R: Proteoglycans in primate arteries. I. Ultrastructural localization and distribution in the intima. J Cell Biol 67: 660–674, 1975.
45. Eisenstein R, Kuettner K: The ground substance of the arterial wall. Part 2. Electron-microscopic studies. Atherosclerosis 27: 37–46, 1976.
46. Riva R, Marchini M, Strocchi R: Proteoglycans and their relationship with the other components of the rabbit aorta wall observed in two different experimental conditions. Acta Histochem 65: 233–242, 1979.
47. Oegema TR, Hascall VC, Eisenstein R: Characterization of bovine aorta proteoglycan extracted with guanidine hydrochloride in the presence of protease inhibitors. J Biol Chem 254: 1312–1318, 1979.
48. Salisbury BGJ, Wagner WD: Isolation and preliminary characterization of proteoglycans dissociatively extracted from human aorta. J Biol Chem 256: 8050–8057, 1981.
49. Schmidt A, Prager M, Selmke P, Buddecke E: Isolation and properties of proteoglycans from bovine aorta. Eur J Biochem 125: 95–101, 1982.
50. Nygren H, Hansson H-A, Linde A: Ultrastructural localisation of proteoglycans in the odontoblast-predentin region of rat incisor. Cell Tissue Res 168: 277–287, 1976.
51. Takagi M, Parmley RT, Denys FR: Ultrastructural localization of complex carbohydrates in odontoblasts, predentin, and dentin. J Histochem Cytochem 29: 747–758, 1981.
52. Hay ED, Meier S: Glycosaminoglycan synthesis by embryonic inductors: neural tube, notochord, and lens. J Cell Biol 62: 889–898, 1974.
53. Cohn RH, Banerjee SD, Bernfield MR: Basal lamina of embryonic salivary epithelia. Nature of glycosaminoglycan and organization of extracellular materials. J Cell Biol 73: 464–478, 1977.
54. Kanwar YS, Fraquhar MG: Anionic sites in the glomerular basement membrane. In vivo and in vitro localization to the laminae rarae by cationic probes. J Cell Biol 81: 137–153, 1979.
55. Kanwar YS, Farquhar MG: Presence of heparan sulfate in the glomerular basement membrane. Proc Natl Acad Sci USA 76: 1303–1307, 1979.
56. Kanwar YS, Hascall VC, Farquhar MG, Partial characterization of newly synthesized proteoglycans isolated from the glomerular basement membrane. J Cell Biol 90: 527–532, 1981.
57. Kanwar YS, Linker A, Farquhar MG: Increased permeability of the glomerular basement membrane to ferritin after removal of glycosaminoglycans (heparan sulfate) by enzyme digestion. J Cell Biol 86: 688–693, 1980.
58. Del Rosso M, Cappelletti R, Vannucchi S, Romagnani S, Chiarugi VP: Selective exposure of mucopolysaccharides is involved in macrophage physiology. Biochim Biophys Acta 586: 512–517, 1979.
59. Vogel KG, Kendall VF: Cell-surface glycosaminoglycans: turnover in cultured human embryo fibroblasts (IMR-90). J Cell Physiol 103: 475–487, 1980.
60. Kjellén L, Pettersson I, Höök M: Cell-surface heparan sulfate: An intercalated membrane proteoglycan. Proc Natl Acad Sci USA 78: 5371–5375, 1981.
61. Simionescu N, Simionescu M: Galloylglucoses of low molecular weight as mordant in electron microscopy. I. Procedure, and evidence for mordanting effect. J Cell Biol 70: 608–621, 1976.
62. Luft JH: The structure and properties of the cell surface coat. Int Rev Cytol 45: 291–382, 1976.
63. Spicer SS, Baron DA, Sato A, Schulte BA: Variability of cell surface glycoconjugates – relation to differences in cell function. J Histochem Cytochem 29: 994–1002, 1981.
64. Danon DL, Goldstein L, Marikovsky Y, Skutelsky E: Use of cationized ferritin as a label of negative charges on cell surfaces. J Ultrastruct Res 38: 500–510, 1972.
65. Simionescu M, Simionescu N, Silbert JE, Palade GE: Differentiated microdomains on the luminal surface of the capillary endothelium. II.

112

Partial characterization of their anionic sites. J Cell Biol 90: 614–621, 1981.

66. Kleinschmidt AK, Zahn RK: Uber Desoxyribonukleinsäure-Molekylen in Protein Mischfilmen. Z Naturforsch (B) 146: 770–779, 1959.

67. Rosenberg L, Hellmann W, Kleinschmidt AK: Macromolecular models of proteinpolysaccharides from bovine nasal cartilage based on electron microscopic studies. J Biol Chem 245: 4123–4130, 1970.

68. Lang D, Mitani M: Simplified quantitative electron microscopy of biopolymers. Biopolymers 9: 373–379, 1970.

69. Thyberg J, Lohmander S, Heinegård D: Proteoglycans of hyaline cartilage. Electron-microscopic studies on isolated molecules. Biochem J 151: 157–166, 1975.

70. Kleinschmidt AK: Electron microscopic studies of macromolecules without appositional contrast. Philos Trans R Soc Lond [Biol] 261: 143–149, 1971.

71. Panessa BJ, McCorkle RA, Hoffman P, Warren JB, Coleman G: Ultrastructure of hydrated proteoglycans using a pulsed plasma source. Ultramicroscopy 6: 139–148, 1981.

72. Fessler JH, Fessler LI: Electron microscopic visualization of the polysaccharide hyaluronic acid. Proc Natl Acad Sci USA 56: 141–147, 1966.

73. Wellauer P, Wyler T, Buddecke E: Electron microscopic and physico-chemical studies on bovine nasal cartilage proteoglycan. Hoppe Seylers Z Physiol Chem 353: 1043–1052, 1972.

74. Yoneda M: Electron microscopic studies of proteoglycans from epiphyseal cartilage of suckling rats. Connect Tissue 9 (2): 1–12, 1977.

75. Buckwalter JA, Rosenberg LR: Electron microscopic studies of cartilage proteoglycans. Direct evidence for the variable length of the chondroitin sulfate-rich region of proteoglycan subunit core protein. J Biol Chem 257: 9830–9839, 1982.

76. Rosenberg L, Hellmann W, Kleinschmidt AK: Electron microscopic studies of proteoglycan aggregates from bovine articular cartilage. J Biol Chem 250: 1877–1883, 1975.

77. Kimura JH, Osdoby P, Caplan AI, Hascall VC: Electron microscopic and biochemical studies of proteoglycan polydispersity in chick limb bud chondrocyte cultures. J Biol Chem 253: 4721–4729, 1978.

78. Heinegård D, Lohmander S, Thyberg J: Cartilage proteoglycan aggregates. Electron-microscopic studies of native and fragmented molecules. Biochem J 175: 913–919, 1978.

79. Faltz LL, Reddi AH, Hascall GK, Martin D, Pita JC, Hascall VC: Characteristics of proteoglycans extracted from the Swarm rat chondrosarcoma with associative solvents. J Biol Chem 254: 1375–1380, 1979.

Author's address:
Department of Histology
Karolinska Institutet
P.O. Box 60400
S-104 01 Stockholm, Sweden

CHAPTER 6

Collagen–proteoglycan interaction

ALESSANDRO RUGGERI and FRANCO BENAZZO

1. Introduction

Collagen and proteoglycans always coexist in the extra-cellular matrix of the connective tissue.

Each of these components has its own specific role in the tissue. Collagen is a structure that gives stiffness and a high degree of tensile strength to the matrix; the highly hydrated proteoglycans confer rigidity to the matrix as well as allowing transport of 'moving' molecules and impeding diffusion of large molecules due to their mesh-like structure. In actual fact, collagen and proteoglycans also interact with proteoglycans, for instance, by influencing collagen fibril formation and growth as well as the assembly of the collagen fibrils and their three-dimensional arrangement according to the specific primary function of the tissue. Taking into account these few considerations, one may easily deduce the biological implications related to collagen–proteoglycan interactions and their possible pathological consequences.

The first experimental studies on collagen–proteoglycan interactions were developed about twenty years ago, when the influence of glycosaminoglycans and proteoglycans on the formation of collagen fibrils *in vitro* was studied (1–3). These studies found renewed interest when the spatial relationship existing between the two components was demonstrated directly in tissues by means of different morphological techniques (mainly electron microscopy) (4–10). However, new and more sophisticated biochemical investigation indicates that proteoglycans present high polydispersity in the same tissue and that their composition varies according to the nature of the tissue examined (11); on the other hand, many types of collagen have been discovered each of which is diversely distributed in the tissues (12). Thus, due to the number of variables of these two matrix components, it is possible that chemical interactions and steric relationships differ from tissue to tissue and according to development and aging.

The aim of this chapter is to present the contribu-

tion of electron microscopy in the visualization of the morphological differences of proteoglycans in tissues and in the knowledge of the interactions between collagen and proteoglycans. Electron microscopy is indeed essential in the observation of the diameter and banding patterns of collagen fibrils reconstituted under the influence of glycosaminoglycans and proteoglycans. Electron microscopy is also necessary in order to reveal the shape, size and three-dimensional array of proteoglycans in relation to collagen in various connective tissues. Lastly, it may be expected that electron microscopy provides new information on the specific site of interaction of proteoglycans within the period of the collagen fibrils and on the participation of these molecules in the aggregation and assembly of fibrils.

2. In vitro studies on collagen–proteoglycan interaction

Even though the authors have no direct experience in *in vitro* interaction between collagen and proteoglycans, the main points are reported in order to give a general view of this subject and some correlations with the other paragraphs of this chapter.

In suitable conditions of temperature and pH, collagen monomers extracted from tissues by weak acid or neutral salt solutions, self-assembly to form D-periodic fibrils. This process passes through a first 'lag period', in which thin aperiodic filaments are formed and a subsequent 'growth period' in which the filaments grow in width and acquire a banding pattern similar to that of native collagen fibrils.

During the 'lag period' the collagen monomers self assemble according to the staggered D-periodicity, offering their hydrophobic and charged amino acid side chains to one another. In this period the filamentous aggregate grows in length and the collagen molecules also seem to be involved in conformational changes in the extra-helical end regions, as

Ruggeri, A and Motta, PM (eds): Ultrastructure of the connective tissue matrix. ISBN-13:978-1-4612-9789-5

well as in a stabilization process. During the 'growth period', on the other hand, the aperiodic filaments grow by lateral accretion of new filamentous intermediate aggregates (for more details, see Chapter 1, p. 23).

Much investigation shows that glycosaminoglycans and proteoglycans influence the formation of fibrils *in vitro*, modifying both the length of the 'lag' and the 'growth period'.

The 'lag period, for example, can be shortened if chondroitin sulphate and keratan sulphate are added to the collagen solution, while it is retarded by heparan sulphate addition (1). At the same time, the 'growth period' distance increases with chondroitin sulphate and hyaluronate, but the degree of aggregation is favored (2). On the other hand, amongst the more complex proteoglycan molecules, dermatan sulphate proteoglycan is able to precipitate soluble collagen monomers (3, 13), while chondroitin sulphate proteoglycan has the capacity of delaying the process. It is also shown by chromatography on agarose-collagen columns that fractions of bovine cornea proteochondroitin sulphate interact with collagen, while no interaction occurs with proteokeratan sulphate fractions (14). Moreover, in similar experimental conditions it is shown that a strong ionic interaction between cartilage proteoglycan and collagen is dependent on the structure of the protein core of the proteoglycan (15). In addition, it is the pool of all the proteoglycans present in tissue which seems to be effective on fibrillogenesis, rather than proteoglycans taken separately (16).

Even though these observations do not lead to conclusions, it seems clear that the self assembly of collagen molecules and their progressive lateral accretion are influenced by the presence of proteoglycans. In general one may assume that it is the glycosaminoglycan side chains that determine the primary banding of proteoglycans to collagen. The interaction is predominately electrostatic in nature (1) and increases according to the charge density of the glycosaminoglycan side chains (17). Among glycosaminoglycans, dermatan sulphate and heparan sulphate interact more strongly with collagen than chondroitin sulphate and keratan sulphate (17). The acid residues of the glycosaminoglycans should bind to the superficial charged groups of collagen aggregates; probably not all groups along the collagen molecule are available, but only clusters of L-arginyl and L-lysil residues (18, 19). The negative charges of the glycosaminoglycans could intervene also by stabilizing the triple helical structure, because the presence of glycosaminoglycans increases the melting point (i.e. the point of conversion from

the triple helical structure to a random coiled form) of collagen from 30°C to 46°C (20). On the other hand, the proteoglycan core protein alone (21) could bind to a reactive site on the fibril different from that of the glycosaminoglycans.

Apart from the specific interactions of both glycosaminoglycan side chains and protein core with collagen, the whole proteoglycan molecule should influence collagen fibril formation by its physical-chemical property of excluded volume. Excluded volume influences the large molecules in solution affecting the equilibrium between the different conformational states and favoring the more compact conformations, i.e. the precipitation of collagen solution in form of collagen fibrils of native type. This mechanism is believed to be very effective for the hyaluronate–collagen interaction, since the former is of high molecular weight and has only one negative charge per disaccharide. It is generally accepted that electrostatic forces and steric exclusion act mutually both under *in vivo* and *in vitro* conditions, but their relative importance still remains difficult to assess (21–23).

3. Proteoglycans visualized in tissues and their relationship with collagen fibrils

Considerable progress was made in ultrastructural research on proteoglycan–collagen relationships directly studied 'in tissues' when the histochemical procedures used in light microscopy were applied to electron microscopy. By means of these methods the electron dense granules, observable in the cartilage matrix (*matrix granules*), were interpreted as aggregates of proteoglycans (6, 24), and similar interpretations were given to thin granules visible in the vessel-wall matrix (25) or to thin filaments detectable among the collagen fibrils of loose and fibrous connective tissues (5).

As reported in Chapter 5, histochemical methods consist in the use of electron dense cationic stains and enzymatic tests. The dye is usually added to the fixator and used on 'blocks' of tissues; in this procedure the cationic stain, in addition to improving the electron density of the proteoglycan particles, represents a factor of fixation. In fact, by linking the dye to the substrate, a more stable complex is formed which becomes less extractable during dehydration and embedding.

From the first histochemical observations, a strict relationship between proteoglycans and collagen fibrils was visible (6, 24); moreover, in tissues where the collagen fibrils were assembled in bundles, a periodic array, with a distance corresponding to the D-period of collagen, was observable (4, 5, 26). Research continued with the introduction of new dyes and staining procedures (8, 27–30) and developed with the study of the ultrastructural differences of proteoglycan particles of various tissues and organs and with the aim of finding more details on the side of attachment of proteoglycans to collagen fibrils (10).

3.1. Proteoglycan particles in different tissues

3.1.1. Cartilage. Hyaline cartilage has been the object of much histochemical research in which various cationic stains have been used.

In general, the appearance of proteoglycan particles is that of highly electron dense, rounded or stellate matrix granules. Frequently, very thin filaments are visible connecting the granules to one another or to neighboring collagen fibrils. The width of the granules is extremely variable (ranging from 5 to 100 nm). Their mean diameter and number varies in relation to the cartilage examined, the anatomical site of the cartilage and the distance from the cell. For example, matrix granules are larger in the deeper layers of articular cartilage than in the superficial ones; moreover they decrease in size and number from territorial to interterritorial cartilage and are bigger in the bovine nasal septum cartilage than in deep layers of articular cartilage of the same animal (Figure 1a, b). Apart from the phenomena of fusion of two or more small granules claimed by some authors (6), the variability of the mean width and number of the granules depends on the differences in size and quantity of the proteoglycan monomers according to their location.

When the same tissues are treated with staining solutions containing $MgCl_2$ in critical electrolyte concentrations ($MgCl_2$ CEC solutions) the proteoglycan particles assume a 'filament-like' or 'leaf-like' aspect instead of a granular aspect (8, 31). According to this procedure, proteoglycan molecules are presumably better fixed and therefore maintain their presumptive extended position. In other words, molecule collapse is at least partially prevented by the dye plus $MgCl_2$ CEC solution treatment. Moreover it is presumed that under these staining conditions fusion does not usually occur (32).

Proteoglycan particles having a characteristic leaf-like shape (Figure 1c, d) were found in every hyaline cartilage examined, whereas filament-like particles (Figure 2a, b) were found to fill fibrous cartilage, such as the more superficial layers of the articular cartilage (33) or the peripheral parts of the annulus fibrosus of the intervertebral disk (34). Both leaf-like and filament-like particles are linear in shape and have a relatively uniform electron density, which allows reliable measurement. In human and bovine nasal septum cartilage, the leaf-like particle width ranges from 20 to 35 nm, whereas in the deep layers of the human and bovine knee joint cartilage, or in the inner part of the annulus fibrosus of the intervertebral disk, it ranges from 20 to 23 nm. Notably different, on the contrary, is the diameter of filament-like particles, which is uniform along the length of the filament and ranges from 8 to 12 nm in articular cartilage surface layers and 12 to 16 nm in peripheral parts of the annulus fibrosus of the intervertebral disk.

As one may deduce from these findings, larger leaf-like proteoglycan particles are found in areas where the collagen fibrils are randomly arranged and widely spaced. In this case the particles appear to be connected to the collagen fibrils only by a short tract of their length (Figure 1d), while the remaining part is expanded in the interfibrillar spaces free of attachment. On the other hand the proteoglycan filaments occupy narrow spaces among neighboring parallel collagen fibrils: the filaments are found to be extensively related and in orthogonal and periodic array to the collagen fibrils. The two ultrastructural aspects respectively countersign the hyaline and fibrous cartilage. As far as the different shape and width of the two kinds of particles are concerned one might assume that leaf-like particles represent large proteoglycan monomers with high molecular weight and numerous and/or long chondroitin and keratan sulphate side chains. The filaments should correspond to small monomers trapped within adjacent collagen fibrils. Actually, the average diameter of the leaf-like particles (20–35 nm) may be correlated with the measurements carried out on rat chondrosarcoma spread monomers having condensed side chains (35). On the other hand the filament-like particles, due to their small and uniform diameter, might correspond to the highly interacting monomers found in the fibrous cartilage close to the bovine articular surface (36) or in the outer layers of the annulus fibrosus of the intervertebral disk (37) and could represent small monomers with few and/or short glycosaminoglycan side chains. Enzyme treatments with testicular hyaluronidase, as well as chondroitinase ABC, support the idea that the filament-like particles are rich in short side chains of keratan sulfate, while the leaf-like particles have high content of long condroitin sulfate chains.

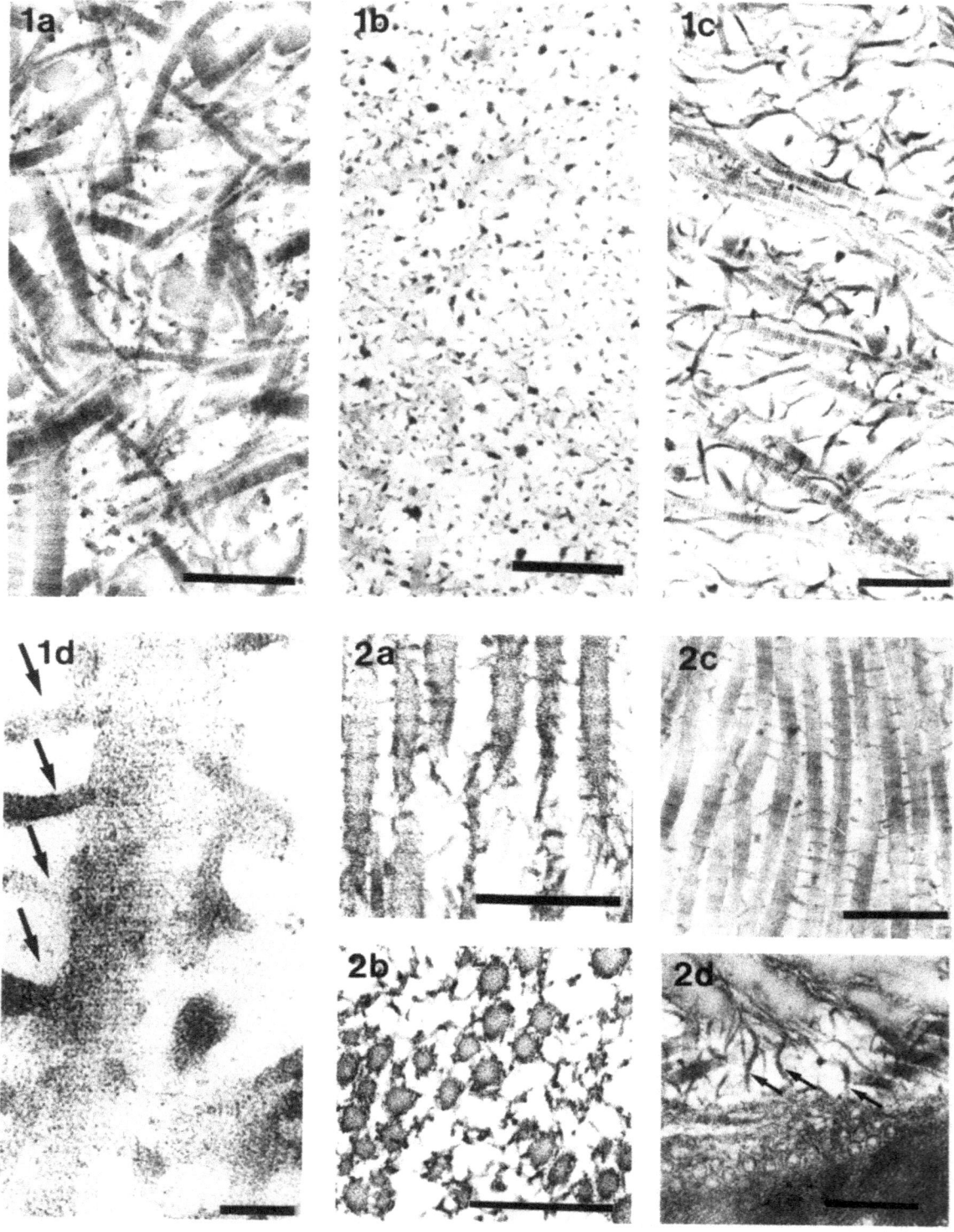

3.1.2. Aorta wall. In the tunica intima and media of the aorta wall proteoglycan granules of varying sizes are found in spaces within muscle cells, elastic fibres and collagen fibril bundles (25). Thin filaments are also seen connecting the granules to the surrounding matrix components. When the specimens are treated with alcian blue diluted in $MgCl_2$ CEC solutions (38), two kinds of proteoglycan particles are visualized: (1) leaf-like particles ranging in width from 20 to 30 nm and scattered at random around the cells, elastic fibres and bundles of collagen fibrils (Figure 2d); (2) 8-nm thick filament-like particles, generally placed within the fibril bundles and in orthogonal, periodic relationship to the collagen fibrils (Figure 2c). These aspects reproduce the same ultrastructural findings of cartilage. The largest proteoglycan particles fill wide areas and only a low percentage of them appear to be linked to the surrounding components; presumably these particles represent large sized monomers involved in a three-dimensional macromolecular network with high retention of water. The thin filaments are in extensive periodic relationship to the collagen fibrils and probably represent small monomers containing dermatan sulphate tightly interacting with the collagen fibrils (39).

3.1.3. Tendon. The tendon is a typical fibrous connective tissue, made up of broad and parallel aligned type I collagen fibrils and a relatively small amount of proteoglycans. Dermatan sulphate proteoglycan, a small monomer with few dermatan sulphate side chains, is the most representative proteoglycan molecule in adult life (40), which interacts very strongly with collagen (17).

The penetration of cationic stains within the tendon is very poor, presumably because of the stiffness of the tissue. In favorable staining conditions (very small blocks of tissue and long incubation time), alcian blue containing $MgCl_2$ CEC solution visualizes very thin filaments, 6–8 nm thick, orthogonally arranged to the collagen fibrils (Figure 3a). These filaments connect the collagen fibrils to one another and are regularly spaced at a distance which corresponds to the D-period of the fibrils. Testicular hyaluronidase digestion only partially abolishes these filaments, which, instead, completely disappear after treatment with chondroitinase ABC (Figure 3b) (41).

Proteoglycans and their relationship to the collagen fibrils have been widely examined in tendons of new born and young rats (10, 30, 40, 42). Scott and co-workers used cupromeronic blue dye (10) diluted in $MgCl_2$ CEC solution and bound to tungstate to increase the electron density of the proteoglycan particles. These authors showed one kind of broad and relatively long rod-like particle with no apparent relationship to the collagen fibrils, and a second kind of thin proteoglycan particle with a filament-like aspect and a prevalent relationship to the fibrils. The latter filaments appear orthogonally and periodically arrayed to the fibrils, or joining two or more of the orthogonal elements at right angles. The orthogonal filaments appear to be attached to the *d* band of the fibrils.

In agreement with the above observations, both thick rod-like particles (averaging 15 nm) and thin filament-like particles (8 nm thick) have been visualized by alcian blue diluted in 0.3 M–$MgCl_2$ solution (Figure 3c) in one-month-old rat Achilles's tendon (42). The two kinds of particles, in this case, appear to be connected to each other. In longitudinal sections thick rod-like particles running parallel to the collagen fibrils and as long as 200 nm minimum are visible. Thin regularly spaced lateral filaments are orthogonally attached to these particles. These thin filaments appear to surround the neighboring collagen fibrils and are regularly spaced at a distance equal to the D-period of the fibrils. In transverse sections, the rod-like particles appear as electron dense dots, and, from these, thin filaments move in the direction of the neighboring collagen fibrils. Both in longitudinal and transverse sections, where the collagen fibrils are not closely packed, rod-like particles and side filaments appear transversely

Figure 1. Bovine knee joint (*a*) and nasal septum (*b*) cartilage fixed conventionally with glutaraldehyde and osmium tetroxide. Electron-dense matrix granules are randomly distributed around the cells and among the collagen fibrils. The average diameter values are lower in articular than in nasal cartilage; *c* and *d* illustrate bovine nasal cartilage fixed with glutaraldehyde, stained over night with 0.05% alcian blue 8 GX diluted in 0.025 M acetate buffer, pH 5.8, containing 0.3 M–$MgCl_2$ (alcian blue plus 0.3 M–$MgCl_2$) and postfixed with osmium tetroxide. Large leaf-like particles, instead of granules, are visualized. Some appear attached to the fibrils at their extremities (see arrow in *d*) while others are randomly distributed in the wide spaces among the fibrils. a, b, c: bars = 0.5 μm; d: bar = 0.05 μm.

Figure 2. Peripheral part of the annulus fibrosus of the bovine intervertebral disk (a, b) and tunica media of rat aorta wall (c, d) fixed with glutaraldehyde, stained with alcian blue plus 0.3 M–$MgCl_2$ and postfixed with osmium tetroxide. Among neighboring, parallel collagen fibrils thin filament-like particles are found in orthogonal and periodic array (a, b, c). In d large leaf-like particles are found (arrows) in the large areas among bundles of collagen fibrils, elastin fibers and muscle cells. Bars = 0.5 μm. Figure 2a reprinted from Marchini et al. (34) with permission of Basic Appl Histochem.

118

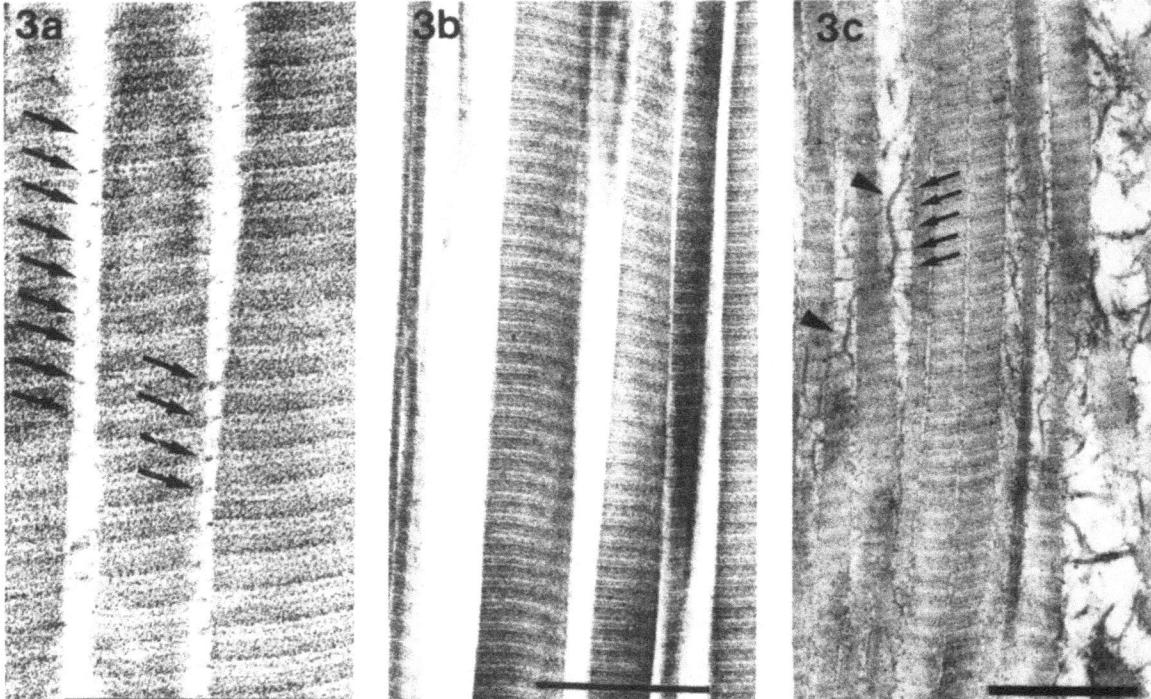

Figure 3. Six-month-old rat tail tendon (a, b) and one-month-old rat Achille's tendon (c) fixed with glutaraldehyde, stained with alcian blue plus 0.3 M–MgCl$_2$ and postfixed with osmium tetroxide. In b fresh specimens were preincubated for 20 h at 37° C with chondroitinase ABC in pH 7.4 phosphate buffer. a: filament-like particles (arrows) visualized by alcian blue are found in orthogonal and periodic array to the collagen fibrils; b: these filaments completely disappear when preincubation with chondroitinase ABC is carried out; c: the orthogonal thin filaments (arrows) appear to be connected at regular intervals to the sides of long rod-like particles (arrow head) running parallel to the collagen fibrils. Where the collagen fibrils are not closely packed rod-like particles and lateral filaments appear transversely placed connecting two or more fibrils. Bars = 0.5 μm.

placed in the empty space and clearly connect two or more fibrils.

From the above observations it appears that tendons of new born and young rats, which contain both chondroitin sulphate and dermatan sulphate proteoglycans (40), show large rod-like and thin filament-like particles. In old rat tendons, in which dermatan sulphate alone is detectable (40), only thin filament-like particles are found. Enzymatic digestion tests are now in progress in our laboratory in order to find possible differences in composition between the two classes of particles.

3.1.4. Other tissues. In the highly hydrated, opaque cornea stroma of eight-day-old chick avian embryo, where the collagen fibrils are sparsely arranged, 30–50 nm thick proteoglycan granules are found in the matrix after Ruthenium red staining. In 12-day-old embryonic avian cornea stroma, where collagen fibrils are more closely packed, 20-nm thick proteoglycan granules in periodic array are visualized (43). When cornea stroma maturation is over, pro-

teoglycans are visualized as very thin filaments (4 nm thick) orthogonally and periodically arrayed to the collagen fibrils (26). It is reasonable to assume that these thin filaments are the ultrastructural representation of dermatan sulphate and/or keratan sulphate proteoglycans, which are both small molecules with only 1–3 glycosaminoglycan side chains per monomer and which have a very small hydrodynamic volume (44). As cornea keratan sulphate proteoglycans show affinity *in vitro* for collagen (14), finer histochemical details could better clarify the chemical nature of these filaments.

Filament-like particles 10 nm thick, very similar to those observable in cornea stroma, have been found by means of alcian blue plus 0.3 M–MgCl$_2$ in six-month-old rat sclera, a tissue rich in dermatan sulphate proteoglycans (45). Filaments showing the same close relationship to the collagen fibrils have been visualized also in other fibrous tissues, such as in rat peritenoneum (Figure 4a) and ligaments.

The matrix secreted by fibroblasts in cultures is characterized by a high content of proteodermatan

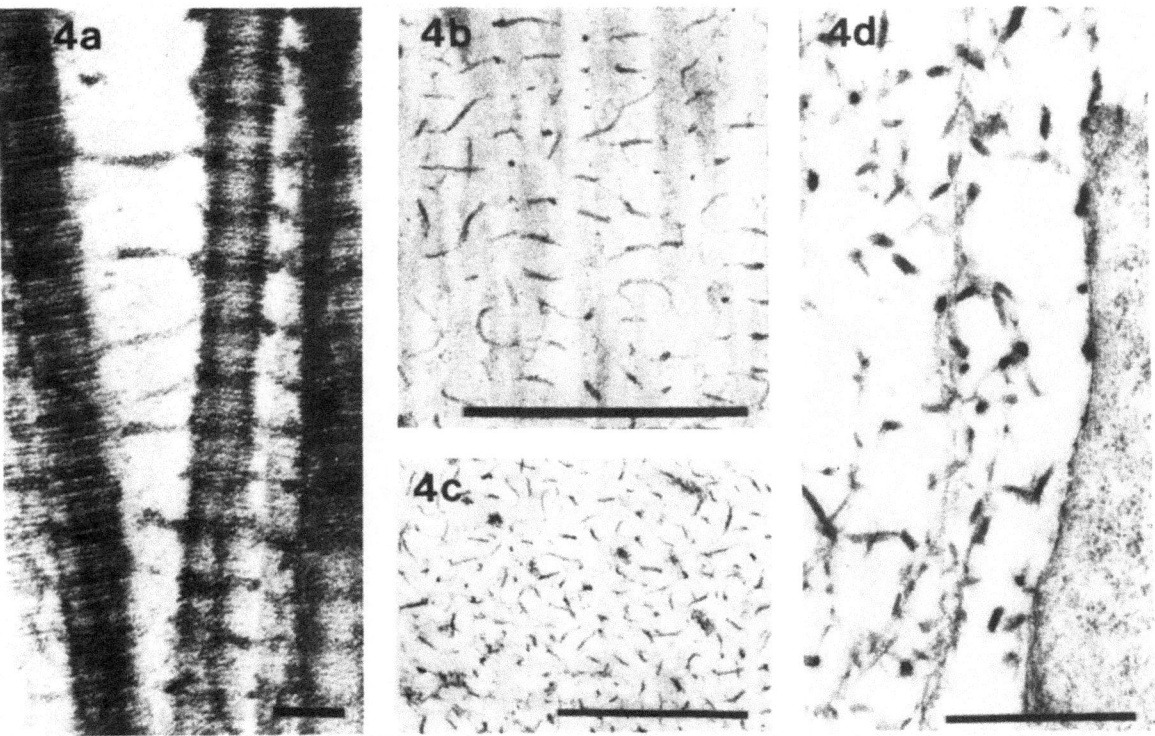

Figure 4. Six-month-old rat tail peritenoneum (a), extracellular matrix of *in vitro* cultured fibroblasts (b, c) and chick embryo presumptive cartilaginous area (d) fixed with glutaraldehyde, stained with alcian blue plus 0.3 M–MgCl₂ and postfixed with osmium tetroxide. Filament-like particles are found where collagen fibrils are fastened in bundles (a, b, c); large rod-like particles are visible where the space among the collagen fibrils is large (d). a: bar = 0.1 μm; b, c, d: bars = 0.5 μm. (Figure 4d reprinted from Ruggeri et al. (8) with permission of Histochem J.)

sulphate and proteoheparan sulphate (46). After treatment with alcian blue plus 0.3 M–MgCl₂ solution (8), thin filament-like particles periodically related to the collagen fibrils are observable (Figure 4b, c).

The extracellular matrix of chick embryo presumptive cartilaginous areas (including perinotochordal matrix) contains chondroitin sulphate proteoglycans (47), prevalently in monomeric, unassociated form (48). On thin sections of traditionally fixed tissue blocks this material is represented by electron dense granules, 20–30 nm wide, mainly attached to aperiodic collagen microfibrils (49). In specimens treated with alcian blue plus 0.3 M–MgCl₂ solution, the same components assume the shape of rod-like and stellate particles ranging in width from 18 to 20 nm and showing a more evident relationship to the collagen microfibrils (Figure 4d).

From the above ultrastructural findings obtained from tissues of different sources one may assume that electron microscopy provides interesting data on the distribution of proteoglycans in tissue and their relationship to collagen fibrils. Two main kinds of particles may be distinguished, which presumably correspond to proteoglycans with different composition and molecular weight: (1) large and electron dense proteoglycan particles placed in large spaces full of water, and (2) thin filament-like particles situated within parallel arrayed collagen fibrils and displaying periodic and orthogonal relationship.

Filament-like particles with size and array similar to those observable in thin sections have also recently been visualized (32, 41, 42) on replicas of fibrous tissues freeze-etched (Figure 5a, b) or submitted to shadow-casting (Figure 5c, d). Replicas of freeze-etched specimens were obtained from six-month-old rat tail tendons fixed with glutaraldehyde, treated with alcian blue diluted in 0.3 M–MgCl₂ solution, washed in phosphate buffer, and then frozen for fracturing. Cleaving was followed by deep-etching, which allowed the proteoglycan molecules placed under the fracture plane to emerge from the surrounding ice. The shadow-casting method was applied to six-month-old rat tail tendons and

Figure 5. Replicas of six-month-old rat tail tendon (a, b) and rat tail peritenoneum (c, d) fixed with glutaraldehyde, stained with alcian blue plus 0.3 M–MgCl₂ and processed for freeze-etching or shadow casting (see text). Filament-like particles (arrows) are found cross-connecting (a) and surrounding (c) neighboring fibrils. These filaments appear to be removed after chondroitinase ABC preincubation (b, d). Bars = 0.1 μm.

their sheaths. The samples were fixed, stained with alcian blue diluted in 0.3 M–MgCl$_2$ solution, washed, postfixed, dehydrated, dried with a critical point apparatus and shadowed. Both kinds of replica show filaments in cross and periodic array to the fibrils and with uniform diameters, averaging 8 nm (detracting the replica's width: 2.5×2 nm). The filaments disappear if fresh samples are submitted to chondroitinase ABC digestion (Figure 5b, d).

These findings, obtained by means of very different procedures, further suggest that in tendons, as well as wherever collagen fibrils are assembled in bundles or in laminae, proteoglycan molecules have such a conformation as to assume a filament-like ultrastructural shape. In other words they should correspond to small dermatan sulphate proteoglycan monomers which have several oligosaccharides, few glycosaminoglycan side chains per molecule (50), and strong interaction with collagen (22). According to these observations, the interaction between collagen and proteoglycans could be schematically represented as drawn in Figure 6a and b, where collagen fibrils, randomly distributed in bundles, are represented.

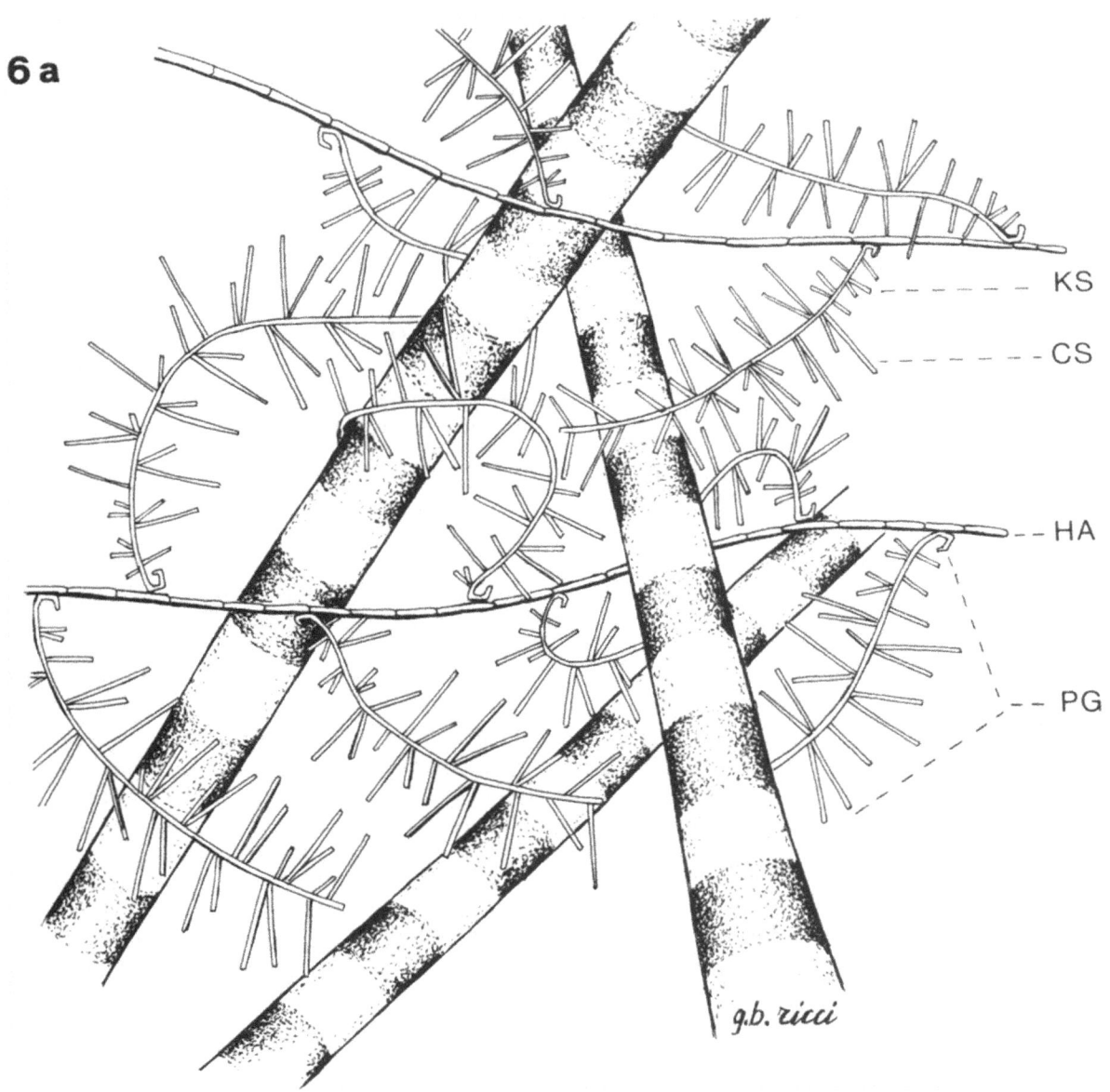

6 a

KS

CS

HA

PG

g.b. ricci

6 b

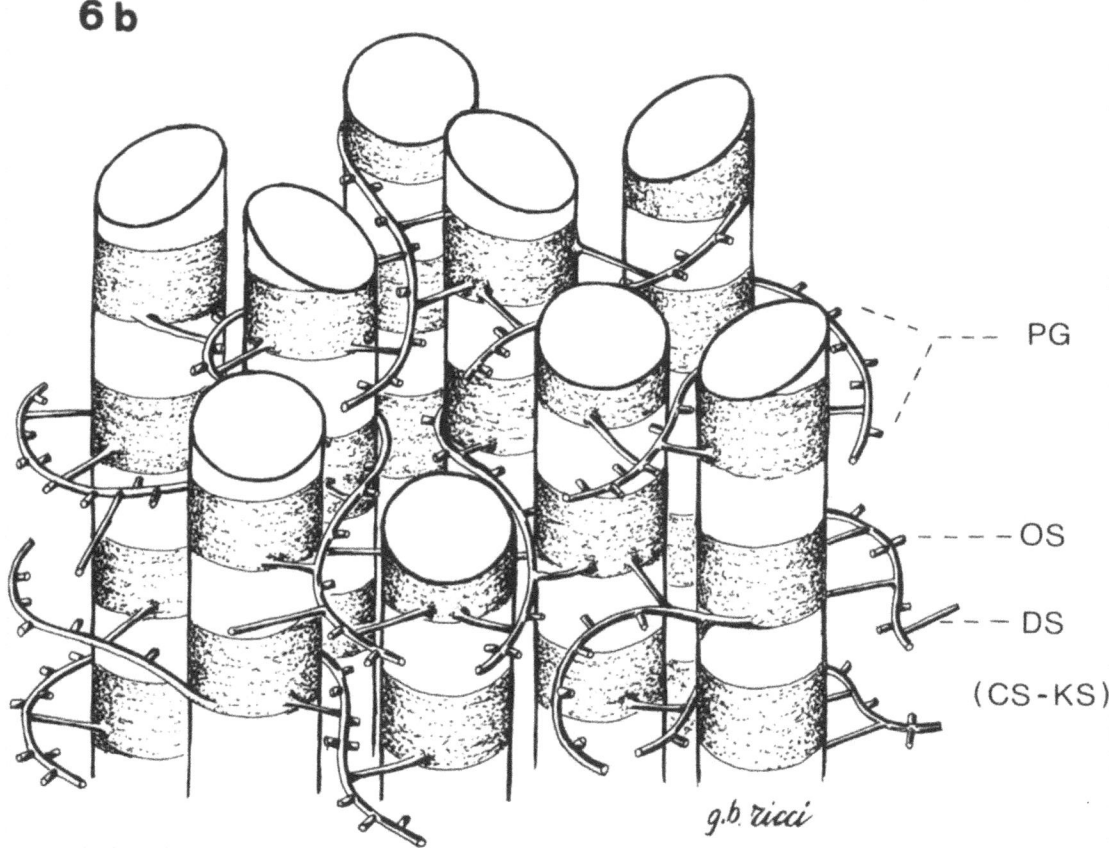

PG

OS

DS

(CS-KS)

g.b. ricci

Figure 6. Tentative schematic representation of proteoglycan monomers and their relationship with collagen fibrils. In a, where the collagen fibrils are irregularly spaced in a network, the proteoglycan monomers are large and randomly arranged; in b, where the collagen fibrils are fastened in bundles, the proteoglycan monomers are small, orthogonally arranged and spaced at a distance that corresponds to the D-period of the fibrils. PG = proteoglycan monomer; HA = hyaluronic acid; CS = chondroitin sulphate chain; DS = dermatan sulphate chain; KS = keratan sulphate chain; OS = oligosaccharide.

3.2. Site of attachment of proteoglycans to collagen fibrils

Within the D-period of the collagen fibrils, the areas with a prevalence of cationic charges suitable for interaction with proteoglycan molecules, are restricted to *a* and *c* bands (19). Attempts to localize the site of this interaction by means of electron microscopy have been made by several authors (4, 40, 51–53). Previous investigation suggested that the *a* band to be a site of proteoglycan interaction with collagen in hyaline cartilage, cornea, synovium and ear cartilage. Recently, Scott and co-workers (40) have shown that young rat tendon dermatan sulphate-rich proteoglycans associate with collagen at the *d* band. It has been suggested by the authors that the proteoglycan monomer extends its glycosaminoglycan side chains from *d* to the roughly equidistant *a*

and *c* bands. These findings need further confirmation. One must consider that several chemical factors and methodologies could influence these results (40). On the other hand the wide range of variability of collagen and proteoglycans in tissues could also entail different sites of interaction.

Other approaches to the study of collagen proteoglycan interactions come from X-ray diffraction analysis (53). Low angle X-ray diffraction patterns of tail tendon collagen fibers stained in blocks with ruthenium red show some diffraction maxima which have been interpreted as site of stain binding with supposed glycosaminoglycan anionic charges. Considering the low specificity conditions adopted for that staining, these data need further confirmation. Nevertheless, it has been recently suggested that the molecular density of proteoglycans themselves should be sufficient to influence the X-ray diffraction pattern (54, 55). Therefore X-ray diffraction analysis should also contribute information on proteoglycan–collagen interaction in future.

4. Functional aspects of collagen proteoglycan interaction

In Section 2 the possible influence of proteoglycans in fibril formation have been reported. In general one may assume that the high density of positive charge in bands *a* and *c* of the collagen attract the charged glycosaminoglycan side chains of the proteoglycans (19). The *a* and *c* bands, situated on either side of the overlap region, correspond to the C-terminus and N-terminus respectively, i.e. to the sites of intermolecular cross-link. It is supposed (40) that the glycosaminoglycan side chains, prevalently attached to these regions, could hinder lateral accretion of the collagen molecules or fibrils, restricting their access to the cross-link sites; following this suggestion, proteoglycans must be displaced during fibril formation.

Among the various macromolecules, it is dermatan sulphate proteoglycan which is believed to promote fibril formation, whereas chondroitin sulphate proteoglycan should limit the process. In fact, to our knowledge, there is an abundance of dermatan sulphate proteoglycans in fibrous tissue containing thick collagen fibrils, while the high percentage of chondroitin sulphate proteoglycans (as in hyaline cartilage) matches with small sized fibrils. In agreement with these observations it has been demonstrated that the stainability of the collagen fibrils with picro-sirius red is related to the presence of proteoglycans in tissues, and a correlation between type of collagen and proteoglycan molecular composition has been found (56). As pointed out in Chapter 4 (p. 89), thickening of the collagen fibrils allows the increase of interfibrillar cross-link density and thus higher tensile strength of the fibrils; on the contrary, small collagen fibrils display a greater interface between themselves and between themselves and the proteoglycans, thus increasing the creep-resistance of the tissue.

The tight interaction of the proteoglycans with collagen could also influence the D-period of the collagen fibrils (i.e. the molecular arrangement). Actually, differences in the D-period length, recently recorded on low angle X-ray diffraction patterns of skin (57, 58), could be due to a different proportion and composition of proteoglycans associated with these tissues (57). It is suggested that 'the glycosaminoglycan binding may alter the charge distributions and water binding in such a way as to permit a somewhat shorter length for the molecule, leading to a shorter D-period'.

In addition to their possible influence on growth and arrangement of the collagen fibrils, proteoglycans should be involved in lining up neighboring collagen fibrils. It has been shown, in fact, that both parallel and anti-parallel neighboring fibrils are aligned across several fibrils by the *a* band, and it has been suggested (19) that this regular interfibrillar bridging is due to the matching of excess positive charge at the *a* band of the collagen fibrils with the negative charges of the proteoglycans.

The hydrodynamic size of the proteoglycans should also settle the adjacent collagen fibrils at a distance consistent with physico-chemical or mechanical properties of the tissues. For example, the small proteoglycans of the cornea, besides possibly controlling the diameter of the collagen fibrils, keep these fibrils regularly spaced in order to maintain the optical transparency of the tissue. In hyaline cartilage, the chondroitin sulphate-rich proteoglycans are complexed in large hydrated aggregates forming together with the collagen fibrils, a wide interlaced meshwork which absorbs comprehensive load.

5. Concluding remarks

Electron microscopy has made a fundamental contribution to the knowledge of the interaction between proteoglycans and collagen, putting in evidence the existence of a specific and periodically repetitive site of attachment of the proteoglycans along the collagen fibril. The examination of the size and density distribution of proteoglycans and collagen in tissue and of their spatial relationship has also helped to explain certain specific differences in composition of the tissues, or parts of tissues, according to their different biological and functional roles.

Thus, for example, the small hydrodynamic size of proteoglycans in tendons, and in fibrous tissues in general, is justified: in fact, due to their size, proteoglycans may be inter-spaced at a distance equivalent to the D-period of the collagen fibril, thus influencing the arrangement 'in phase' of the fibrils themselves and their close packing. On the other hand, the presence in vessel walls both of small and large proteoglycans may also be justified: the small proteoglycans are situated within bundles of collagen, thus maintaining the collagen fibrils close to each other, whereas the large proteoglycans occupy large spaces between bundles of collagen, elastic fibers and muscle cells, thus absorbing remarkable volume excursion. Lastly, the gradual changes in composition which proteoglycans undergo during development and aging may be considered in relation to the progressive reduction in hydrodynamic volume of the proteoglycans which occurs simultaneously with tissue stiffening.

Even though electron microscopy has made progress in this field, at present the techniques are still inadequate. There is the necessity of improvement in *in situ* visualization of proteoglycans in order to reach the same degree of visualization of proteoglycans as that obtained for extracted proteoglycans and to collect more detailed ultrastructural findings regarding the site and mode of interaction of proteoglycans and collagen. However, the possibility of new methods of research in this field is less remote than a few years ago. As accurate extraction of both specific genetical types of collagen and monomers of proteoglycans of various compositions is now possible, it may be assumed that in future *in vitro* study followed by ultrastructural observation of the interactions of these two components may supply more precise and detailed data. It is believed that this investigation, will be developed in strict relationship with X-ray diffraction, neutron diffraction, and optic diffraction analysis, applied to histochemically stained or enzymatically treated tissues. It is also believed that improvement in the preparation of samples for freeze-fracturing, which are being stud-

ied at present in some laboratories, gives a more detailed visualization of the site of interaction of proteoglycan molecules with collagen. Furthermore it is believed that it would be useful to support ultrastructural histochemical findings with ultrastructural immunochemistry which is in rapid evolution.

The interaction between collagen and proteoglycan is only a small part of the complex system of interactions which involves all the other structural components of the extracellular matrix, such as elastic fibers, fibronectin, laminin and other structural glycoproteins. Some ultrastructural data on the interaction of proteoglycans and collagen with elastin and of these three components with the structural elements of the basal membranes are already known. At present only biochemical and immunochemical data are available regarding the interaction of fibronectin with the other structural components of the matrix, but it is believed that an essential contribution will also soon be made by electron microscopy.

Acknowledgements

The personal investigations mentioned in this paper have been supported by grants from the Italian National Research Council.

References

1. Wood GC, Keech MK: The formation of fibrils from collagen solution. 1. The effect of experimental conditions: kinetic and electron-microscope studies. Biochem J 75: 588–598, 1960.
2. Mathews MB, Decker L: The effect of acid mucopolysaccharides and acid-mucopolysaccharides-proteins on fibril formation from collagen solution. Biochem J 109: 517–527, 1968.
3. Toole BP, Lowther DA: Dermatan sulfate-protein: isolation from and interaction with collagen. Arch Biochem Biophys 128: 567–578, 1968.
4. Smith JW, Peters TJ, Serafini-Fracassini A: Observations on the distribution of the proteinpolysaccharide complex and collagen in bovine articular cartilage. J Cell Sci 2: 129–136, 1967.
5. Meyers DB, Highton TC, Rayns DG: Acid mucopolysaccharides closely associated with collagen fibrils in normal human synovium. J Ultrastruct Res 28: 202–213, 1969.
6. Anderson HC, Sajdera SW: The fine structure of bovine nasal cartilage. Extraction as a technique to study proteoglycans and collagen in cartilage matrix. J Cell Biol 49: 650–663, 1971.
7. Módis L, Módis-Süveges I, Conti G: Quantitative polarisations-optische Untersuchungen an Mucopolysacchariden. Acta Histochem (Jena) 12 (suppl): 169–176, 1972.
8. Ruggeri A, Dell'Orbo C, Quacci D: Electron microscopic visualization of proteoglycans with Alcian Blue. Histochem J 7: 187–197, 1975.
9. Junqueira LCU, Bignolas G, Mourão PAS, Bonetti SS: Quantitation of collagen-proteoglycan interaction in tissue section. Connect Tissue Res 7: 91–96, 1980.
10. Scott JE, Orford CR: Dermatan sulphate-rich proteoglycan associates with rat tail-tendon collagen at the d band in the gap region. Biochem J 197: 213–216, 1981.
11. Kennedy JF: Proteoglycans-biological and chemical aspects. In

Human Life. Kennedy JF (ed), Amsterdam, Elsevier, 1979 pp 145–194.
12. Miller EJ, Rhodes RK: Preparation and characterization of the different types of collagen. Methods Enzymol 82: 33–64, 1982.
13. Tsiganos CP, Muir H: Studies on protein-polysaccharides from pig laryngeal cartilage. Heterogeneity, fractionation and characterization. Biochem J 1113: 885–894, 1969.
14. Speziale P, Bardoni A, Balduini L: Interactions between bovine cornea proteoglycans and collagen. Biochem J 187: 655–659, 1980.
15. Greenwald RA, Schwartz LE, Cantor JO: Interaction of cartilage proteoglycans with collagen-substituted agarose gels. Biochem J 145: 610–605, 1975.
16. Muthiah P, Kuhn K: Studies on proteoglycans obtained from bovine knee joint and ear cartilages: sequential extraction, fractionation, characterization and their effect on fibril formation. Biochim Biophys Acta 304: 12–19, 1973.
17. Öbrink B: A study of the interactions between monomeric topocollagen and glycosaminoglycans. Eur J Biochem 33: 387–400, 1973.
18. Öbrink B, Laurent TC, Carlsson B: The binding of chondroitin sulphate to collagen. FEBS Lett 56: 166–169, 1975.
19. Doyle BB, Hukins DWL, Hulmes DJS, Miller A, Woodhead-Galloway J: Collagen polymorphism: its origins in the amino acid sequence. J Mol Biol 91: 79–99, 1975.
20. Gelman RA, Blackwell J: Collagen-mucopolysaccharides interactions at acid pH. Biochim Biophys Acta 342: 254–261, 1974.
21. Oegema TR, Laidlow J, Hascall VC, Dziewatkowski DD: The effect of proteoglycans on the formation of fibrils from collagen solutions. Arch Biochem Biophys 170: 698–709, 1975.
22. Öbrink B: The influence of glycosaminoglycans on the formation of fibers from monomeric tropocollagen in vitro. Eur J Biochem 34: 129–137, 1973.
23. Öbrink B: Polisaccharide-collagen interactions. In: Structure of Fibrous Polimers. Atkins EDT (ed), Woburn, 1975 pp 81–92.

24. Matukas VJ, Panner BJ, Orbison JL: Studies on ultrastructural identification and distribution of protein-polysaccharides in cartilage matrix. J Cell Biol 32: 365–373, 1967.
25. Wight TN, Ross R: Proteoglycans in primate arteries. I. Ultrastructural localization and distribution in the intima. J Cell Biol 67: 660–674, 1975.
26. Smith JW, Frame J: Observations on the collagen and proteinpolysaccharide complex of rabbit corneal stroma. J Cell Sci 4: 421–436, 1969.
27. Shepard N, Mitchell N: The localization of proteoglycans by light and electron microscopy using safranin O. A study of epiphyseal cartilage. J Ultrastruct Res 54: 451–460, 1976.
28. Shepart N, Mitchell N: Simultaneous localization of proteoglycan by light and electron microscopy using toluidine blue O. A study of epiphyseal cartilage. J Histochem Cytochem 24: 621–629, 1976.
29. Ruggeri A, Dell'Orbo C, Quacci D: Electron microscopic visualization of proteoglycans with ruthenium red. Histochem J 9: 249–252, 1977.
30. Scott JE: Collagen–proteoglycans interactions – localization of proteoglycans in tendon by electron microscopy. Biochem J 187: 887–891, 1980.
31. Ruggeri A, Reale E, Dell'Orbo C: Relationships between connective tissue proteoglycan and fibrous proteins, as shown by electron microscopy. Ups J Med Sci 82: 74, 1977.
32. Aureli G, Castellani AA, Balduini C, Bonucci E, Cetta G, DeLuca G, Rizzotti M, Ruggeri A: Structural and metabolic aspects of macromolecules in the connective tissue. Basic Appl Histochem 25: 217–254, 1981.
33. Shepard N, Mitchell N: The localization of articular cartilage proteoglycan by electron microscopy. Anat Rec 187: 463–476, 1977.
34. Marchini M, Strocchi R, Castellani PP, Riva R: Ultrastructural observations on collagen and proteoglycans in the annulus fibrosus of the intervertebral disc. Basic Appl Histochem 23: 137–148, 1979.
35. Hascall VC, Hascall GK: Proteoglycans. In: Cell biology of Extracellular Matrix. Hay ED (ed), New York, 1981, pp 39–63.
36. Poole AR, Pidoux I, Reiner A, Tang L-H, Choi H, Rosenberg L: Localization of proteoglycan monomer and link protein in the matrix of bovine articular cartilage: an immunohistochemical study. J Histochem Cytochem 28: 621–635, 1980.
37. Stevens RL, Dondi PG, Muir H: Proteoglycans of intervertebral disc. Absence of degradation during the isolation of proteoglycans from the intervertebral disc. Biochem J 179: 573–578, 1979.
38. Riva R, Marchini M, Strocchi R: Proteoglycans and their relationships with the other components of the rabbit aorta wall observed in two different experimental conditions. Acta Histochem 65: 233–242, 1979.
39. Wight TN, Hascall VC: Proteoglycans in primate arteries. III. Characterization of the proteoglycans synthesized by arterial smooth muscle cells in culture. J Cell Biol 96: 167–176, 1983.
40. Scott JE, Orford CR, Hughes E: Proteoglycan–collagen arrangements in developing rat tail tendon: an electron microscopical and biochemical investigation. Biochem J 195: 573–581, 1981.
41. Reale E, Ruggeri A, Benazzo F, Marchini M: Proteoglycan–alcian blue complexes on thin sections and on freeze-etching and shadow-casting replicas. In: VII European Symposium on Connective Tissue Research. Prague, 1980, pp 152-153.
42. Benazzo F, Reale E, Ruggeri A, Marchini M, Castelli C: Further observations on the relationships of proteoglycans with collagen fibrils. Basic Appl Histochem 26 (Suppl): 75, 19 .
43. Hay ED: Fine structure of embryonic matrices and their relation to the cell surface in ruthenium red-fixed tissues. Growth 42: 399–423, 1978.
44. Hascall VC, Kimura JH: Proteoglycans: isolation and characterization. Methods Enzymol 82: 769–800, 1982.
45. Borcherding MS, Black LJ, Sitting RA, Bizzel JW, Breen M, Weinstein HG: Proteoglycans and collagen fibre organization in human corneoscleral tissue. Exp Eye Res 21: 59–70, 1975.
46. Carlstedt J, Coster L, Malmström A: Isolation and characterization of dermatan sulphate and heparan sulphate proteoglycans from fibroblast culture. Biochem J 197: 217–225, 1981.
47. Ruggeri A: Ultrastructural histochemical and autoradiographic studies on the developing chick notochord. Z Anat Entwickl Gesch 138: 20–33, 1972.
48. Vasan NS, Lash JW: Heterogeneity of proteoglycans in developing chick limb cartilage. Biochem J 164: 179–183, 1977.
49. Földes I, Modis L, Antalffy J, Ádám I: Ultrastructure of extracellular matrix of embryonic chick limb bud cartilage. Acta Biol Acad Sci Hung 31: 81–95, 1980.
50. Heinegard D, Inerot S, Paulsson M, Sommarin Y, Wieslander J: Proteoglycan variability in health and disease. In: VII Meeting of the Federation of European Connective Tissue Societies. Copenhagen, 1982, p 20.
51. Nemetscheck T: Unterschiede in der intensität der Querstreifen von Kollagen und der Auswascheffekt. Z. Naturforsch 20: 77–79, 1965.
52. Meyers DB: Electron microscopic autoradiography of 35 SO4-labelled material closely associated with collagen fibrils in mammàlian synovium and ear cartilage. Histochem J 8: 191–199, 1976.
53. Lam R, Claffey WJ, Geil PH: Small angle X-ray diffraction studies of mucopolysaccharides in collagen. Biophys J 24: 612–628, 1978.
54. Berthet C, Hulmes DJS, Miller A: Structure of collagen in cartilage of intervertebral disk. Science 199: 547–549, 1978.
55. Meek KM, Elliott GF, Hughes RA, Nave C: The axial electron density in collagen fibrils from human corneal stroma. Current Eye Res 2: 471–477, 1983.
56. Junqueira LCU, Toledo OMS, Montes GS: Correlation of specific sulfated glycosaminoglycans with collagen types I, II and III. Cell Tissue Res 217: 171–175, 1981.
57. Stinson RH, Sweeny PR: Skin collagen has an unusual d-soacing. Biochim Biophys Acta 621: 158–161, 1980.
58. Brodsky B, Eikenberry F: Characterization of fibrous forms of collagen. Methods Enzymol 82: 127–174, 1982.

Authors' addresses:
A. Ruggeri
Università di Bologna
Istituto di Anatomia Umana Normale
Via Irnerio, 48
40126 Bologna, Italy

F. Benazzo
Clinica Ortopedica
Università
27100 Pavia, Italy

CHAPTER 7

The ultrastructural organization of the elastin fibre

IVONNE PASQUALI-RONCHETTI and CLAUDIO FORNIERI

1. Introduction

Elastin is the highly hydrophobic protein that confers elasticity to all the connective tissues; it is therefore widely spread in the body and is particularly abundant in organs and tissues subjected to specific and periodic stress. As to the amount of elastin in mature tissues of different sources see the review by Ayer (1).

Although some recent studies have obtained information on the organization of elastin molecules and their aggregates in lower animals (2–4), the distribution, shape and size of these aggregates are better known in higher vertebrates and, particularly, in mammals.

From the studies so far conducted, it can be concluded that elastin material forms fibres, bundles and lamellae, whose shape and size are extremely heterogenous, mainly depending on the organ and, therefore, on the strength and direction of the forces working on the aggregating material during development. The eventual importance of other connective tissue components in the orientation of the fibre is treated elsewhere in this book (Chapter 2).

As it can be roughly appreciated on histological sections and confirmed by electron microscopy, all the elastin fibres of a given organ and tissue are interconnected, giving rise to a three-dimensional reticulum system, whose organization is peculiar to each tissue and organ (1, 5, 6). In human skin the elastin system is mainly represented in the deep dermis, where compact, long and branched fibres form a loose network, which is connected through smaller elastin bundles to the papillar dermis and to the basement membrane (5); in the tunica propria of seminiferous tubules a fine network of thin fibres has been shown to surround the tubules and to undergo modification in various pathological conditions (6); in elastic vessels concentric layers of elastin fibres, interconnected by fine fibrils running at right angles, have been described (7); in beef ligamentum nuchae

all the elastin fibres are oriented roughly parallel to the axis of the ligament and are connected by rare branchings (8); on the contrary, in the lung a completely disordered interbranched meshwork of elastin fibres has been described.

2. Biophysical properties of mature elastin

Elastin is characterized by a high degree of reversible elasticity, which is not affected by the vigorous chemical and physical methods used during its isolation and purification from the other connective tissue components. On the contrary, the elasticity of this material is greatly modified by dehydration (9).

Several physico-chemical parameters have been measured on both native and purified elastin, with the aim of understanding the molecular basis of the elastic deformation. From a number of these studies it emerges that the mechanical and thermodynamic properties of elastin are consistent with the kinectic theory of rubber elasticity, and therefore the molecular organization of elastin has been compared to that of rubber (10, 11). This hypothesis has been partially confirmed by some studies conducted with other methodological approaches (9, 12, 13); however, the majority of chemical, physical and electron microscopical data seem to suggest that elastin has a completely different molecular organization or, at least, is a more complicated and multiphase system (9, 11, 14–19).

In favour of a random rubber-like molecular organization are the data by Hoeve and Flory, who emphasized the importance of both swelling and temperature on the thermoelastic properties of elastin (20). They found that these properties were typical for a ideal rubber by assuming that elastin would not change volume with temperature. Similar results were obtained also by Dorrington and McCrum from studies on the thermal behaviour of elastin in water (10).

Ruggeri, A and Motta, PM (eds): Ultrastructure of the connective tissue matrix. ISBN-13:978-1-4612-9789-5

Following the rubber theory, elastin would consist of a three-dimensional network of polypeptide chains, that in the relaxed state adopt a random conformation. Upon stretching, the chains would be constrained to extend in the direction of the applied force; this would result in a decrease of their conformational entropy, generating the elastic resorting force. Similarities between elastin and natural polymers were also found by the presence in both cases of interchain crosslinks, connecting, in one case, long polypeptide chains and, in the latter, long isoprene chains. Moreover, the correlation time for the rotation of the backbone carbon atoms in elastin was found to be of the same order as that of natural elastomers (16).

However, by combining physico-chemical and biochemical data, it appeared that elastin system was more complicated than a simple three-dimensional array of homogeneously charged isoprenoid chains. First of all, the polypetide chain in elastin was shown to be about 70% non-polar, and the hydrophobic amino acids not homogeneously distributed along the molecule (14, 21). Therefore, by taking into account the importance of water for the mechanical properties of this protein, it was suggested that hydrophobic regions in the molecule might play a very important role and that hydrophobic interactions, better than simple molecule conformation, could be responsible for the elasticity (9, 22, 23).

Models have been proposed, in which hydrophobic regions of the polypeptide chain were packed into spherical globules, 5 nm in diameter, in order to minimize contact with the surrounding water (22). By stretching, the globules would be deformed with a decrease in entropy of the water-globule system and generation of a retractive force (22).

Other models based on hydrophobic interactions were proposed by Gray and co workers (14), who assumed a helical conformation for the polypeptide chains responsible for the hydrophobic forces accounting for elasticity.

Therefore, also confirming that the deformation mechanism is essentially entropic, other forces, mainly hydrophobic, seem to contribute to the elasticity of elastin, to its temperature-dependent swelling and to the organization of the molecules in a three-dimensional network (9, 23).

In the formulation of the latter models, chemical data derived from fragmentation of purified elastin have also been taken into account. In fact, from purified mature elastin segments have been isolated and characterized that indicate the presence of repetitive polypeptide sequences along the elastin molecule. Some of these sequences are characterized by

proline, valine and glycine, and therefore are highly hydrophobic; some by a high content of alanine and lysine (14, 21, 24, 26). These latter areas have been demonstrated to be the locus of the desmosine intermolecular, and perhaps intramolecular, crosslinks (25, 26) and regions of some order in the molecule (14, 25, 27). Actually, the alanine-rich sequences have lower segmental mobility than the proline–valine-rich regions and have been suggested to have an ordered alfa-helical configuration (14, 25).

A rather pronounced order in elastin fibres has also been described by others on the basis of ultrastructural and biophysical studies. Gotte and co workers (15) observed that purified elastin, when studied by negative staining electron microscopy, consists of rather parallel fibrils, 5 nm thick. They reached the conclusion that elastin molecules are organized into filament ropes, roughly parallel to the long axis of the fibre, and whose diameter changes upon stretching (15). The filament organization of purified and fragmented elastin has been confirmed by many authors on the basis of electron microscopical observations (18). Similar conclusions were reached by Serafini-Fracassini by X-ray low angle scattering on stretched fibres (17) and by Pasquali Ronchetti and co workers by freeze-fracture electron microscopy on natural fibres (19).

Although a fibrillar model is widely compatible with the physico-chemical characteristics of elastin material described above, it is not completely clear what is the contribution of the various treatments, such as stretching, fixation, staining and dehydration, on the ultimate appearance of this material in electron microscopical studies, and, on the other hand, what are the interferences on the various phases of the system of factors, such as solvent, temperature and stretching, as far as the other biochemical and biophysical studies are concerned.

3. Biochemical characterization of elastin

Tropoelastin is the soluble precursor of mature elastin. It is a 72,000 molecular weight protein, which is mainly, but not only (28, 29), synthesized by fibroblasts and smooth muscle cells, either *in vivo* and *in vitro* (30, 31). There are reports indicating that tropoelastin is synthesized as proelastin, with higher molecular weight and richer in hydrophilic amino acids than tropoelastin, and containing histidine, methionine and cystine residues (32, 33). These findings have not been confirmed by others (34). There is also indication that two genetically distinct tropoelastin molecules are synthesized in higher verte-

128

brates (30, 35, 36), and that tropoelastins synthesized in different organs have slightly different amino acid compositions and crosslinking properties (37–39). In any case, in the extracellular space the tropoelastin molecules are linked together, through the lysine-derived crosslinks desmosines, to form an insoluble polymer: the mature elastin (30, 40, 41). Tropoelastin can be isolated and purified from organs of animals kept under cupper deficient diets (42) or treated with chemicals that induce lathyrism (33, 43). This is a single polypeptide chain of about 800 residues, whose amino acid composition is given in Table 1.

The amino acid composition of insoluble elastin, after its isolation and purification from various organs, is reported in Table 2. Apart from small variations, which might depend on the different sources and on the different methodological procedures for isolating and purifying elastin, both tropoelastin and mature elastin exhibit an amino acid composition characterized by having more than 60% neutral amino acids. Proline, valine and alanine content is unusually high; glycin accounts for about one third, as in collagen. However its distribution in elastin is random and not in every third position as in a typical collagen sequence. Tropoelastin has a higher lysine content than mature elastin, whereas in mature elastin lysine-derived crosslinks can be evidenced.

Although the amino acid sequence of the whole molecule is unknown, several polypeptide segments have been characterized. By chemical or enzymatic treatments, fragments from both mature elastin and tropoelastin have been isolated, and these fragments are repeatedly represented in the molecule. Among these, sequences such as Lys-Ala-Ala-Lys and Lys-Ala-Ala-Ala-Lys have been identified and considered the locus where desmosine crosslinks are formed, through a process of oxidative deamination of lysyl residues. This reaction is catalyzed by the enzyme lysyl oxidase in the extra-cellular matrix (26, 44, 45). The pyridine rings so formed from four lysine residues might link covalently from two up to four tropoelastin molecules, giving rise to the three-dimensional network of mature elastin (26, 44). Moreover, as already mentioned, the physico-chemical behaviour of the alanine-rich segments suggests a helical conformation of these regions and, therefore, some order in the elastin molecule (27).

Other polypeptides, isolated from both mature elastin and tropoelastin, are characterized by a high content of glycine, proline and valine (14, 21). Also in this case repetitive sequences in forms of tetra-, penta- and hexapeptides have been recognized in the whole molecule (21). These segments are highly hydrophobic and are considered to be responsible for the hydrophobic forces involved in the mechanism of elasticity (9, 11). Some authors have suggested that they play an important role in the molecular and supermolecular organization of the whole molecule (46, 47).

Table 1. Tropoelastin content expressed as residues/1,000 (References in parentheses)

| | Chick | | | Pig | | |
	Aorta (87) *	(32) *	Lung (93)°	Aorta (39) *	(89)°	Ear (39) *
Lysine	40	39	43	38	39	30
Histidine	–	–	2	2	1	2
Arginine	3	8	8	7	4	7
Hydroxyproline	12	10	12	10	4	20
Aspartic acid	9	4	8	10	3	6
Threonine	13	11	13	16	9	15
Serine	16	7	12	15	8	13
Glutamic acid	15	14	18	18	11	21
Proline	132	124	120	107	116	101
Glycine	315	338	313	320	326	327
Alanine	169	178	180	230	274	222
Valine	182	175	163	122	120	123
Isoleucine	19	17	21	15	11	18
Leucine	49	54	58	49	38	45
Tyrosine	11	12	15	16	10	13
Phenylalanine	18	20	14	29	21	29

* Animals kept on a Beta-aminopropionitrile-containing diet.
° Animals kept on a Cu-deficient diet.

Table 2. Elastin content expressed as residues/1,000 (References in parentheses)

	Chick embryo			Human			Calf		
	Aorta (43)	Skin (43)	Lung (43)	Aorta (88)	Skin (92)	Lung (90)	Aorta (91)	Lig.Nu. (38)	Ear (38)
Lysine	3.9	7.5	6.1	6.0	5.7	10.5	11.0	3.3	7.3
Histidine	0.2	1.0	0.5	–	0.7	1.9	1.5	0.5	0.7
Arginine	4.6	5.0	3.9	6.5	7.1	9.3	15.5	5.8	14.3
Hydroxyproline	11.1	10.5	13.3	11.0	8.6	12.2	35.7	8.1	11.7
Aspartic acid	2.3	4.5	4.0	6.5	5.5	9.5	15.5	5.8	21.4
Threonine	5.6	5.1	4.4	8.5	10.3	11.1	11.3	9.3	12.9
Serine	4.4	6.0	4.0	5.7	9.4	9.8	16.1	8.7	14.8
Glutamic acid	12.5	18.1	16.7	21.5	20.1	23.1	27.6	15.4	35.7
Proline	139.1	136.5	131.3	117.0	119.5	117.2	116.8	115.2	113.7
Glycine	367.2	351.2	361.2	308.0	306.9	312.2	327.9	328.1	307.3
Alanine	177.5	175.2	180.5	227.0	247.8	237.0	206.7	227.0	204.3
Valine	162.4	165.5	164.5	148.0	122.4	118.6	113.4	131.6	105.0
Methionine	tr	tr	tr	tr	0.8	–	–	–	–
Isoleucine	19.8	20.0	19.4	24.6	24.0	23.7	21.5	23.9	19.9
Leucine	54.1	55.6	53.6	57.1	59.8	59.1	46.8	59.4	62.2
Tyrosine	12.9	14.7	13.3	21.8	23.3	21.2	7.1	5.9	16.0
Phenylalanine	20.0	20.9	20.3	23.9	23.6	22.6	24.0	29.3	29.3
Isodesmosine	1.2	1.3	1.3	1.1	1.3	1.36	– 5.4*	4.9*	
Desmosine	1.2	1.4	1.7	1.9	2.1		–	10.1*	10.4*

* Expressed as lysine equivalent.

4. Structure of mature elastin

4.1. Optical and electron microscopical stains

As already mentioned, by light microscopy the overall organization of the elastin system can be visualized in the various organs. This can be achieved by virtue of the affinity of elastin for stains which do not react, or react very poorly, with the other connective tissue components. One of these stains, used since the beginning of this century (48), is based on the formation of a complex between elastin and ferric hematoxylin in the presence of iodine. Some improvements have been introduced by Lillie et al. (49) and more recently by Churukian and Schenk (50), who adopted iron gallein and obtained a more controlled reaction. Resorcin-fucsin has been also widely used to visualize elastin material; however, it has been recently shown that type III collagen is stained as well and that the stain particularly binds to the surface of elastin fibres, where a microfibrillar component distinct from elastin is highly represented (51, 52).

Böch described a method for obtaining a rather specific staining of elastin based on permanganate oxidation and mild methylation of the tissue respectively before and after aldehyde-fucsin staining (52). Some of these optical stains were used in electron microscopy studies. Elastin does not bind the usual electron microscopical stains, such as uranyl acetate and lead citrate, and therefore remains completely electron lucent (Figure 1). However, in spite of the poor electron density, the elastin fibres are easily recognized in conventionally stained thin sections, when their diameter is wider than 100 nm. On the contrary, with this staining procedure very thin fibres cannot be seen and the internal organization of the fibre can hardly be studied. Verhoeff's iron hematoxylin followed by lead citrate has been shown to improve the contrast of the elastin fibres (53). The use of tetraphenylporphine sulfonate complexed with gold and silver has been described to give specific and strong electron density to elastin (54). Recently, tannic acid, either added to the fixation media (55) or used on sections (56), has been shown to greatly improve the contrast of elastin, which appears as a homogeneous electron dense material (Figure 2).

The introduction of this simple staining procedure greatly contributed to the identification in several organs of small elastin fibres, down to 35–50 nm in diameter, which are barely detectable in routinely stained thin sections. More recently, aqueous solutions of palladium chloride followed by lead citrate were found to give good contrast to elastin and attempts were made to understand the mechanism of

130

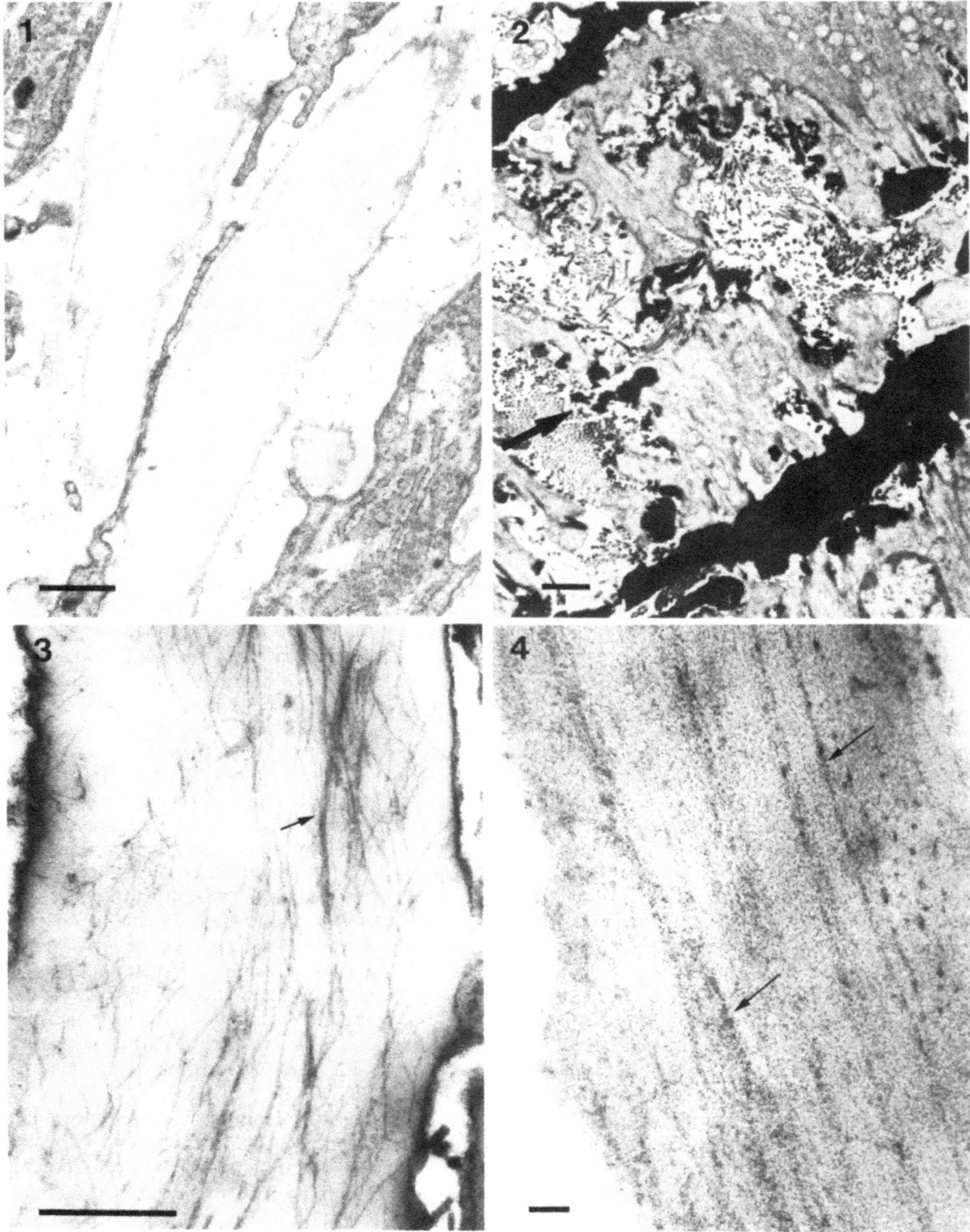

binding. It was shown that chemically purified elastin strongly bound palladium chloride and that the reaction was not affected by glutaraldehyde, whereas OsO$_4$ fixation reduced the amount of palladium bound to elastin; however, it was also found that palladium binding increased the susceptibility of elastin to elastase digestion, so indicating that during staining 'a significant alteration of elastin has occurred' (57).

By combining conventional with specific elastin staining both at light and electron optical levels, the different types of elastin fibres were described and defined as oxytalan, elaunin and true elastin. These could be recognized as consecutive steps in normal elastogenesis (12, 58, 59) or, as in skin, as different quantitative distribution of elastin and associated microfibrils in different regions of the same organ (5).

4.2. Electron microscopical studies

Since the application of electron microscopy to the study of biological material, elastin has been studied either *in situ* or after isolation with the aim of understanding the molecular basis of its mechanical properties. However, due to its low reactivity and to the difficulty of obtaining purified small fragments, understanding of the organization of elastin did not proceed as well as for other biological structures.

In the early 1940s it was shown that elastin fibres do not exhibit specific structural features, such as collagen. It also became clear that all elastin tissues have an essentially similar organization (60, 61). One of the first observations was that fragmented aortic elastica could swell by acid treatment, revealing, by electron microscopy, a fine fibrillar organization towards molecular dimension (1). However, a better understanding of the structure of the fibre was reached by selective and controlled digestion of elastin material. By metal shadowing with chromium, isolated elastin fibres were shown to split, upon digestion with elastase, into finer elements, lying roughly parallel to one another, and it was suggested that, independently from the tissue, the elastin material consisted of 20-nm wide fibrils, bound together by a cement substance, that could be dissolved by elastase (61). By prolonged elastase digestion, these fine fibrils were also observed to be reduced to granular material and it was concluded that elastin 'must be dual in nature consisting of thin fibres, in close association with a dense non-fibrous component', which 'seems to function as cement material' (61).

These results, obtained by Hall and co-workers on fragmented elastin fibres, were largely confirmed on fixed and sectioned material. Actually, fixed, plastic embedded and thin-sectioned elastin fibres revealed the coexistence of an amorphous electron trasparent component filling the spaces among electron dense strands, oriented in the direction of the fibre (62). In 1952 Dempsey could observe in thin sections that had been deplastified and shadowed with chromium, without previous chemical extraction of the amorphous component, the presence within the elastin fibre of a three-dimensional network of 20-nm thick fibrils (63). This led to the hypothesis of the existence of elastin material within the fibre in a dense and a rarefied form (60).

Subsequent combined electron microscopical and biochemical studies confirmed the coexistence of two components within the elastin fibre, and definitely clarified their different chemical nature. The fine quasi-parallel fibrils more resistant to elastase and capable of being positively stained with uranyl and lead salts, were identified as a glycoprotein material associated with, but distinct from, elastin, whereas true elastin was recognized as the amorphous cementing material. These conclusions were

Figure 1. Thin section of aorta of a seven-day-old chick. The material was fixed in glutaraldehyde and osmium tetroxide, embedded in epoxy resins and stained with uranyl acetate and lead citrate. The elastin fibres appear amorphous and electron transparent. Inside the elastin matrix, electron dense strands of residual microfibrils are seen. Bar = 1 μm.
Figure 2. Thin section of tonaca media of abdominal rabbit aorta. The specimen was fixed and embedded as in Figure 1; the sections were stained with tannic acid and lead citrate. Elastin is homogeneously electron dense. With this procedure small elastin fibres can be identified among smooth muscle cells (arrows). Bar = 1 μm.
Figure 3. Ultracryosection of elastin fibre of bovine ligamentum nuchae. The specimen was quickly frozen in liquid nitrogen, and cryosectioned at −80°C. The sections were stained with uranyl acetate and lead citrate. Elastin material is electron lucent, whereas microfibrillar residues inside the fibre are stained and reveal their non-homogeneous and roughly reticular organization, with prevalent orientation in the direction of the fibre (arrows). Bar = 1 μm.
Figure 4. Ultracryosection of an elastin fibre from human aorta chemically treated in order to purify elastin from all the other constituents of the vessel. The purified elastin fibres were rehydrated, frozen in liquid nitrogen and cryosectioned at −80°C. The sections were stained with uranyl acetate and lead citrate. Microfibrillar elements still persist inside the fibre (arrow). Purified elastin was provided by L. Gotte. Bar = 100 nm.

132

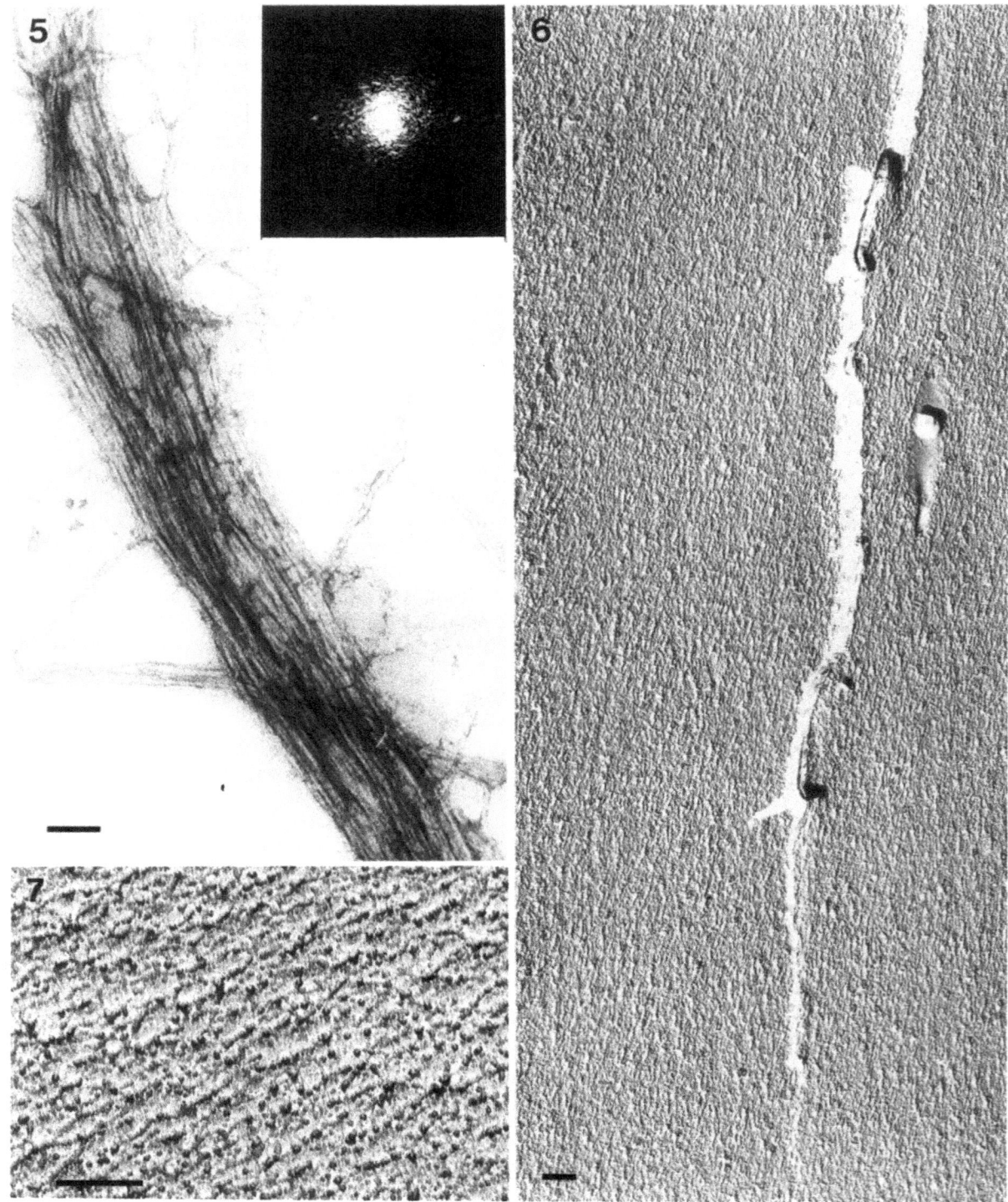

mainly reached by the observation that the microfibrillar and amorphous components exhibit different staining properties on thin sections (12, 62) and that the relative amount changes during fibrogenesis (12, 64, 65). Furthermore the two components reveal different enzymatic sensitivity (66, 67) and, when at least partially purified, different chemical composition (68–69, 70).

These conclusions, reached more than ten years ago, are still valid. By transmission electron microscopy on thin sections, elastin material, independently of the tissue, always exhibits the same organization, as shown in Figure 1. It consists of two structurally distinct components: microfibrillar and amorphous. The microfibrillar component heavily binds uranyl acetate and lead citrate, becoming electron dense, whereas the amorphous component remains completely electron transparent. The microfibrillar component has been shown to be more abundant in embryonic tissues (12, 64–66), to appear first in tissue culture during fibrogenesis (65) and to become gradually less represented in the fibre as this grows by progressive increase of the amorphous component. In the mature fibre the microfibrillar component is therefore scarce and is recognized on thin sections either as long microfibrils, 12 nm wide, roughly parallel to the long axis of the fibre and located at the periphery of the elastin fibre, or as electron dense ill-defined strands inside the fibre (Figure 3). As already mentioned, the microfibrillar component has not been fully purified and characterized; however, there is good evidence that it is a glycoprotein (68–70). Some recent reports suggest it has some affinity with collagen (70). On the contrary, the amorphous component has been identified as the highly hydrophobic protein elastin (68). The relationships between the two components during fibrogenesis are extensively discussed elsewhere in this book by Serafini-Fracassini (see Chapter 8).

At present it is completely unknown whether the glycoprotein component plays some functional and/or structural role in the fibre. It is perhaps worthwhile to point out that in some organs, such as in human aorta, it is particularly abundant and is tightly bound to elastin. This appears from the presence of uranyl and lead positive strands of material inside the fibre after exhaustive purification of elastin with guanidinium, dithioerythritol and collagenase (Figure 4).

The organization of the elastin fibre was also studied by scanning electron microscopy; however, the degree of resolution reached was in the majority of cases very low (7, 71, 72). Evidence for a filament nature in the 120 nm size range has been provided on critical-point dried purified fetal elastin fibres (73). It was suggested that each mature elastin fibre is formed by a number of thin elastin fibrils, bound together by a microfibrillar protein, which fills in the spaces between fibrils and which can be removed by guanidine-mercaptoethanol-collagenase treatment in immature fibre, whereas it remains trapped by elastin fibrils in mature elastin (73).

Evidence for a discrete submicroscopic organization of elastin was also obtained with other electron microscopical approaches. By negative staining, chemically purified elastin fibres fragmented by sonication and/or by mechanical treatments consisted of sets of well-oriented filaments (15, 17, 18) (Figure 5). The mean diameter of the filaments, revealed by optical diffraction, was around 5 nm. Smaller filaments were also observed by some authors (18) and suggested to represent the elemental organization of the elastin molecules or to be due to some stretching, and therefore thinning, of wider elements. In some reports a regular periodicity of about 4 nm along the filament axis has been described and taken as evidence for a helical organization of the filaments (15). In all cases, contamination by glycoproteins or other constituents was rouled out by amino acid analysis, and gross molecular damage was excluded as unsonicated samples exhibited identical filament structure (15).

A fibrillar substructure for elastin was also supported by low angle X-ray scattering data. On stretched purified elastin, equatorial reflections, corresponding to 5-nm spacings, were identified and

Figure 5. Purified and fragmented elastin from beef ligamentum nuchae. The specimen was negatively stained on the grid with 1% uranyl acetate and 20 mM oxalic acid pH 6.8. The elastin appears as bundles of filaments, mainly parallel, with a diameter around 5 nm. Purified elastin was provided by D. Volpin. Bar = 100 nm.

Figure 6. Freeze-fracture image of elastin fibres from beef ligamentum nuchae. The elastin bundles were stretched up to 100%, chemically fixed in glutaraldehyde and osmium tetroxide and quickly frozen in liquid nitrogen. The fracture plane runs parallel to the axis of the fibres and reveals a network of longitudinal filaments. Bar = 100 nm.

Figure 7. Freeze-fracture surface of an elastin fibre stretched up to 100% before freezing. On the fracture surface a platinum/carbon film was evaporated during rotation from an angle of 32°. The fracture plane runs parallel to the fibre axis and reveals the filament network. Each filament appears made up of globular subunits about 6 nm in diameter. Bar = 100 nm.

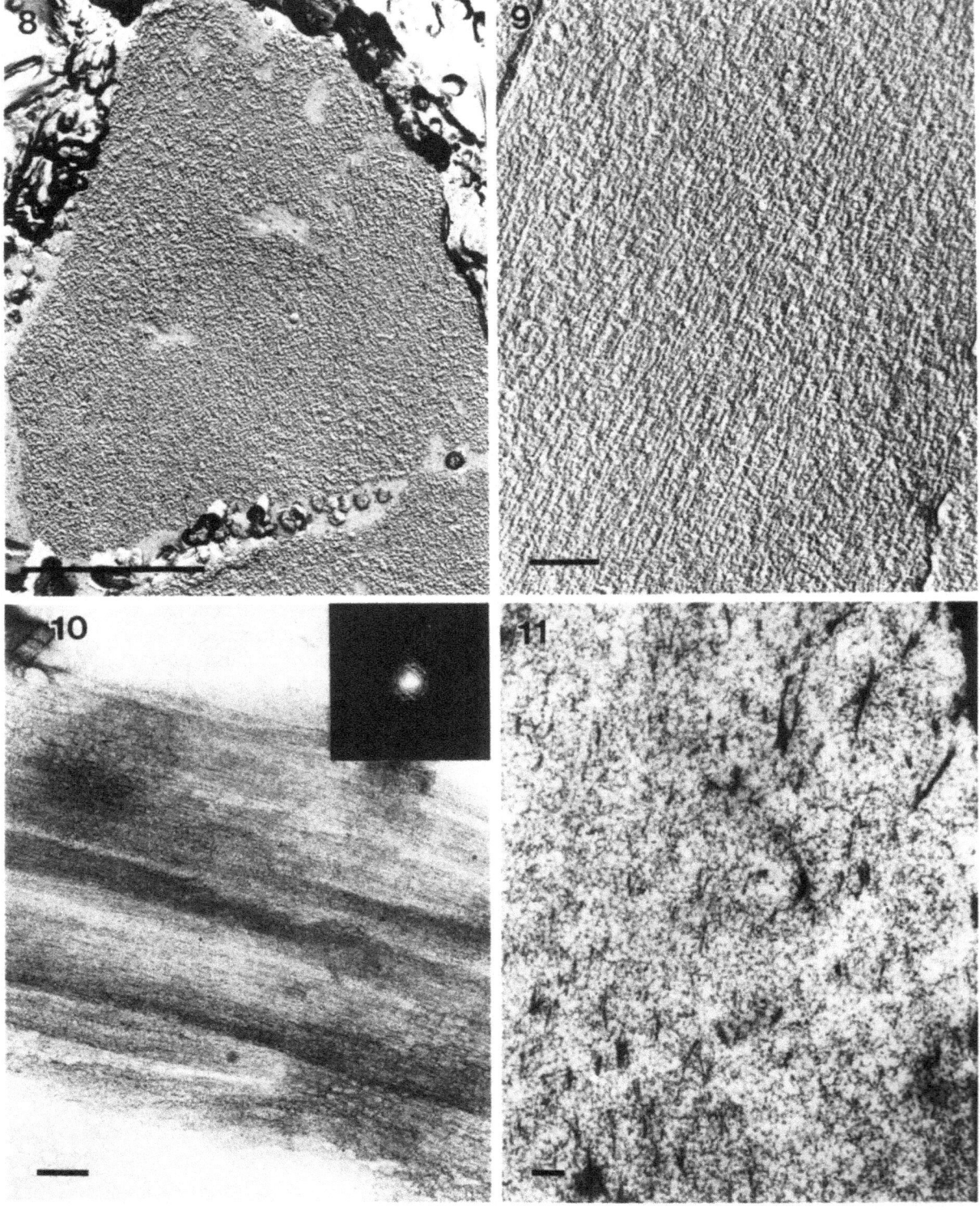

interpreted as primary elastin filaments (17). It was suggested that in relaxed fibre the filaments are randomly arranged and that by stretching they are oriented in the direction of the applied force.

In spite of such good evidences for a filament organization of the elastin molecules, it is clear that artefacts cannot be ruled out in negatively stained specimens. Dehydration, stretching phenomena, interferences with the negative stains or even reaggregation of molecular fragments could greatly interfere with the final appearance of a protein material, which is characterized by high hydrophobicity and peculiar deformation properties.

Some of these possible interferences have been recently overcome by using cryotechniques in electron microscopy. By freeze-fracture native ox ligamentum nuchae was shown to consist of elastin fibres exhibiting a disordered fine granular structure when frozen in relaxed state. However, by stretching, the fibres appeared to be made of discrete filaments forming a three-dimensional network (Figure 6). Upon stretching up to twice their length, the fibres were solved into sets of quasi-parallel filaments (Figure 9). The mean diameter of the filaments was around 6–7 nm. The filaments exhibited a beaded structure with a periodicity of about 6 nm (Figure 7). These results were not affected by previous chemical fixation nor by glycerol antifreeze treatment. The specimens were rapidly frozen in the hydrated state and relevant artefacts could only derive from plastic deformation during fracturing (19). As shown in Figures 6, 7, 8 and 9, fibrillar elements form a three-dimensional network, whose meshes seem to be filled by water or by some completely amorphous material. Actually, two possibilities have to be considered: (a) the fibrillar elements represent supermolecular arrangement of elastin molecules and the meshes are filled by water or the fibrils are glycoprotein material and the meshes are

filled by 'amorphous elastin'. The first hypothesis seems the most realistic, considering the high amount of fibrillar component revealed by this technique and the scarce amount of glycoprotein component within the same elastin fibre shown on conventional and on frozen thin sections (compare Figures 1, 3 and 11 with Figures 6 to 9). In favour of the first hypothesis are also physical and chemical data that elastin is a highly hydrated material, containing more than 60% water (16), and that water is present in a multiphase state (9).

The fibrillar organization of elastin revealed by freeze-fracture is very similar to that observed on negatively stainend purified and fragmented elastin (17–19) (Figure 5). In the latter case, flattening and probable deformation of the fibre, as well as lateral aggregation and stretching of the filaments due to dehydration, could have induced collapse of the three-dimensional array, with visualization of sets of quasi-parallel filaments. The discrepancies in the thickness of the filaments between negative stained and freeze-fractured samples (about 5 and 7 nm respectively) could also be due to dehydration and/or to superposition phenomena in negatively stained specimens.

The filament organization of native elastin was also confirmed on cryo-ultrathin sections of beef ligamentum nuchae (74). Thin sections of untreated frozen ligament bundles showed that elastin fibres had a compact homogeneous structure like that described in conventionally fixed, embedded and sectioned material. They exhibited also the same staining properties as in thin conventional section: uranyl acetate and lead citrate stained the glycoprotein strands inside the fibre, whereas tannic acid markedly increased the electron density of the whole fibre to that in Figure 2. However, cryoultrasections of elastin bundles, either stretched and chemically fixed or swollen in glycerol before freezing, revealed

Figure 8. Elastin fibre from beef ligamentum nuchae, treated as in Figure 6, but fractured across the fibre. In cross fracture the elastin filaments do not form any regular arrangement. Bar = 1 µm.

Figure 9. Freeze-fractured elastin fibre from bovine ligamentum nuchae. Fracture parallel to the axis of the fibre. The elastin bundle was stretched up to 180%, chemically fixed with glutaraldehyde and tannic acid before freezing. The fracture plane was shadowed with platinum/carbon from 42°. Filaments arranged in the direction of the force applied can be very well appreciated. Bar = 100 nm.

Figure 10. Ultracryosection of an elastin fibre from beef ligamentum nuchae. The bundle was stretched up to 150% of its original length, fixed with glutaraldehyde and osmium tetroxide and quickly frozen in liquid nitrogen. Cryosections were performed at −80° C and then stained with 1% uranyl acetate and 20 mM oxalic acid pH 6.8. The fibre consists of filaments aligned in the direction of the fibre. Optical diffraction analysis revealed a mean filament thickness of 4.8 nm. Bar = 100 nm.

Figure 11. Elastin fibre from beef ligamentum nuchae, fixed in glutaraldehyde for three days and then in 1% osmium tetroxide for two days, dehydrated and embedded in durcupan. The sections were stained with tannic acid and lead citrate. Elastin appears as a network formed by 5–7-nm thick electron opaque segments. The microfibrillar residues are still evident inside the fibre. It has to be noted that the fibres were about twice in diameter as compared to those processed by conventional fixation schedule; without excluding some chemical extraction, their structure seemed to be less affected by subsequent dehydration. Bar = 100 nm.

by negative staining a filament organization of the fibre practically identical to that seen on purified and fragmented elastin using the same staining technique (19, 74) (Figure 10).

Therefore, cryotechniques in electron microscopy seem, at present, to confirm that elastin molecules are organized in supermolecular filament structures, which form a three-dimensional network and exhibit some orientation in the direction of the fibre (19, 74).

Such a fibrillar organization was also suggested by some recent results obtained in our laboratory on conventional embedded and sectioned samples. As already observed by Cliff in 1971, osmium tetroxide is able to solubilize elastin after prolonged exposition. However, by well controlled treatments with osmium tetroxide, it was possible to 'fix' the elastin structure inside the fibre without inducing tremendous damage. As shown in Figure 11 the overall architecture of the fibre is maintained, as well as the glycoprotein strands inside the fibre. Elastin material appears in the form of a fine fibrillar network. Osmium tetroxide has been shown to induce breaking of polypeptide chains (75–77) and to be particularly active on pyridine-containing molecules (77). However its action is progressive with time, and a condition can be achieved in which extraction is not so dramatic. We found that this very precise moment depends on several factors, such as osmium concentration, temperature and specimen thickness, and that it can be rougly judged by the loss of elasticity of the elastin bundle. Following these criteria, images like that in Figure 11 can be easily found. The similarities between this image and those by freeze-fracture untreated material are rather impressive (compare Figures 6, 7 and 11).

In favour of a discrete filament assembly of the elastin molecules, there are also a number of studies on the aggregation properties and structure of the soluble precursor of elastin, tropoelastin, and of chemical fragments of mature elastin. It has been observed that alfa-elastin, a water soluble polydisperse system obtained by oxalic acid treatment from mature elastin, could form reversible precipitates by heating. These precipitates, called coacervates, were shown to exhibit physical properties similar to those of original elastin (78). Similar coacervates were later obtained from purified tropoelastin (46, 79). By negative staining electron microscopy, both types of coacervates were shown to consist of bundles of about 5-nm thick filaments, identical to those of native elastin observed with the same technique (17, 18, 46, 79, 81). It was argued that intramolecular alternating hydrophobic and hydro-

philic regions in the molecule could lead to the filament assembly *in vitro* and that intermolecular hydrophobic interactions could be responsible for the aggregation and alignment of the filaments in the coacervate (17, 18, 46).

The importance of the hydrophobic interactions in the coacervation process was demonstrated by Urry and co-workers (82–85) who studied the physicochemical and structural properties of synthetic polypeptides with the same repeating hydrophobic sequences found in natural elastin (14, 21). They found strict relationships between coacervates of natural products and of synthetic hydrophobic polypeptides from both ultrastructural (80–83) and physico-chemical points of view (47, 83–85).

This was taken as evidence of the importance of the hydrophobic sequences for the conformation of the tropoelastin molecules and for their aligning during fibrogenesis. It was in fact suggested that a process of coacervation could be a key step in elastogenesis, by favouring molecular approaching and alignment of the protein chains prior to crosslinking (46, 47, 83, 85).

These observations together with the fact that during coacervation of both natural and synthetic polypeptide chains an increasing order was measured by circular dicroism (84), were taken as support of the hypothesis that elastin fibre is not a randomly arranged system of proteins, with random crosslinks, but that some intramolecular as well as intermolecular order has to be admitted in the natural polymer (47, 83, 84), as also suggested by negative staining and freeze-fracture electron microscopy on native elastin. This hypothesis is also confirmed by recent chemical studies proposing two crosslinks, desmosine and lysinonorleucine, separated by 35 amino acid residues, in the natural molecule and therefore implying some asymmetry at molecular level (86); and by the finding that in avian tropoelastin the polypeptide sequence between residues 17 and 54 is similar to that found in collagen (87).

5. Conclusions and perspectives

Scanning and transmission electron microscopy studies played a fundamental role in clarifying the disposition of the overall elastin system in the different tissues. Furthermore, combined biochemical and ultrastructural studies have stated that the elastic fibre is a dual system of glycoprotein and elastin protein and that rather precise relationships between these two components seem to regulate fibrogenesis.

Electron microscopy has also made an important contribution to the study of the organization of elastin molecules within the fibre. However, at present, different electron microscopical approaches give different data on the molecular organization of the fibre. The results, in fact, are contradictory and apparently difficult to reconcile with those obtained with other physico-chemical approaches. In thin sections of resin embedded fibres, elastin molecules appear to be arranged in an amorphous isotropic three-dimensional array; on replicas of frozen hydrated specimens the elastin molecules seem to form supermolecular filament structures that give rise to a highly anisotropic system by stretching; by negative staining, elastin can be visualized as a set of 5-nm thick filaments roughly parallel to the long axis of the fibre. To a certain extent these different data can be reconciled by considering the eventual interferences of the various methodological approaches on a protein system with such peculiar hydrophobic and elastic properties (9, 11, 16, 23). However, efforts have to be made to improve specimen preparation and image analysis in order to fully understand elastin organization in fresh hydrated fibres.

From mechanochemical and other biophysical studies, elastin appears as a highly disordered and random polymer. There is agreement that within the molecule the crosslinking regions present an ordered alfa-helix or extended helix conformation (14, 27); there is also good evidence that some hydrophobic portions of the molecule exhibit a beta-turn and beta-band organization (85). However, even if the elastin molecule consists of segments with different mobilities (15), the overall elastin network seems to be formed by highly mobile and disordered chains (13, 25). These conclusions seem to be in contrast with the electron microscopical data that suggest an ordered filament organization for the elastin molecules; however, as already observed by several authors, the beaded filaments observed by electron microscopy in any case represent supermolecular assemblies of molecules, wherein the single polypeptide chain, or portions of it, may well display a random conformation.

We believe that electron microscopy can be of great help in clarifying elastin molecular organization as well as many other problems, such as elastin secretion from cells, fibrogenesis and pathological alterations. For all these problems, however, appropriate methodological approaches have to be studied. As far as the organization of the elastin molecules within the fibre is concerned, it is essential to improve specimen preparation for high resolution electron microscopy, to preserve hydration and possibly to avoid chemicals, such as fixatives and stains. This could be achieved in the future by the observation of frozen-hydrated specimens, by low-dose techniques and by STEM.

References

1. Ayer JP: Elastic tissue. Intern Rev Connect Tissue Res 2: 33–100, 1980.
2. Spina M, Garbisa S, Field JM, Serafini Fracassini A: The salmonid elastic fibril. An investigation of some chemical and physical parameters. Arch. Biochem Biophys 192: 430–437, 1979.
3. Sage H, Gray RW: Studies on the evolution of elastin. I.Phylogenetic distribution. Comp Biochem Physiol [B] 64: 313–327, 1979.
4. Sage H, Gray RW: Studies on the evolution of elastin. III. The ancestral protein. Comp Biochem Physiol [B] 68: 473–480, 1981.
5. Cotta Pereira G, Guerra Rodrigo F, David Ferreira JF: Comparative study between the elastic system fibers in human thin and thick skin. Biol Cell 31: 297–302, 1978.
6. Wiedmer BJ, Vogel A, Hedinger C: Elastic fibers in the tunica propria of the seminiferous tubules. Virch Arch [Cell Pathol] 27: 267–277, 1978.
7. Smith P: A comparison of the orientation of the elastin fibers in the elastic laminae of the pulmunary trunk and aorta rabbits using the scanning electron microscope. Lab Invest 35: 525–529, 1976.
8. Kewly MA, Williams G, Steven FS: Studies of elastic tissue formation in the developing bovine ligamentum nuchae. J Pathol 124: 95–101, 1977.
9. Gosline JM: The temperature dependent swelling of elastin. Biopolymers 17: 697–707, 1978.
10. Dorrington K, McCrum NG: Elastin as a rubber. Biopolymers 16: 1201–1222, 1977.
11. Gosline JM, French J:Dynamic mechanical properties of elastin. Biopolymers 18: 2091–2103, 1979.
12. Greenle TK, Ross R, Hartman JL: The fine structure of elastin fibers J Cell Biol 30: 59–71, 1966.
13. Torchia DA, Piez KA: Mobility of elastin chains as determined by ^{13}C-nuclear magnetic resonance. J Mol Biol 76: 419–424, 1973.
14. Gary RW, Sandberg LB, Foster JA: Molecular model for elastin structure and function. Nature (Lond) 246: 461–466, 1973.
15. Gotte L, Giro GM, Volpin D, Horne RW: The ultrastructural organization of elastin. J. Ultrastruct Res 46: 23–33, 1974.
16. Ellis GE, Packer KJ: Nuclear spin relaxation studies of hydrated elastin. Biopolymers 15: 813–832, 1976.
17. Serafini-Fracassini A, Field JM, Spina M, Stephens WGS, Delf B: The molecular organization of the elastin fibrils. J Mol Biol 100: 73–84, 1976.
18. Cleary EG, Cliff WJ: The substructure of elastin. Exp Mol Pathol 28: 227–246, 1978.
19. Pasquali Ronchetti I, Fornieri C, Baccarani Contri M, Volpin D: The ultrastructure of elastin revealed by freeze-fracture electron microscopy. Micron 10: 89–99, 1979.
20. Hoeve CAJ, Flory PJ: The elastic properties of elastin. J Am Chem Soc 80: 6523–6526, 1958.
21. Foster JA, Bruenger E, Gary WR, Sandberg LB, Isolation and amino acid sequences of tropoelastin peptides. J Biol Chem 248: 2876–2879, 1973.
22. Weis-Fogh T, Andersen SO: New molecular model for the long-range elasticity of elastin. Nature (Lond) 227: 718–721, 1970.
23. Gosline JM: Hydrophobic interaction and a model for the elasticity of elastin. Biopolymers 17: 677–695, 1978.
24. Partridge SM, Elsden DF, Thomas J: Constitution of the cross-linkages in elastin. Nature (Lond) 197: 1297–1298, 1963.
25. Lyerla JR, Torchia DA: Molecular mobility and structure of elastin deduced from solvent and temperature dependence of ^{13}C magnetic resonance relaxation data. Biochemistry (USA) 14: 5175–5183, 1975.
26. Baig KM, Vlaovic M, Anwar RA: Amino acid sequences C-terminal to

the cross-links in bovine elastin. Biochem J 185: 611–616, 1980.

27. Foster JA, Bruenger E, Rubin L, Imberman M, Kagan H, Mechan RP, Franzblau C: Circular dicroism studies of an elastin crosslinked peptide. Biopolymers 15: 833–841, 1976.

28. Haeger P, Rosenbloom J: Biosynthesis of tropoelastin by elastic cartilage. Connect Tissue Res 8: 21–26, 1980.

29. Cantor JO, Keller S, Parshley MS, Darnule TV, Darnule AT, Carreta JM, Turino JM, Mandl I: Biosynthesis of crosslinked elastin by an endothelial cell culture. Biochem Biophys Res Commun 95: 1381–1386, 1980.

30. Narayanan AS, Sandberg LB, Ross R, Layman DL: The smooth muscle cell. III. Elastin synthesis in arterial smooth muscle cell culture. J Cell Biol 68: 411–419, 1976.

31. Snider R, Faris B, Verbitzki V, Moscaritolo R, Salcedo LL, Franzblau C: Elastin biosynthesis and cross-link formation in rabbit aortic muscle cell. Biochemistry (USA) 20: 2614–2617, 1981.

32. Foster JA, Mecham RP, Franzblau C: A high molecular weight species of soluble elastin. Biochem Biophys Res Commun 72: 1399–1406, 1976.

33. Foster JA, Mecham RP, Rich CB, Cronin MF, Levine A, Imberman M, Salcedo LL, Proelastin: synthesis in cultured smooth muscle cells. J Biol Chem 253: 2797–2803, 1978.

34. Bressan GM, Prockop DJ: Synthesis of elastin in aortas from chick embryos. Conversion of newly secreted elastin to cross-linked elastin without apparent proteolysis of the molecule. Biochemistry (USA) 16: 1406–1412, 1977.

35. Kucich U, Christner P, Rosenbloom J, Weinbaum G: An analysis of the organ and species immunospecificity of elastin. Connect Tissue Res 8: 121–126, 1981.

36. Barrineau LL, Rich CB, Przybyla A, Foster JA: Differential expression of aortic and lung elastin genes during chick embryogenesis. Dev Biol 87: 46–51, 1981.

37. Field JM, Rodger GW, Hunter JC, Serafini-Fracassini A, Spina M: Isolation of elastin from bovine auricular cartilage. Arch Biochem Biophys 191: 705–713, 1978.

38. Keith DA, Paz MA, Gallop PM: Differences in Valyl-Proline sequence content in elastins from various bovine tissues. Biochem Biophys Res Commun 87: 1214–1217, 1979.

39. Foster JA, Rich CB, DeSa MD: Comparison of aortic and ear cartilage tropoelastins isolated from lathyritic pigs. Biochim Biophys Acta 626: 383–389, 1980.

40. Francis G, John R, Thomas J: Biosynthetic pathway of desmosines in elastin. Biochem J 136: 45–55, 1973.

41. Miyoshi M, Kanamori M, Rosenbloom J: Synthesis of elastin. A rapid formation of lysine-derived crosslinks by chick embryo aorta. J Biochem 79: 1235–1243, 1976.

42. Buckingham K, Heng-Khoo CS, Dubick M, Lefevre M, Cross C, Julian L, Rucker R: Copper deficiency and elastin metabolism in avian lung. Proc Soc Exp Biol Med 166: 310–319, 1981.

43. Pasquali Ronchetti I, Fornieri C, Castellani I, Bressan GM, Volpin D: Alterations of the connective tissue components induced by beta-aminopropionitrile. Exp Mol Pathol 35: 42–56, 1981.

44. Foster JA, Rubin L, Kagan HM, Franzblau C, Bruenger E, Sandberg LB: Isolation and characterization of cross-linked peptides from elastin. J Biol Chem 249: 6191–6196, 1974.

45. Jurikova M, Franzblau C, Faris R, Deyl Z, Adam M: Some lysine-containing peptides from the elastase digest of elastin and their relation to lysinonorleucine cross-link. Biochim Biophys Acta 386: 239–243, 1975.

46. Cox BA, Starcher BC, Urry DW: Coacervation of tropoelastin results in fiber formation. J Biol Chem 249: 997–998, 1974.

47. Rapaka RS, Okamoto K, Urry DW: Coacervation properties in sequential polypeptide models of elastin. Synthesis of H(Ala-Pro-Gly-Gly)n Val O Me and H(Ala-Pro-Gly-Val-Gly)n Val O Me. Intern J Pept Prot Res 12: 81–92, 1978.

48. Verhoeff FH: Some new staining methods of wide applicability, including a rapid differential stain for elastic tissue. J Am Med Assoc 50: 876–877, 1908.

49. Lillie RD, Pizzolato P, Donaldson PT: Hematoxylin substitutes: gallein as a biological stain. Stain Technol 49: 339–346, 1974.

50. Churukian CJ, Schenk EA: Iron gallein elastic method. A substitute for Verhoeff's elastic tissue stain. Stain Technol 51: 213–217, 1976.

51. Puchtler H, Meloan SN, Pollard GR: Light microscopic distinction between elastin pseudo-elastica (Type III collagen) and interstitial collagen. Histochemistry 49: 1–14, 1976.

52. Böck P: Staining of elastin and pseudoelastica (elastic fiber microfibrils, type III and type IV collagen) with paraldehyde fucsin. Mikroskopie 33: 332–341, 1977.

53. Brissie RM, Spicer SS, Hall BS, Thompson NT: Ultrastructural staining of thin sections with iron hematoxylin. J Histochem Cytochem 22: 895–907, 1974.

54. Albert EN, Fleischer E: A new electron dense stain for elastic tissue. J Histochem Cytochem 18: 697–708, 1970.

55. Mizuhira V, Nakamura H, Fujioka T: New staining method for the elastic fibers using tannic acid-glutaraldehyde mixture. J Electron Microsc (Tokyo) 21: 240–245, 1972.

56. Kajikawa K, Yamaguchi T, Katsuda S, Miva A: An improved electron stain for elastic fibers using tannic acid. J Electron Microsc (Tokyo) 24:287–289, 1975.

57. Morris SM, Stone JP, Rosenkrans WA, Calore JD, Albright JT, Franzblau C: Palladium chloride as a stain for elastin at the ultrastructural level. J Histochem Cytochem 26: 635–644, 1978.

58. Fullmer HM: A comparative histochemical study of elastic, pre-elastic and oxytalan connective tissue fibers. J Histochem Cytochem 8: 290–295, 1960.

59. Albert EN: Developing elastic tissue. Am J Pathol 69: 89–94, 1972.

60. Dempsey EW, Lansing AI: Elastic tissue. Intern Rev Cytol 3: 436–453, 1954.

61. Hall DA, Reed R, Tunbridge RE: Electron microscope studies of elastic tissue. J Exp Cell Res 8: 35–48, 1955.

62. Gross J: The structure of elastic tissue as studied with the electron microscope. J Exp Med 89: 699–708, 1949.

63. Dempsey EW: The chemical characterization and submicroscopic structure of elastic tissue. Science 116: 520, 1952.

64. Haust MD, More RH, Bencosme SA, Balis JU: Elastogenesis in human aorta: an electron microscopic study. Exp Mol Pathol 4: 508–524, 1965.

65. Ross R: The smooth muscle cell. II. Growth of smooth muscle in culture and formation of elastic fibers. J Cell Biol 50: 172–186, 1971.

66. Kadar A, Gardner DL, Bush V: Susceptibility of the chick embryo aorta to elastase: an electron microscope study. J Pathol 4: 261–266, 1971.

67. Bodley HD, Wood RL: Ultrastructural studies on elastic fibers using enzymatic digestion of thin sections. Anat Rec 172: 71–88, 1972.

68. Ross R, Bornstein P: The elastic fiber. I. The separation and partial characterization of its macromolecular components. J Cell Biol 40: 366–381, 1969.

69. Lamberg SI, Poppke DC, Williams BR: Isolation of elastic tissue microfibrils derived from cultured cells of calf ligamentum nuchae. Connect Tissue Res 8: 1–8, 1980.

70. Sear CHJ, Kewley MA, Jones CJP, Grant ME, Jackson DS: The identification of glycoproteins associated with elastin tissue microfibrils. Biochem J 170: 715–718, 1978.

71. Crissman RS, Ross JN Jr, Davis T: Scanning electron microscopy of an elastin fiber network which forms the internal elastic lamina in canine saphenous vein. Anat Rec 198: 581–593, 1980.

72. Meyer W, Neurand K, Radke B: Elastic fiber arrangement in the skin of the pig. Arch Dermatol Res 270: 391–401, 1981.

73. Kewley MA, Stevens FS, Williams G: The presence of fine elastin fibrils within the elastin fibre observed by scanning electron microscopy. J Anat 123: 129–134, 1977.

74. Fornieri C, Pasquali Ronchetti I, Edman AC, Sjöström M: Contribution of cryotechniques to the study of elastin ultrastructure. J Microsc 125: 87–93, 1982.

75. Cliff WJ: The ultrastructure of aortic elastica as revealed by prolonged treatment with OsO_4. Exp Mol Pathol 15: 220–229, 1971.

76. Porter KR, Kallman F: The properties and effects of osmium tetroxide as a tissue fixative with special reference to its use for electron microscopy. Exp Cell Res 4: 127–141, 1953.

77. Deetz JS, Behrman EJ: Kinetics of the reaction of some tryptophan derivatives with the osmium tetroxide-pyridine reagent. J Org Chem 45: 135–140, 1980.

78. Partridge SM: Diffusion of solutes in elastin fibres. Biochim Biophys Acta 140: 132–141, 1967.

79. Bressan GM, Castellani I, Giro GM, Volpin D, Fornieri C, Pasquali Ronchetti I: Banded fibers in tropoelastin coacervates at physiological temperatures. J Ultrastruct Res 82: 335–340, 1983.

80. Volpin D, Pasquali Ronchetti I, Urry DW, Gotte L: Banded fibres in high temperature coacervates of elastin peptides. J Biol Chem 251:

6871–6873, 1976.

81. Volpin D, Pasquali Ronchetti I: The ultrastructure of high temperature coacervates from elastin. J Ultrastruct Res 61: 295–302, 1977.

82. Volpin D, Urry DW, Pasquali-Ronchetti, Gotte L: Studies by electron microscopy on the structure of coacervates of synthetic polypeptides of tropoelastin. Micron 7: 193–198, 1976.

83. Long MM, Rapaka RS, Volpin D, Pasquali Ronchetti I, Urry DW: Spectroscopic and electron micrographic studies on the repeat tetra-peptide of tropoelastin. Val-Pro-Gly-Gly. Arch Biochem Biophys 210: 445–452, 1980.

84. Volpin D, Urry DW, Cox BA, Gotte L: Optical diffraction of tropo-elastin and alfa-elastin coacervates. Biochim Biophys Acta 439: 253–258, 1976.

85. Urry DW, Long MM: On the conformation, coacervation and function of polymeric models of elastin. Adv Exp Biol Med 79: 685, 714, 1977.

86. Mecham RP, Foster JA: A structural model for desmosine crosslinked peptides. Biochem J 173: 617–625, 1978.

87. Smith DW, Sandberg LB, Leslie HB, Wolt TB, Minton ST, Myers B, Rucker RB: Primary structure of a chick tropoelastin peptide: evidence for a collagen-like amino acid sequence. Biochem Biophys Res Commun 103: 880–885, 1981.

88. Sikes BC, Partridge SM: Isolation of a soluble elastin from lathyritic chicks. Biochem J 130: 1171–1172, 1972.

89. John R, Thomas J: Chemical composition of elastins isolated from aortas and pulmonary tissues of humans of different ages. Biochem J 127: 261–269, 1972.

90. Mecham RP, Foster JA: Characterization of insoluble elastin from copper deficient pigs. Its usefulness in elastin sequence studies. Biochim Biophys Acta 577: 147–158, 1979.

91. Keller S, Mandl I, Turino GM: Determination of the relative amount of elastin in lung tissues. Biochem Med 25: 74–80, 1981.

92. Schwartz E, Adamany AM, Blumenfeld OO: Isolation and character-ization of the internal elastic lamina from calf thoracic aorta. Exp Mol Pathol 34: 299–306, 1981.

93. Pasquali Ronchetti I, Volpin D, Baccarani Contri M, Castellani I, Peserico A: Pseudoxanthoma Elasticum: biochemical and ultrastruc-tural studies. Dermatologica 163: 307–325, 1981.

94. Buckingham K, Heng-Khoo CS, Dubick M, Lefevre M, Cross C, Juline L, Rucker R: Copper-deficiency and elastin metabolism in avian lung. Proc Soc Exp Biol Med 166: 310–319, 1981.

Authors' address:
Istituto di Patologia Generale
Via Campi, 287
41100 Modena, Italy

CHAPTER 8

Elastogenesis in embryonic and post-natal development

AUGUSTO SERAFINI-FRACASSINI

1. Introduction

Despite increasing emphasis on the subject, at the time of this writing the overall process of biogenesis of the elastic fibre is not completely understood. This is, in part at least, a reflection of the persistence of uncertainties concerning the interpretation of several aspects of the conformation and supramolecular organization of elastin. It is, however, to be ascribed primarily to the difficulty of applying classical biochemical techniques to the analysis of a macromolecule that in the extracellular matrices forms, as discussed fully elsewhere in this book (Chapter 1), extensive polypeptide networks stabilized by covalent cross-links. A major turning point in this field of investigation has therefore been the isolation of a soluble form of the protein (1), designated as tropoelastin (2), from the aorta of copper-deficient pigs. This discovery has supplied a substrate suitable for sequence analysis and has also provided an invaluable tool for the study of many facets of the molecular biology of elastin.

2. Elastogenic cells

A correlation within developing connective tissue matrices between either the rate of accumulation of elastic fibres or the rate of synthesis of the elastin precursor and concomitant cytological events, explicitly indicative of biosynthetic activity, has been widely used for the identification of cells performing an elastogenic function. It should be noted, however, that cells undergoing rapid differentiation, particularly when located in a tissue which displays a considerable degree of cellular heterogeneity, cannot always be satisfactorily categorized by the exclusive use of morphological criteria. Moreover, it is not always feasible to establish with a sufficient degree of certainty whether a particular morphological pattern, visualized under the electron microscope,

unequivocally represents a specific functional condition within the living tissue. Because of these difficulties, some ambiguities still persist in the identification of the cell(s) responsible for elastin biosynthesis within some connective tissue matrices. In the aorta of a few species, of which man is an example, it has been demonstrated quite convincingly that the smooth muscle cell is the only cell capable of elastogenic function (3, 4). However, in the chick, formation of elastic fibres appears to occur prominently in direct contact with three different types of medial cells, viz. undifferentiated mesenchymal cells, fibroblasts and smooth muscle cells (5). In the latter species, the deposition of the collagenous component occurs at an early stage of embryonic development and is well advanced by the eighth day, when the rate of elastin synthesis increases abruptly. As a result, elastic fibres become readily detectable in the extracellular milieu by the next day (6). Thereafter, as the aorta develops, the relative rates of elastin and collagen biosynthesis rapidly diverge, with the former predominating (7). On the ninth day most of the mesenchymal cells are already differentiated into fibroblasts (8). These, in turn, begin to transform into smooth muscle cells on about the 11th day, when the concentration of elastin in the tissue is about 8%, on a dry weight basis. This process starts from the deep layers of the tunica media and gradually proceeds upwards. At the time of hatching, when the concentration of elastin has increased to 40%, the three cell types can still be identified and all exhibit morphologies suggestive of protein biosynthetic activity. The undifferentiated mesenchymal cells are now found only underneath the endothelium, while the fibroblasts are interdispersed among the smooth muscle cells which predominate. Takagi's view of multiple cellular involvement in elastogenesis (8) is not shared by others (9) who believe that this function is served only by the smooth muscle cells. Recently, both in the chick and in the hamster, an elastogenic function has been

Ruggeri, A and Motta, PM (eds): Ultrastructure of the connective tissue matrix. ISBN-13:978-1-4612-9789-5

tentatively assigned, on the basis of immuno-electron microscope evidence, also to endothelial cells (10).

The lung is another organ where elastin plays a major role in the generation of the functional properties. In the rat, alveolar septum formation occurs mainly between the 4th and 15th day of post-natal life (11). Budding of the new septa from the wall of the large, thick-walled saccules that at birth form the immature lung is associated with the formation, within the developing structures, of a network of elastic fibres which frequently appear to originate at the tip of the septa. Concurrently, interstitial fibroblasts undergo major structural and functional changes (12, 13) leading to the formation of two types of cells which display distinct distributions within the tissue (14). Those cells which are located at the tip of the septa have several of the morphological characteristics of the myofibroblast, viz. an extensive rough endoplasmic reticulum, multiple membrane-bound vesicles associated with a well-developed Golgi complex and prominent cytofilaments. Each of these cells is in close connection with several elastic fibres. On the other hand, those cells which are situated at the base of the septa do not have any specific relationship with elastic fibres, exhibit only a few organelles and their cytoplasm is filled with osmiophilic material. When the alveolar wall is fully developed, both types of cells are no longer recognizable having acquired dormant morphological characteristics.

Elastin is also the predominant component of the ligamentum nuchae of large ruminants. In this tissue, elastic fibres become detectable in appreciable amount at the beginning of the seventh month of foetal life (15). At about this time, the rate of accumulation of elastin markedly increases and reaches a peak between the 225th day of gestation and the end of the first post-natal month, after which it falls gradually to adult levels (15, 16). Because of this, the concentration of elastin in the tissue soon exceeds that of collagen, which represents the major fibrous protein component during the earlier stages of development. The predominant cell type in the collagenous phase of ligament growth is represented by the primary fibroblast, characterized by a well-developed ergastoplasm, a prominent Golgi apparatus, osmiophilic bodies and a few mitochondria (15). The ensuing phase of intense elastogenic activity is instead associated with the differentiation of a second type of fibroblast which has lost most of the ergastoplasm and many of the cell organelles and is characterized by the presence of both widely dilated cisternae, within the endoplasmic reticulum, and abundant coated vescicles (17).

A similar modification of the fibroblast population in relation to elastogenesis has been reported to occur in the course of wound healing (18, 19).

The ability of chondroblasts to carry out *in vitro* elastin biosynthesis has been demonstrated recently (20, 21).

3. The biosynthetic precursor

As pointed out earlier, an important contribution to the elucidation of the nature of the immediate biosynthetic precursor of insoluble elastin came from the investigation of connective tissue disorders brought about in growing animals by copper deficiency. In these, the most striking pathological changes affect the wall of the large blood vessels, in particular that of the aorta, and consist of a degeneration of the elastic lamellar system. Concomitantly, within the tissue there is a reduction in the concentration of insoluble elastin and the accumulation of a soluble form of the protein (tropoelastin) which cannot be detected in normal animals (22). The amino acid composition of tropoelastin was found (1) to be virtually identical to that of insoluble elastin, with the exception of a much higher lysine content and the absence of the polyfunctional amino acids, which in the latter protein are responsible for the cross-linking of the polypeptide chains (Table 1). As a precursor–product relationship between lysine and polyfunctional amino acids had already been ascertained (23), the biochemical defect induced by copper deficiency was clearly identified as the inhibition of the mechanism responsible for the conversion of tropoelastin to cross-linked elastin, in the course of fibre formation. Equivalent forms of tropoelastin have been subsequently isolated from chick (24, 25) and bovine (26) aorta.

Tropoelastins are soluble in cold salt solutions, but undergo reversible coacervation within the temperature range 23–37° C, dependent upon concentration (2). Tropoelastin from porcine aorta has a molecular weight of 74,000 (27) and comprises some 850 amino acid residues. A large proportion of its primary structure is known (28), although the order in which the sequenced fragments are arranged in the parent molecule has not yet been established. In spite of this, there is sufficient evidence to suggest that the macromolecule is constituted by several domains, each comprising from about 50 to 100 amino acid residues, joined by intervening short alanine-rich runs of sequence which normally incorporate two lysyl residues separated by one to three alanines.

142

Table 1. Amino acid composition of porcine aortic tropoelastin and insoluble elastin

	Tropoelastin (27)		Insoluble elastin (2)
	Residues per 1,000 amino acid residues	Residues per mole (70,000 mol.wt.)	Residues per 1,000 amino acid residues
Hydroxyproline	9.9	9	14.5
Aspartic acid	3.7	3	8.8
Threonine	14.4	13	7.4
Serine	11.6	10	8.1
Glutamic acid	16.9	15	20.9
Proline	104.0	92	93.8
Glycine	326.0	287	328.9
Alanine	230.0	203	233.3
Cysteine	0.0		0.0
Valine	132.0	116	124.9
Methionine	0.0		1.7
Isoleucine	16.0	14	19.6
Leucine	45.2	40	57.4
Tyrosine	16.3	14	16.9
Phenylalanine	26.5	23	32.3
Lysine	43.3	38	7.6
Histidine	0.0		1.1
Arginine	4.3	4	6.6
Isodesmosine/4*	0.0		8.1
Desmosine/4*	0.0		8.3

* Expressed as lysine equivalents.

The lysine residue located C-terminally in each pair is followed, apparently with equal frequency, by either a further alanine or by a large hydrophobic residue (29–31). The effect of such an arrangement on the biogenesis of the cross-links will be discussed later.

The large, lysine-free domains are mainly responsible for the generation of the mechanical properties of the protein. They are predominantly hydrophobic, and an interesting feature of their primary structure is the appreciable presence of runs of sequence in which 4-, 5- or 6-residue units are significantly repeated. These units are the tetrapeptide Val-Pro-Gly-Gly, the pentapeptide Val-Pro-Gly-Val-Gly and the hexapeptide Ala-Pro-Gly-Val-Gly-Val (28). The secondary structure of model polypeptides of these repeat sequences have been extensively studied by Urry and his associates (32). All three polypeptides were found to adopt a type II β-turn conformation. Moreover, in the polypentapeptide, which is elastomeric and capable of forming anisotropic structures upon coacervation, the succession of β-turns generates a β-spiral which contains about three repeating pentamers per turn. The

acquisition of the latter conformation segregates lipophilic and hydrophilic groups in discrete spiral bands that alternate at the external surface of the structure. This arrangement is probably responsible for chain alignment in the formation of anisotropic aggregates during coacervation and fibre formation, with hydrophobic interactions providing the necessary driving force for assembly (32).

Cell- and organ-culture systems together with cell-free translating polysomes and elastin mRNA have been shown to synthesize a 70,000-dalton elastin precursor molecule which exhibits many of the properties of tropoelastin (33–37). However, recently two translational proteins have been identified (38–40). These, which are referred to as tropoelastin *a* and *b*, display a high degree of homology and have an M_r of 73,000 and 70,000 respectively. The *b* form behaves on PAGE like authentic tropoelastin. Tropoelastin *a*, on the other hand, contains a larger amount of cysteine than tropoelastin *b*, and its extraction from tissues necessitates cleavage of disulphide bonds. In the chick and pig lung, expression of each of these two precursor proteins accounts, throughout the developmental period, for about 50% of both the elastin mRNA population and the total elastin synthetic rate. In the aorta, on the other hand, form *b* is preferentially synthesized to the extent that on the 16th day of development it represents 96% of the soluble elastin translated (40). As different tissues appear to synthesize closely similar, if not identical, tropoelastins (41), the differential expression of homologous forms of the precursor could be envisaged to represent a mechanism by which distinct types of insoluble elastin could be produced to meet specific functional requirements.

Elastin is translated in a pretropoelastin form (42). The signal sequence of pretropoelastin *b* contains 24 amino acid residues, possesses N-terminally an initiator methionyl residue and, as in all other peptides of this type, its central domain is largely hydrophobic. It ends with two glutamines followed immediately by the N-terminal sequence of the definitive tropoelastin.

4. Control of gene expression

Nothing can be said at present concerning the actual mechanism responsible for the stimulation of elastin gene expression in the course of development or in response to pathological conditions. It is, however, obvious that it must be under strict regulation as in each anatomical location elastin biosynthesis is operative, at a significant level, only for a specific and relatively short period of foetal and/or post-natal life.

Bearing on the question of the control of the rate of collagen and elastin synthesis, the involvement of hormonal regulation has been investigated and the administration of hydrocortisone found to cause in the aorta a relative depression in the synthesis of collagen and an absolute increase in that of tropoelastin (7).

Both in the case of this hormone-induced de-repression of the elastin genome and in the course of normal embryogenesis, a good positive correlation has been demonstrated between the rate of tropoelastin synthesis and the corresponding level of elastin mRNA, a result which suggests that in the elastogenic cells, like in many other eukariotic systems, the major control of gene expression is at transcriptional level (36).

It should also be borne in mind that both in developing elastic ligaments (16) and in tissue cultures of foetal fibroblasts (43) there is a temporal dissociation between cell proliferation and elastogenesis, in that maximal synthesis occurs concomitantly with a diminution in cell growth. Although several other explanations of the phenomenon are equally plausible, nevertheless this could be interpreted as the expression of the influence exerted by cell–cell and matrix–cell interactions on the process of cell differentiation (44).

5. Post-translational modifications

The hydroxylation of some of the prolyl residues which are already incorporated into peptide linkage is catalyzed by the enzyme prolyl hydroxylase, an oxygenase present in the microsomal fraction. At variance with collagen, hydroxylation does not appear to be an essential feature of elastin biogenesis as the inhibition of the hydroxylase activity does not result in a depression of tropoelastin secretion (34). Moreover, only about 25% of the prolyl residues that in elastin are followed by glycine, and are therefore susceptible to enzymic recognition, do actually undergo modification (45). A possible explanation of this phenomenon is offered by the observation that hydroxylation of Urry's synthetic polypentapeptide interferes with chain association (46). A high hydroxyproline content in elastin could, therefore, impair the process of fibre formation.

Not much is known at present of the mechanism by which tropoelastin molecules are transported intracellularly. Various types of vesicles have been identified in elastogenic cells and some of these have been tentatively associated with the transfer of tropoelastin and the precursors of other connective tissue macromolecules (6, 15). Recently, immunological staining of sections of embryonic aorta using elastin-specific antibodies has produced images which suggest that in both smooth muscle and endothelial cells tropoelastin is indeed secreted via vescicular structures (10). These are probably the dense-core vesicles described by Thyberg and associates (47, 48). The involvement of microtubules in the secretory process has also been demonstrated (34).

During, or immediately after extrusion into the extracellular milieu, probably via fusion of the above mentioned vesicles with the plasma membrane, tropoelastin undergoes a further modification that initiates the process of cross-link formation (49). It consists of the conversion of some of the lysyl residues located in the alanine-rich sequences into residues of α-amino adipic δ-semialdehyde (allysine), by oxidative deamination at C-6. The reaction is catalyzed by the enzyme lysyl oxidase, a Cu^{2+}-dependent amine oxidase (50). The enzyme $M_r \sim 32,000$ is soluble only in chaotropic solutions. Where exposed to other aqueous environments, it rapidly aggregates to a series of oligomeric species that exceed 1×10^6 molecular weight (51). Connective tissue matrices appear to contain multiple forms of the enzyme which are closely related with respect to compositional, immunological and functional parameters (51, 52). The significance of these isotypes is not fully understood since all of them appear to utilize both collagen and elastin as substrates (52).

During active growth, levels of lysyl oxidase activity in different matrices appear to be subjected to tissue-specific, age-dependent fluctuations that overlap those displayed by the rates of collagen and elastin biosynthesis (53).

The formation of an effective enzyme-substrate complex appears to require complementarity of distribution of electrostatic charges between the anionic oxidase and cationic sites on elastin (54). Moreover, it has been shown that *in vitro* the activity of the enzyme is affected by the conformation of the substrate, insoluble elastin and coacervated tropoelastin being more effectively modified than monomeric tropoelastin (55). It could therefore be postulated that *in vivo* the binding of the enzyme is induced or facilitated by the formation of small clusters of lysyl residues generated by the juxtaposition of polyalanyl sequences belonging to adjacent elastin chains during fibre assembly. This type of binding suggests the possibility of a charge-based, substrate-directed modulation of the enzymic ac-

144

tivity in both normal and pathological conditions (54).

Increasing evidence leads to the conclusion that the amino acid sequence vicinal to the susceptible lysines does not play a major role in enzyme recognition (56). However, it seems that in elastin those lysyl residues which are followed by a large hydrophobic side chain are not deaminated (29, 31).

Peptidyl aldehydes are very reactive and condense spontaneously with one another or with the ε-amino group of an unmodified lysyl residue to form the structurally related polyfunctional amino acids reported in Table 2. Although the exact sequence of events in this aspect of biosynthesis is still uncertain, a likely pathway (24) is illustrated in Figure 1 where two alanine-rich sequences belonging to adjacent tropoelastin molecules are shown in exact register.

Table 2. Concentration of cross-links in three bovine tissues (73)

Amino acid	Ligamentum nuchae	Aorta	Cartilage
Aldol condensation product/2	4.3	3.7	5.7
Dehydrolysinonorleucine/2	0.0	0.2	0.0
Lysinonorleucine/2	2.2	2.2	2.1
Dehydromerodesmosine/3	0.2	0.7	0.7
Merodesmosine/3	0.5	0.6	tr.
Isodesmosine/4	5.4	5.6	4.9
Desmosine/4	10.1	9.7	10.4

Values are expressed as lysine equivalents per 100 amino acid residues.

Lysyl residues originally in positions A, B and C are deaminated, while that in D is unmodified, followed by a tyrosine. The aldehyde function of each allysine can react with that of any other such residue forming an aldol condensation product which in the figure appears to join arbitrarily A and C. Alternatively, each of these functions, for example that in position B, may form a Schiff-base with the ε-amino group in D. Both reaction products, the aldol condensate and the $\Delta^{6,7}$-dehydrolysinonorleucine – mainly in the form of its stable derivative lysinonorleucine – are present in elastin where they function as cross-links. On the other hand, they may react with each other and produce, depending upon the direction of the initial condensation of the allysines, a residue of either dihydrodesmosine or dihydroisodesmosine. These species in turn are oxidized without additional enzymic involvement to form stable desmosine and isodesmosine (55, 57). Table 2 shows the concentrations of the polyfunctional amino acids in elastins isolated from three bovine tissues. It should be noted that several of the lysyl residues which undergo modification in tropoelastin are still unaccounted for on the basis of the cross-links identified so far (58).

6. Morphogenesis of the elastic fibre

No large pool of premature elastin exists in the extracellular matrices because the rapid formation of cross-links results in the production of insoluble elastin (59), which is characterized by a very slow rate of metabolic turnover.

Electron microscopic investigations of several developing connective tissue matrices (6, 15, 60, 61, 62) have repeatedly indicated that the formation of elastic fibres involves the orderly deposition of the newly synthesised elastin on a pre-existing supportive structure. This is provided by the so-called microfibrillar component which is morphologically constituted by fibrils that exhibit an affinity for lead, osmium and uranyl acetate (60) and have a thickness of about 11 nm. In the latter respect, it should be noted that significant differences in their mean diameter have been recorded in various species together with progressive age-related changes that take place during embryonic and post-natal development (63).

The microfibrils are among the first fibrous structures which can be recognised during the morphogenesis of elastic tissues and are found grouped in small bundles in proximity to elastogenic cells (Figure 2a). At a later stage, elastin has been shown to appear in the form of an amorphous material in several discrete loci within each microfibrillar

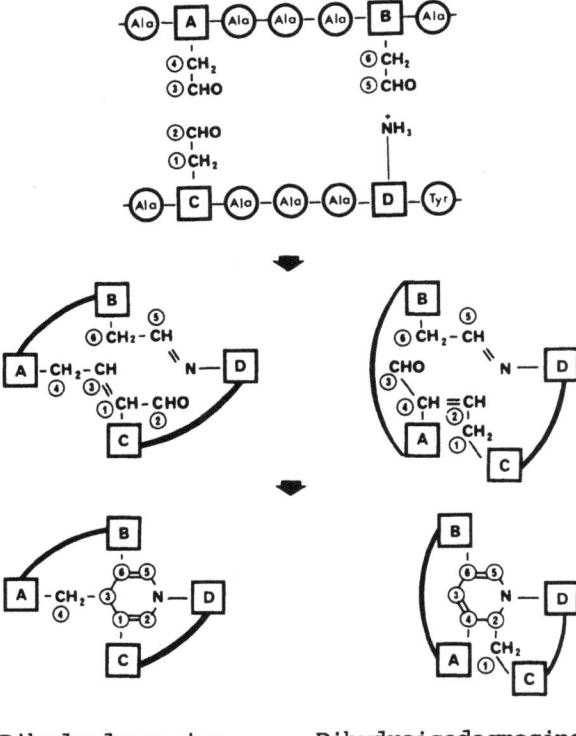

Dihydrodesmosine   Dihydroisodesmosine

Figure 1. Schematic representation of a pathway leading to the formation of dihydrodesmosine and dihydroisodesmosine (modified from 29).

bundle (Figure 2b). These amorphous areas gradually coalesce (Figure 2c) and generate the central core of the newly formed elastic fibre (60, 61).

In this process and during the subsequent stages of fibre development, the majority of the microfibrils appear to be progressively displaced at the outer aspect of the structure (Figures 3a, b), a position they retain in the mature tissue. Concomitantly, the elastin component rapidly loses its affinity for electron-dense metal ions (17), probably as a result of the establishment of a tight macromolecular packing (64).

One of the most salient features of the morphogenesis of elastic tissues is the close interrelationship which is often observed between elastogenic cells and immature elastic fibres, the latter being situated in infoldings of the cell surface (5, 14, 60, 61) which are occasionally characterized by a dense thickening of the plasma membrane (65). As far as embryonic bovine ligamentum nuchae is concerned, in the author's experience there is no uniform pattern of interaction. Extensive cell-fibre contacts appear to be restricted to minute loci which are spread throughout the tissue and separated by large do-

mains where relationships are less frequent and more superficial. In the former areas, fibroblasts enclose completely a very high proportion of the developing fibres (Figure 3c) and occasionally produce spiral arrangements of their plasma membrane (Figure 4a) which are suggestive of a rotational movement of the cell in relation to the axis of the fibre. The gap between the plasma membrane and the fibre is often just sufficient to accomodate the microfibrillar layer. Frequently, where a close contact is established, numerous vescicles can be seen associated with the plasma membrane (Figure 4b). These are probably the expression of either endocytosis or simply intense local stimulation. Moreover, cells exhibit, albeit to a variable extent, cytoplasmic polarization arising from the preferential localization of the endoplasmic reticulum and cell organelles in an area near to the site of interaction (Figure 4b).

Although this evidence is not in itself explicit, it is however conducive to the suggestion that elastogenesis progresses unevenly throughout the tissue and that the regions of extensive cell-fibre interaction represent sites of intense biosynthetic activity. The existence in these areas of an intimate contact between the plasma membranes and the surface of the growing elastic fibres would significantly reduce the problem associated with the extracellular transport of tropoelastin from the site of synthesis to that of fibre formation; it will in fact be recalled that this macromolecule undergoes coacervation *in vitro* when exposed to physiological conditions. It would also provide a micro-environment suitable for the formation of a complex between elastin and lysyl oxidase, an enzyme that is equally highly prone to aggregation. In such a system, the microfibrillar component could be envisaged to constitute a substratum suitable for the attachment of fibroblasts to the surface of the developing elastic fibres. In this role it would therefore be akin to those glycoproteins which are instrumental in establishing cell–collagen interactions (66). Cell-adhesion could in turn initiate a biosynthetic episode resulting in the deposition of a finite amount of newly synthesized elastin on the pre-existing structure. Electron microscopic images like that in Figure 3c suggest that this could be in the form of fine fibrils, each provided with a microfibrillar envelope, which can be seen to occupy, in relation to the core of the fibre, a satellite position. Fibrils of such small dimensions (~100 nm wide) have been visualized also by scanning electron microscopy (67). Figures 3a and 3b suggest that they are subsequently incorporated into the fibre central core.

146

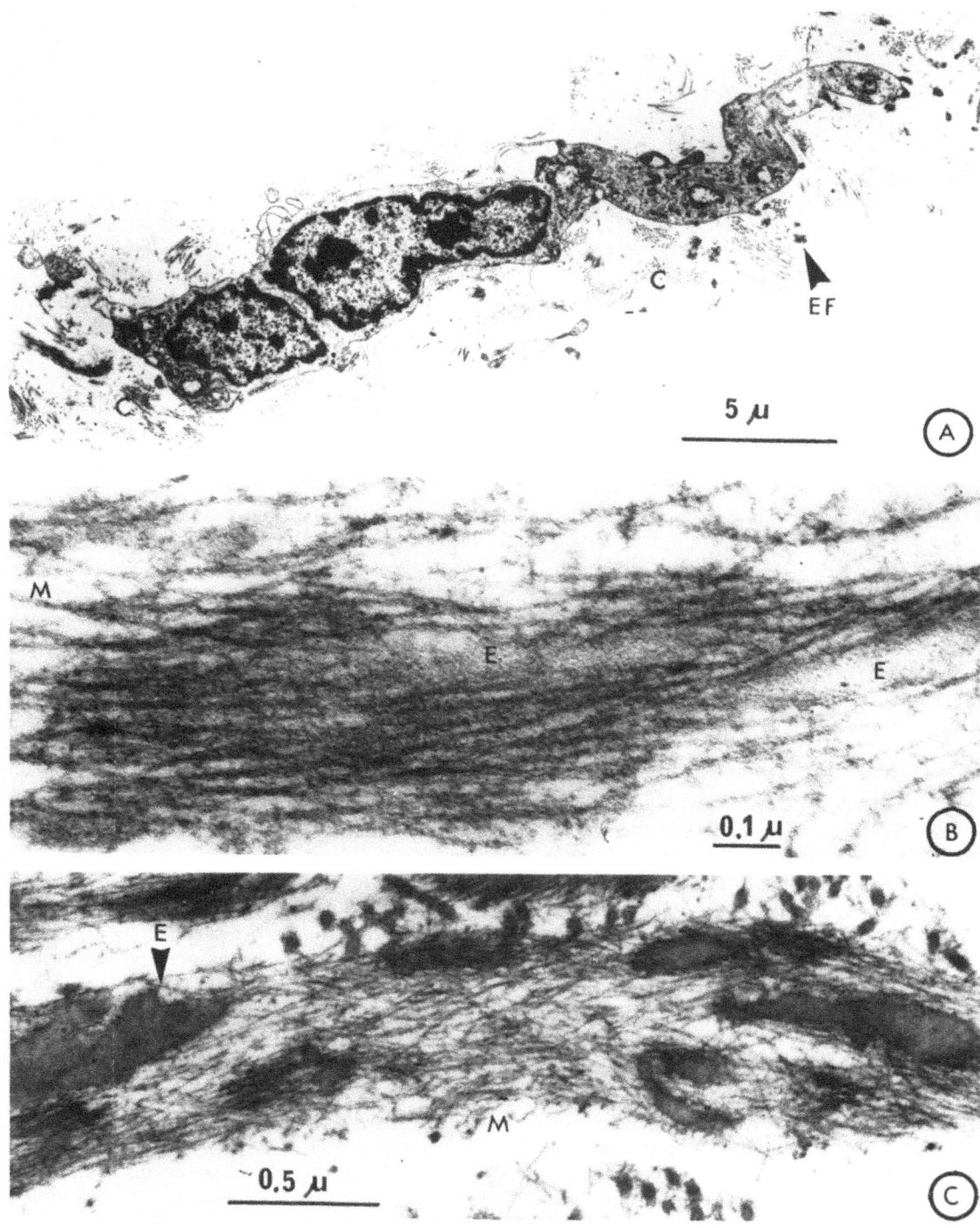

Figure 2. (a) Transverse section of four-month foetal bovine ligamentum nuchae, stained with uranyl acetate-lead citrate. This electron micrograph shows a fibroblast surrounded by collagen fibres (C) and immature elastic fibres (EF) (× 4800). (b) Longitudinal section of a four-month foetal bovine ligamentum nuchae, stained with uranyl acetate-lead citrate. The immature elastic fibre shown in this micrograph comprises a bundle of microfibrils (M) and two amorphous areas of newly synthesized elastin (E) (× 99,000). (c) Longitudinal section of a six-month foetal bovine ligamentum nuchae, stained with uranyl acetate-lead citrate. The elastic fibre shown is constituted predominantly of microfibrils (M). The central core areas (E) are wider than those in Figure 2b, but still largely independent (× 46,500). Note in both (b) and (c) the relatively dark staining of the elastin component.

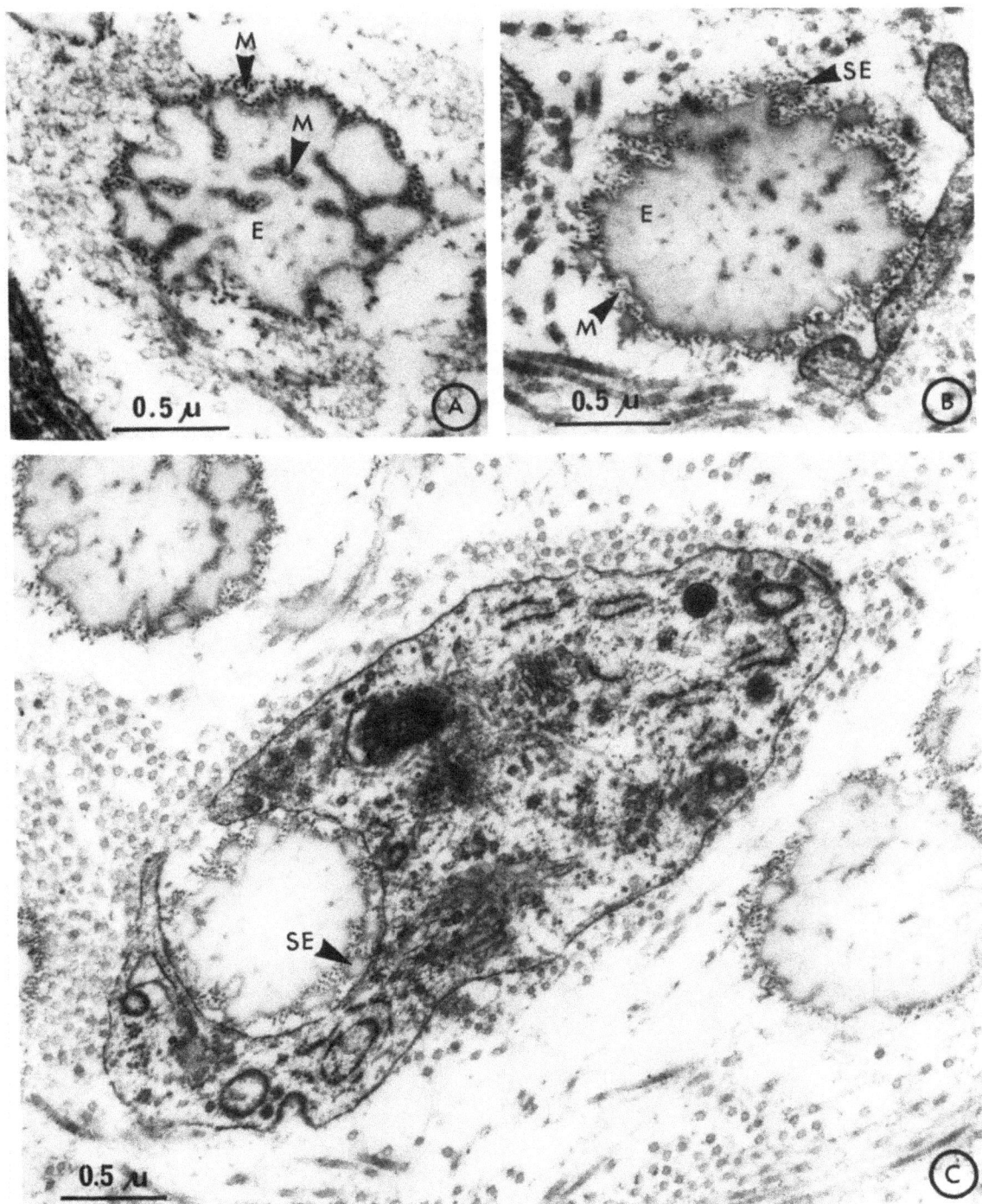

Figure 3. (a) This electron micrograph demonstrates the transverse section of an elastic fibre in a seven-month foetal bovine ligamentum nuchae, stained with uranyl acetate-lead citrate. The elastin component forms a clearly identifiable central core (E) with the microfibrils (M) in both peripheral and interstitial position (× 37,000). (b) The elastic fibre shown in this micrograph was located in proximity to that in (a). The central core (E) is more homogeneous and the microfibrils (M) are almost entirely segregated at the outer aspect of the structure. Small satellite elastic fibres (SE) can be seen in a peripheral location (× 37,000). Note the lighter staining of the elastin component in both (a) and (b) when compared with that in Figures 2b and 2c. (c) The fibroblast in this transverse section of a seven-month foetal bovine ligamentum nuchae completely encloses an elastic fibre. Note the small satellite elastic fibres (SE) in close contact with the plasma membrane. Stained with uranyl acetate-lead citrate (× 31,800).

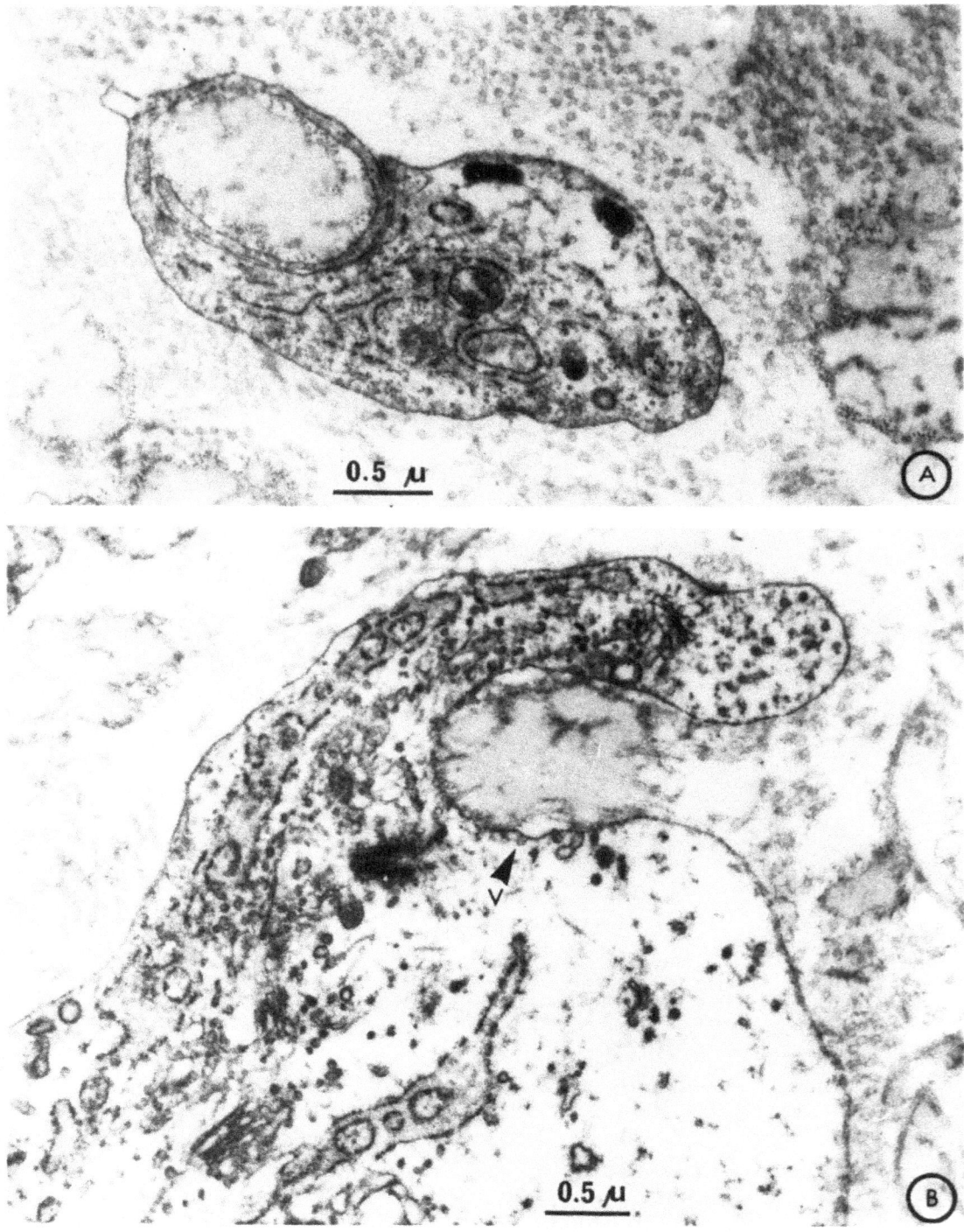

Figure 4. (a) This electron micrograph shows the transverse section of a fibroblast wrapped around an elastic fibre. Seven-month foetal bovine ligamentum nuchae, stained with uranyl acetate-lead citrate ($\times$ 28,600). (b) Transverse section of a fibroblast in seven-month foetal bovine ligamentum nuchae, stained with uranyl acetate-lead citrate. This electron micrograph demonstrates the preferential distribution of the endoplasmic reticulum and cell organelles in the area adjacent to the infolding of the plasma membrane that contains an elastic fibre. Note the vesicles (V) in the area of contact ($\times$ 24,700).

Concerning the interactions between elastogenic cells and a pre-formed support, it should be noted that the presence of elastic fibres appears to be conditional for foetal fibroblasts to produce cross-linked elastin *in vitro* (43).

Microfibrils disaggregate in chaotropic solutions containing a reducing agent and the view has developed that they are constituted by disulphide-linked glycoproteins (68). Ideas concerning the number and nature of these constituents are at this stage tentative. Tissue cultures of foetal bovine ligamentum nuchae have been reported to produce two PAS-positive proteins, designated MFP1 and MFP2, which are precipitated by an anti-microfibrillar protein antiserum (69, 70). MFP1 is a collagenous glycoprotein ($M_r \sim 150,000$) which is susceptible to digestion with bacterial collagenase, while MFP2 is non-collagenous and has an M_r of 300,000. On the other hand, bovine ligamentum nuchae, when extracted with chaotropic solutions under reducing conditions,

yields, among other proteins, a glycoprotein of 34,000 molecular weight which, upon dialysis, aggregates to form fibrils with a uniform diameter of about 11 nm and therefore of a size identical to that exhibited by native microfibrils (71). This glycoprotein was also found to act as a substratum for growth of fibroblasts and other cell types *in vitro* with fibronectin mediating the attachment and spreading of the cells (72).

The organization of the newly synthesized unbranched fibres into the larger units which are characteristic of the adult tissues constitutes the final stage in the morphogenesis of elastic matrices. In bovine ligamentum nuchae the foetal elastic fibres, which are about 1.5 μm wide, fuse, perhaps under the effect of stress, during post-natal development and form branched fibres that may ultimately reach a diameter of 14 μm (17). The development of the lamellar system in the tunica media of the aorta is equally a post-natal event (6).

References

1. Smith DW, Weissman N, Carnes WH: Cardiovascular studies on copper deficient swine. 12. Partial purification of a soluble protein resembling elastin. Biochim Biophys Res Commun 31: 309–315, 1968.
2. Sandberg LB, Weissman N, Smith DW: The purification and partial characterisation of a soluble elastin-like protein from copper deficient porcine aorta. Biochemistry 8: 2940–2945, 1969.
3. Haust MD, More RH, Bencosme SA, Balis JU: Elastogenesis in human aorta: an electron microscope study. Exp Mol Pathol 4: 508–524, 1965.
4. Narayanan AS, Sandberg LB, Ross R, Layman DL: The smooth muscle cell. 3. Elastin synthesis in arterial smooth muscle cell culture. J Cell Biol 68: 411–419, 1976.
5. Takagi K, Kawase O: An electron microscopic study of elastogenesis in embryonic chick aorta. J Electron Microsc (Tokyo) 16: 330–339, 1967.
6. Karrer HE, Cox J: Electron microscopic study of developing chick embryo aorta. J Ultrastruct Res 4: 420–454, 1960.
7. Eichner R, Rosenbloom J: Collagen and elastin synthesis in the developing chick aorta. Arch Biochem Biophys 198: 414–423, 1979.
8. Takagi K: Electron microscopical and biochemical studies of the elastogenesis in embryonic chick aorta. 1. Fine structure of developing embryonic chick aorta. Kumamoto Med J 22: 1–14, 1969.
9. Kadar A, Gardner DL, Bush V: The relation between the fine structure of smooth-muscle cells and elastogenesis in the chick-embryo aorta. J Pathol 104: 253–260, 1971.
10. Damiano V, Tsang A, Kucich U, Weinbaum G, Rosenbloom J: Immuno electron microscopic studies on cells synthesizing elastin. Connect Tissue Res 8: 185–188, 1981.
11. Burri PH, Dbaly T, Weibel ER: The postnatal growth of the rat lung. Anat Rec 178: 711–730, 1974.
12. Burri PH: The postnatal growth of the rat lung. 3. Morphology. Anat Rec 180: 77–98, 1974.
13. Kaufmann SL, Burri PH, Weibel ER: The postnatal growth of the rat lung. 2. Autoradiography. Anat Rec 180: 63–76, 1974.
14. Vaccaro C, Brody JS: The role of the interstitial fibroblast. Anat Rec 192: 467–480, 1978.
15. Fahrenbach WH, Sandberg LB, Cleary EG: Ultrastructural studies on early elastogenesis. Anat Rec 155: 563–576, 1966.
16. Cleary EG, Sandberg LB, Jackson DS: The changes in chemical composition during development of the bovine nuchal ligament. J Cell Biol 33: 469–479, 1967.
17. Kewley MA, Williams G, Steven FS: Studies of elastic tissue formation in the developing bovine ligamentum nuchae. J Pathol 124: 95–101, 1978.
18. Williams G: The pleural reaction to injury: a histological and electron-optical study with special reference to elastic-tissue formation. J Pathol 100: 1–7, 1970.
19. Williams G: The late phases of wound healing: histological and ultrastructural studies of collagen and elastic-tissue formation. J Pathol 102: 61–68, 1970.
20. Quintarelli G, Starcher BC, Vocaturo A, Di Gianfilippo F, Gotte L, Mecham RP: Fibrogenesis and biosynthesis of elastin in cartilage. Connect Tissue Res 7: 1–19, 1979.
21. Heeger P, Rosenbloom J: Biosynthesis of tropoelastin by elastic cartilage. Connect Tissue Res 8: 21–25, 1980.
22. Weissman N, Shields GS, Carnes WH: Cardiovascular studies on copper-deficient swine. 4. Content and solubility of the aortic elastin, collagen and hexosamine. J Biol Chem 238: 3115–3118, 1963.
23. Miller EJ, Martin GR, Mecca CE, Piez KA: The biosynthesis of elastin cross-links. The effect of copper deficiency and a lathyrogen. J Biol Chem 240: 3623–3627, 1965.
24. Rucker RB, Goettlich-Riemann W, Tom K: Properties of chick tropoelastin. Biochim Biophys Acta 317: 193–201, 1973.
25. Foster JA, Shapiro R, Voynow P, Crombie G, Faris B, Franzblau C: Isolation of soluble elastin from lathyritic chicks. Comparison to tropoelastin from copper deficient pigs. Biochemistry 14: 5343–5347, 1975.
26. Whiting AH, Sykes BC, Partridge SM: Isolation of salt-soluble elastin from ligamentum nuchae of copper-deficient calf. Biochem J 141: 573–575, 1974.
27. Smith DW, Brown DM, Carnes WH: Separation and properties of salt-soluble elastin. J Biol Chem 247: 2427–2432, 1972.
28. Sandberg LB, Leach CT, Leslie JG, Torres RA, Alvarez VL: Structural studies of porcine aortic tropoelastin. Front Matrix Biol 8: 69–77, 1980.
29. Foster JA, Rubin L, Kagan HM, Franzblau C, Bruenger E, Sandberg LB: Isolation and characterisation of cross-linked peptides from elastin. J Biol Chem 249: 6191–6196, 1974.
30. Gerber GE, Anwar RA: Structural studies on cross-linked regions of elastin. J Biol Chem 249: 5200–5207, 1974.
31. Gerber GE, Anwar RA: Comparative studies of the cross-linked regions of elastin from bovine ligamentum nuchae and bovine, porcine and human aorta. Biochem J 149: 685–695, 1975.
32. Urry DW: Sequential polypeptides of elastin: structural properties and molecular pathologies. Front Matrix Biol 8: 78–103, 1980.
33. Smith DW, Carnes WH: Biosynthesis of soluble elastin by pig aortic

tissue in vitro. J Biol Chem 248: 8157–8161, 1973.

34. Uitto J, Hoffman H-P, Prockop D: Synthesis of elastin and procollagen by cells from embryonic aorta. Differences in the role of hydroxyproline and the effects of proline analogs on the secretion of the two proteins. Arch Biochem Biophys 173: 187–200, 1976.

35. Ryhänen L, Graves PN, Bressan GM, Prockop DJ: Synthesis of an elastin component of molecular weight about 70,000 by polysomes from chick embryo aortas. Arch Biochem Biophys 185: 344–351, 1978.

36. Burnett W, Eichner R, Rosenbloom J: Correlation of functional elastin messenger ribonucleic acid levels and rate of elastin synthesis in the developing chick aorta. Biochemistry 19: 1106–1111, 1980.

37. Rosenbloom J, Harsch M, Cywinski: Evidence that tropoelastin is the primary precursor in elastin biosynthesis. J Biol Chem 255: 100–106, 1980.

38. Foster JA, Rich CB, Fletcher S, Karr SR, Przybyla A: Translation of chick aorta elastin messenger ribonucleic acid. Comparison to elastin synthesis in chick aorta organ culture. Biochemistry 19: 857–864, 1980.

39. Foster JA, Rich CB, Fletcher S, Karr SR, Desa MD, Oliver T, Przybyla A: Elastin biosynthesis in chick embryonic lung tissue. Comparison to chick aortic elastin. Biochemistry 20: 3528–3535, 1981.

40. Barrineau LL, Rich CB, Przybyla A, Foster JA: Differential expression of aortic and lung elastin genes during chick embryogenesis. Dev Biol 87: 46–51, 1981.

41. Foster JA, Rich CB, Desa MD: Comparison of aortic and ear cartilage tropoelastins isolated from lathyritic pigs. Biochim Biophys Acta 626: 383–389, 1980.

42. Karr SR, Foster JA: Primary structure of the signal peptide of tropoelastin b. J Biol Chem 256: 5946–5949, 1981.

43. Mecham RP, Lange G, Madaras J, Starcher B: Elastin synthesis by ligamentum nuchae fibroblasts: effects of culture conditions and extracellular matrix on elastin production. J Cell Biol 90: 332–338, 1981.

44. Hay ED: Extracellular matrix. J Cell Biol 91: 205–223s, 1981.

45. Gray WR, Sandberg LB, Foster JA: Molecular model for elastin structure and function. Nature (Lond) 246: 461–466, 1973.

46. Urry DW, Sugano H, Prasad KU, Long MM, Bhatnagar RS: Prolyl hydroxylation of the polypentapeptide model of elastin impairs fibre formation. Biochem Biophys Res Commun 90: 194–198, 1979.

47. Thyberg J, Hinek A: Fine structure of rabbit ear chondrocytes in vitro and after autotransplantation. Cell Tissue Res 180: 341–356, 1977.

48. Thyberg J, Hinek A, Nilsson J, Friberg U: Electron microscopic and cytochemical studies of rat aorta. Intracellular vescicles containing elastin- and collagen-like material. Histochem J 11: 1–17, 1979.

49. Kagan, HM, Hewitt NA, Salcedo LL, Franzblau C: Catalytic activity of aortic lysyl oxidase in an insoluble enzyme-substrate complex. Biochim Biophys Acta 365: 223–234, 1974.

50. Siegel RC: Lysyl oxidase. Int Rev Connect Tissue Res 8: 73–118, 1979.

51. Kagan HM, Sullivan KA, Olsson TA, Cronlund AL: Purification and properties of four species of lysyl oxidase from bovine aorta. Biochem J 177: 203–214, 1979.

52. Stassen FCH: Properties of a highly purified lysyl oxidase from embryonic chick cartilage. Biochim Biophys Acta 438: 49–60, 1976.

53. Brody JS, Kagan H, Manalo A: Lung lysyl oxidase activity: relation to lung growth. Am Rev Respir Dis 120: 1289–1295, 1979.

54. Kagan HM, Tseng L, Simpson DE: Control of elastin metabolism by elastin ligands. Reciprocal effects on lysyl oxidase activity. J Biol Chem 256: 5417–5421, 1981.

55. Narayanan AS, Page RC, Kuzan F, Cooper CG: Elastin cross-linking in vitro. Studies on factors influencing the formation of desmosines by lysyl oxidase action on tropoelastin. Biochem J 173: 857–862, 1978.

56. Kagan HM, Tseng L, Trackman PC, Okamoto K, Rapaka RS, Urry DW: Repeat polypeptide models of elastin as substrates for lysyl oxidase. J Biol Chem 255: 3656–3659, 1980.

57. Davis NR, Anwar RA: On the mechanism of formation of desmosine and isodesmosine cross-links of elastin. J Am Chem Soc 92: 3778–3782, 1970.

58. Franzblau C. Elastin. In: Comprehensive Biochemistry, vol 26c. Florkin M, Stolz EH (eds), Amsterdam, Elsevier 1971, pp 659–712.

59. Miyoshi M, Kanamori M, Rosenbloom J: Synthesis of elastin. A rapid formation of lysine-derived crosslinks by chick embryo aorta. J Biochem 79: 1235–1243, 1976.

60. Greenlee TH Jr, Ross R, Hartman JL: The fine structure of elastic fibres. J Cell Biol 30: 59–71, 1966.

61. Greenlee TK Jr, Ross R: The development of the rat flexor digital tendon, a fine structure study. J Ultrastruct Res 18: 354–376, 1967.

62. Jones AW, Barson AJ: Elastogenesis in the developing chick lung: a light and electron microscopical study. J Anat 110: 1–15, 1971.

63. Cleary EG, Fanning JC, Prosser I: Possible roles of microfibrils in elastogenesis. Connect Tissue Res 8: 161–166, 1981.

64. Serafini-Fracassini A. The electron microscopy of fibrous proteins of connective tissue. In: Electron Microscopy of Proteins, vol 2. Harris JR (ed), London Academic Press, 1982, pp 195–231.

65. Hinek A, Thyberg J: Electron microscopic observations on the formation of elastic fibres in primary cultures of aortic smooth muscle cells. J Ultrastruct Res 60: 12–20, 1977.

66. Kleinman HK, Klebe RJ, Martin GR: Role of collagenous matrices in the adhesion and growth of cells. J Cell Biol 88: 473–485, 1981.

67. Kewley MA, Steven FS, Williams G: The presence of fine elastin fibrils within the elastin fibre observed by scanning electron microscopy. J Anat 123: 129–134, 1977.

68. Ross R, Bornstein P: The elastic fibre. 1. The separation and partial characterization of its macromolecular components. J Cell Biol 40: 366–381, 1969.

69. Sear CHJ, Kewley MA, Jones CJP, Grant ME, Jackson DS: The identification of glycoproteins associated with elastic-tissue microfibrils. Biochem J 170: 715–718, 1978.

70. Sear CHJ, Grant ME, Jackson DS: The nature of the microfibrillar glycoproteins of elastic fibres. A biosynthetic study. Biochem J 194: 587–598, 1981.

71. Serafini-Fracassini A, Ventrella G, Field MJ, Hinnie J, Onyezili NI, Griffiths R: Characterization of a structural glycoprotein from bovine ligamentum nuchae exhibiting dual amine oxidase activity. Biochemistry 20: 5424–5429, 1981.

72. Knox P, Wells P, Serafini-Fracassini A: A non-collagenous glycoprotein from elastic tissue acts as substratum for growth of cells in vitro. Nature, (Lond) 295: 614–615, 1982.

73. Field JM, Rodger GW, Hunter JC, Serafini-Fracassini A, Spina M: Isolation of elastin from bovine auricular cartilage. Arch Biochem Biophys 191: 705–713, 1978.

Author's address:
Department of Biochemistry and Microbiology
University of St. Andrews
St. Andrews, Fife KY16 9AL, U.K.

CHAPTER 9

Pathobiology and aging of elastic tissue

ANNA KÁDÁR

1. Introduction

Elastic recoil has an ample amount of importance in the structural integrity and function of blood vessels, mainly of the arteries. The elasticity of various tissues – like blood vessels, pulmonary tissue, skin, elastic cartilage – is due to the presence of elastic fibers.

Elastic fibers are composed of two ultrastructurally and biochemically distinct compounds. One of the components is *elastin*, which has a so-called 'amorphous' appearance in electronmicrographs (1, 4), possesses an amino acid composition (approximately 96% non polar amino acids) with specific cross linkages (2) and behaves as a rubber-like elastomer. The other component is the microfibril, which has a fibrillar structure in electron micrograph. It contains a high amount of polar amino acids and presumably is composed of glycoproteins (3).

During the development of elastic fiber, microfibrils appear first in the extracellular matrix, acting as a framework upon which the soluble elastin becomes polimerized into insoluble elastin (4).

2. Definitions

The terms 'elastic fiber' and 'elastin' are not uniformly used in the literature. Therefore, hereafter I will reserve the expression 'elastin' for the 'amorphous'-looking cross-links containing preotein component, and 'elastic fiber' shall designate that morphological-ultrastructural entity which is composed of elastin and microfibrils (4).

Further details on the elastic fiber, including the literature up to 1979, can be found in a recent monograph (4).

3. The ultrastructure and development of normal elastic fiber

During the last 25 years a number of important discoveries and publications have been issued in the field of elastic fiber research. Considerable achievements were the isolation and crystallization of elastase (5), the purification of elastin (6) and the discovery of the specific cross-links (desmosin, isodesmosin, lysinonorleucin, merodesmosin) by different groups (7). More recently, progress was made by sequencing the elastin molecule (8) by isolation and identification of glycoprotein microfibrils (1, 4, 9) by the purification of non-cross-linked soluble elastin and by characterization of proelastin 'a' and 'b' (10).

Electron microscopy offered new insight into the research of elastic tissue, although no systemic study is available on the development of elastic tissue in humans. The ultrastructure of normal human aorta was published only recently (11).

4. Pathological formation of elastic fibers in experiments

In the biosynthesis of elastin cross-links, four lysine molecules are involved. The oxidative deamination of three lysine molecules out of four requires an enzyme, a lysil oxidase (2).

The copper-containing lysil oxidase activity can be inhibited by restricting copper from the diet of developing animals (4).

The active agent of the Lathyrus Odoratus the beta-amino-propionitrile (BAPN) also inhibits the normal function of lysil oxidase resulting in the lack of formation of cross-links and consequently of insoluble elastin. Lathyrism is a state which develops in experimental animals fed by BAPN causing severe damage in connective tissue, generally, not just in elastic tissue.

In both experimental conditions – in copper defi-

Ruggeri, A and Motta, PM (eds): Ultrastructure of the connective tissue matrix. ISBN-13:978-1-4612-9789-5

152

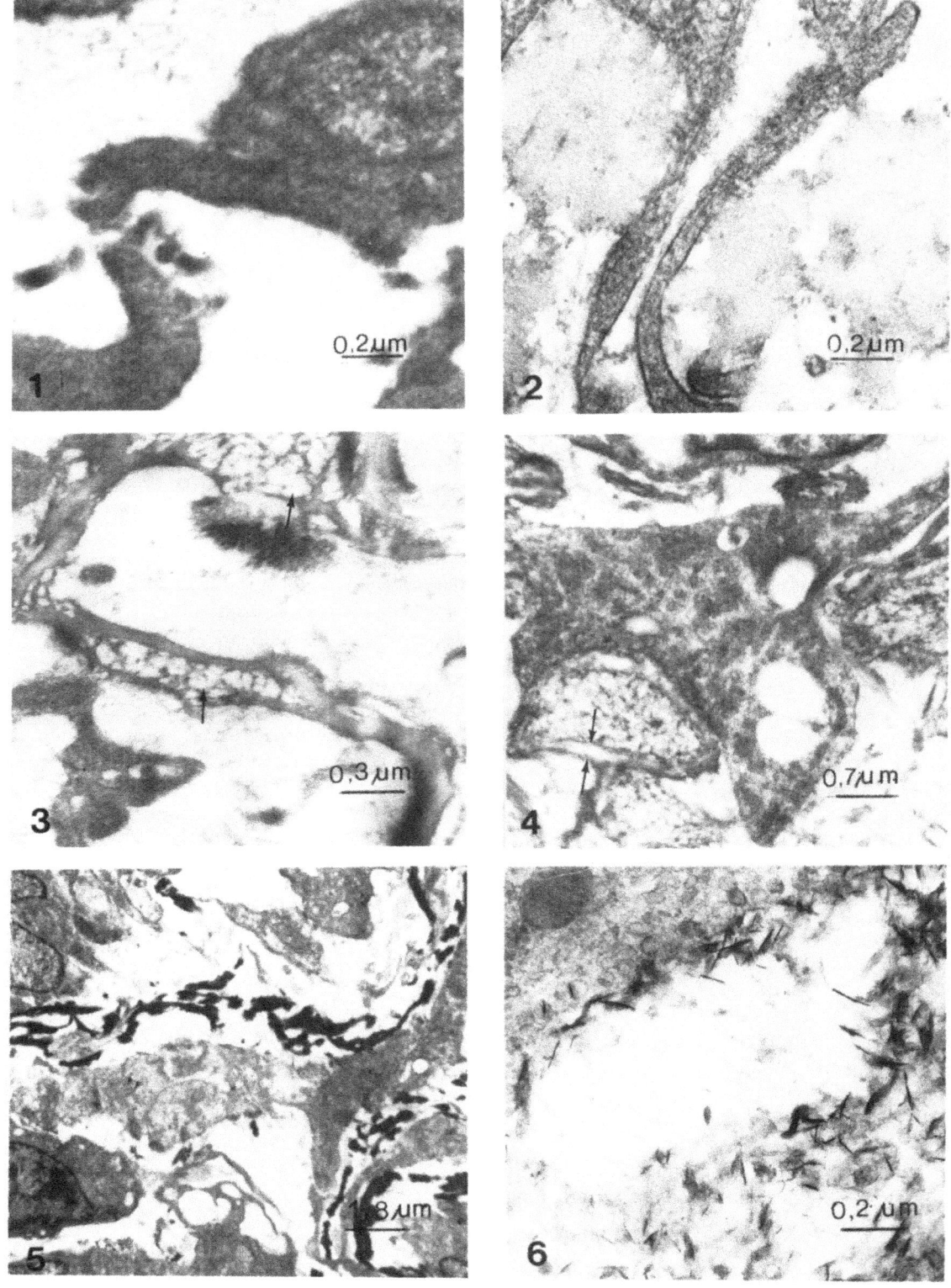

ciency and in lathyrism – the metabolic disturbance of connective tissue runs in a similar manner causing an accumulation of soluble elastin – proved biochemically (10) – and severe disturbance of aggregation of insoluble elastin as revealed by the electron microscope (4, 12).

4.1. Experimental results in copper deficiency

A very high mortality rate was caused by internal hemorrage and ruptures of the main blood vessels among chicks kept on a copper deficient diet. Copper depletion causes lesions mainly in the cardiovascular system: the disturbance of the development of the vascular elastic tissue may lead to the rupture of the large elastic-type vessels or of the heart itself, invariably resulting in death. Elastic fibers were damaged in paraffin sections, while further investigations revealed that, around the ruptures, a material stained similarly to that of an elastin-like substance had been deposited, which proved to be phosphotungstic acid (PTA) positive microfibrils (13).

These studies facilitated the preparation of a soluble precursor to elastin and underlined the view that *copper* plays an important role in the *synthesis of elastin*. Thus no desmosine is formed in the absence of copper resulting in the delay and inadequacy of elastic fiber formation (14). Cross-linkages similar to those found in the elastic fiber are responsible for the stabilization of the collagen molecule; therefore copper depletion also effects the synthesis of collagen.

To clarify the biosynthesis of elastin, investigations were focused on the identification of a pro-elastin or tropoelastin. It is in this sense that copper deficiency and lathyrogen are to be understood as efficient experimental conditions.

At first, Sandberg et al. (14) isolated a soluble elastin from pigs kept on a copper deficient diet. Later tropoelastin was identified in aortic organ and cell cultures (15, 16) and in the lathyritic chick aorta.

Copper deficiency causes a delay of the elastic fiber formation in the chick embryo. Elastic aggregates are small and irregularly scattered (Figure 1)

and continue to bind PTA strongly even in newborns compared to normal elastic fiber (Figure 2), and elastin polymerization on the microfibrils cannot take place.

As a result several structural variations of pathological elastic fiber formation can be noted in newborn chicks (Figure 3). When this pathologically structured fiber is treated with elastase, its 'honeycomb'-like structure is resistant against elastase digestion, while areas where cross-linking had already been formed and which contained amorphous-like elastin could to a certain degree be digested with elastase (Figure 4) (4). Elementary units of elastic fibers in chick embryos were demonstrated with the electron microscope on the sixth day of embryonic development (4). The aorta of the copper depleted embryos contained no fibers even on the tenth day of embryonic life. The described microcyst formation is a characteristic lesion in copper deficiency, and it is the consequence of a focal deterioration of elastic laminae in the vascular injuries, judged from the presence of scars in the media.

The increased resistance of the abnormal fiber structures to elastase and the fact that the fiber developed after copper depletion is in every respect similar to the postdigestion patterns of the normal fiber to the 'honeycomb' structure unequivocally suggests the absence of substrate specific elastin of the fibers synthesized under the condition of copper depletion.

The prolonged existence of the PTA-positivity and the great number of microfibrils represent the increase of the GAG containing substance, viz. structural glycoproteins.

Both copper deficiency and the lathyrogen diet may inhibit an elastase inhibitor which is most likely found in the blood. In this way, the digestive effect of elastase is released, leading to the eventual disintegration of elastic fiber. This hypothesis deserves special attention for the occurence of elastase in the aortic wall.

In summary: the morphological signs of elastic damage caused by copper deficiency consist of irregularly shaped and twisted, elastin-poor elastic fibers, which retain their PTA binding capacity, and an

←

Figure 1. Irregular, small elastic aggregates in copper-depleted animal strongly binds PTA. PTA: × 50,000.
Figure 2. Normal elastic fibers are large and electron lucent. Ua Pb: × 51,000.
Figure 3. Honeycomb-like elastic fiber (arrows) in copper-deficient newborn chick. PTA: × 27,500.
Figure 4. The amorph cross-linked elastin (arrows) in copper-depleted newborn chick is susceptible to elastase. PTA + elastase: × 14,000.
Figure 5. Irregular, electron dense elastic aggregates revealing a spotted pattern in a BAPN-treated newborn chick. PTA: × 5,400.
Figure 6. Ca-apatite on microfibrils in extracellular space of intima proliferation. Ua Pb: × 44,000.

accumulation of microfibrils. At the site of elastic fibers in the aorta wall the formation of the glycoprotein containing microfibrillar frame results in a 'honeycomb'-like structure. Applying the cleaving effect of elastase to the pathological fibers, microfibrillar frame appears resistant and only the amorphous elastin situated at the margin is digestible with elastase. In certain areas it is quite difficult to observe the polymerized cross-linked elastin.

4.2. The role of beta-amino propionitrile

Lathyrism has been known to be a perilous human disease for about 100 years, occuring primarily in India and Spain where people ate chickling peas (Lathyrus Sativus) during great famines. Efforts to isolate the toxic factor from this pea type has proven futile, but the seeds of another species, the Lathyrus Odoratus, when taking up 50% of the diet under experimental conditions, lead to bone deformities and aortic rupture. Later the toxic factor isolated from this particular lathyrus seed was to become the beta-amino-propionitrile (BAPN) and called the 'lathyrus factor'.

BAPN, in addition to causing bone deformity in young animals, is also responsible for various lesions of the aorta, especially ruptures. Although it is generally accepted that BAPN assaults the connective tissue fiber, its precise point of attack is still unknown. BAPN causes the inhibition of collagen and elastic fiber formation followed by a change of the ground substance.

As a direct result of BAPN at the site of elastin fibers among the smooth muscle cells of the aortic wall, a disorganized, filamentous-granular substance and irregular aggregates are noticeable binding PTA even in newborn animals (Figure 5).

BAPN practically causes the disappearance of intercellular substances, while the smooth muscle cell layers are placed closely together in most instances permitting only tiny intercellular spaces (4).

BAPN inhibits the elastin polymerization on the micrifibrils via the inhibition of the usual enzyme process, thus preventing lysine oxidation into aldehyde. It is most likely that the copper-containing lysyl oxidase essential to lysine oxidation is directly and competitively blocked (17). In lathirysm and/or in copper deficiency, the disturbed metabolic process of the connective tissue is analogous to each other, raising the possibility of similar cross-linkage occurences in the collagen as well as in the elastin.

The lack of lysyl oxidase and the subsequent failure of desmosine and isodesmosine building are fairly representative of the effect of lathyrogen on elastic fiber formation. It seems that BAPN plays no decisive role in the inhibition of tropoelastin formation but inhibits polymerization. BAPN not only restrains polymerization, it also brings a considerable amount of instability to the newly formed, though still reversible, cross-links (18). In the course of connective tissue maturation these interim linkings continue to stabilize chemically, becoming fully irreversible at the end. This mechanism could explain the almost instant effect of lathyrogen in embryos or in young animals.

The underlying mechanism in these conditions – similar to those of certain human diseases – is the inhibition of lysyl oxidase through direct mechanism or competitive inhibition. The hindering of the exertion of the effect of lysyl oxidase leads to the inhibition of the formation of cross-links (desmosine, isodesmosine, lysinonorleucin, merodesmosin) and subsequently to impaired elastic fiber formation due to the lack of insoluble elastin.

5. The importance of elastin fibers in various human pathological conditions

The pathobiology of the elastic tissue either as an independent substance or as an integral part of the connective tissue has not been very widely discussed in the preceding literature. It is evident that an impressive number of connective tissue diseases are now known and can be related to specific sites of matrix macromolecule synthesis and degradation. The elastic tissue disorders need a conceptual reconsideration from the point of view of experimental and human pathology. Transmission and scanning electron microscopy offered new possibilities for studying the structure and development of elastic fibers.

Elasticity, a unique physical property of elastin, renders elastic fibers to maintain the structural integrity and functional adaptability of the medium sized and large blood vessels. The intrinsic elastic nature of the elastomer seems to de dependent on its primary structure (8, 19).

It seems fairly certain that changes in the structure and composition of elastin and subsequently of elastic fibers play an important role in many human 'connective tissue' diseases such as those related to the cardiovascular system and are present in aging processes as well (20).

Changes in elastin may lead to fragmentation of elastic fibers, calcification and lipid deposition with loss of elasticity. Consequently these changes act as